Environmental impact assessment

Environmental impact assessment

A comparative review

Second Edition

Christopher Wood

PEARSON
Prentice
Hall

Harlow, England • London • New York • Boston • San Francisco • Toronto • Sydney • Singapore • Hong Kong
Tokyo • Seoul • Taipei • New Delhi • Cape Town • Madrid • Mexico City • Amsterdam • Munich • Paris • Milan

Pearson Education Limited
Edinburgh Gate
Harlow
Essex CM20 2JE
England

and Associated Companies throughout the world

Visit us on the World Wide Web at:
www.pearsoned.co.uk

First published 1995
Second edition 2003

ISBN-10: 0-582-36969-x
ISBN-13: 978-0-582-36969-6

British Library Cataloguing-in-Publication Data
A catalogue record for this book can be obtained from the British Library

Library of Congress Cataloging-in-Publication Data

Wood, Christopher, 1944-
    Environmental impact assessment : a comparative review / Christopher Wood.- 2nd ed.
        p. cm.
    Includes bibliographical references and index.
    ISBN 0-582-36969-X (alk. paper)
        1. Environmental impact analysis. I. Title.

    TD194.6.W66 2002
    333.7'14--dc21

                                                                        2002025163

10 9 8 7 6 5 4
07 06

Typeset in 11/12pt Adobe Garamond by 3

Printed in Malaysia VVP

In memory of

Norah Wood

1915–1996

Contents

Boxes

Figures

Tables

Preface

Environmental impact assessment (EIA) is now over 30 years old. Born in the United States, it was initially ignored, then (in turn) caused great disturbance and antagonism, began to change people's lives for the better, settled down and learned from experience, became respectable and, eventually, was extensively imitated all over the world. As concern about the environment has grown, EIA has been widely seen as a panacea to environmental problems. It is not. EIA is an anticipatory, participatory, integrative environmental management tool which has the ultimate objective of providing decision-makers with an indication of the likely consequences of their decisions relating to new projects or to new programmes, plans or policies. Effective EIA alters the nature of decisions or of the actions implemented to reduce their environmental disbenefits and render them more sustainable. If it fails to do this, EIA is a waste of time and money.

Interest in EIA has burgeoned and there are now over 100 EIA systems in existence worldwide. While the various EIA systems all differ in detail, their basic principles are similar and demonstrate many common problems. Different jurisdictions have used different means to try to solve these problems and to improve the effectiveness of their EIA systems. There is growing interest in learning from the experience of others whose EIA systems have elements worthy of emulation. South Africa, simultaneously a developed and a developing federal country, has been substituted for the state of California and for the treatment of developing countries in this edition. The book now contains accounts of four federal and three unitary national (and no state) jurisdictions.

While there is a huge literature on EIA, there is no other book which presents an accessible comparative step-by-step review of international EIA procedures and practice. This second edition, by reviewing seven EIA systems in detail (the United States, UK, the Netherlands, Canada, Commonwealth of Australia, New Zealand and South Africa) is intended to fill the gap.

As in the first edition, each EIA system is evaluated against a set of criteria which enables comparisons to be made easily. The EIA process is analysed step by step in a succession of chapters. The structure of this edition is identical to that of the first but the text has been greatly modified and extended. Each chapter contains a discussion of appropriate evaluation criteria and methods for the relevant step in the EIA process and then goes on to an analysis of its treatment in each of the seven EIA systems. The evaluation criteria are intended to be generally applicable and can be (and have been) used to analyse other EIA systems. Only marginal changes have been made to the evaluation criteria in this edition. Numerous tables and diagrams are employed to summarise the comparative findings.

EIA is interdisciplinary and involves large numbers of different practitioners but is most closely associated with the professionals concerned with siting new development. The second edition of this book should therefore again be of interest to practitioners involved in government, development and land-use planning, landscape design, the environment, law and engineering. Hopefully, academics and researchers will also continue to find this edition helpful. In addition, the book is intended for those taking specific undergraduate and postgraduate EIA courses, and short EIA training courses, and for students of land-use planning, law, geography, environmental studies, development studies and engineering.

Acknowledgements

Much of this book was written while I was on sabbatical leave from the School of Planning and Landscape, University of Manchester. I wish to thank all my colleagues at Manchester, and especially Carys Jones, of the EIA Centre, for their support. Jeremy Carter and Stephen Jay provided much appreciated research assistance. I owe a great deal to Abigail Shaw, who patiently typed many drafts of the book, and to Elaine Jones and Mary Howcroft who also helped with its presentation. Thanks are due to the Pearson team, and especially to Jill Birch, for their encouragement and professionalism.

My sabbatical leave in South Africa and my visits to Australia, New Zealand, the Netherlands, Canada and the United States were partially funded by the British Council and the University of Manchester, to whom I am most grateful. I wish to record my thanks to Professor Richard Fuggle, Michael Meadows and others at the Department of Environmental and Geographical Science, University of Cape Town, for making me welcome there. I am very grateful to Professor John Arnold for providing me with a 'secret' office at Manchester Business School while I was writing in Manchester.

In the first edition of this book I acknowledged the assistance of Ron Bass, Dinah Bear, Ray Clark, Shannon Cunniff, Anne Miller, Ken Mittleholtz, Joseph Montgomery, David Powers, Russell Train and Nicholas Yost with the United States sections. I am also very grateful to Robert Bartlett, Dinah Bear, Larry Canter, Ray Clark, Bill Dickerson, John Echeverria, Anne Miller and Bob Smythe for their help with the second edition.

In the UK, I would like to thank David Aspinwall, Jim Burns, Kim Chowns, Roger Smithson and John Zetter for their patience, comments, responses and advice about EIA and SEA (strategic environmental assessment). Special thanks are due to Kim Chowns and Roger Smithson for commenting on a draft of the version on the UK.

In addition to those whose assistance in the Netherlands I acknowledged in the first edition of this book, I would like to thank Jos Arts, Jan Jaap de Boer, Jessica Hoitink, Joop Marquenie, Michiel Odijk, Sibout Nooteboom, Jules Scholten, Job van der Berg, Hans van Maanen and Rob Verheem for their comments, responses and advice. Jos Arts, Jan Jaap de Boer and Sibout Nooteboom were kind enough to comment on a draft of the account of the Dutch EIA system.

I acknowledged the help of David Barnes, Gordon Beanlands, Karen Brown, Bill Couch, Karen Finkle, Kathy Fischer, Robert Gibson, Stephen

Hazell, Patrice LeBlanc, Jon O'Riordan, Stephane Parent, Ray Robinson, Barry Sadler and Graham Smith in writing the account of Canadian EIA in the first edition. In addition, I am grateful to Robert Connelly, Keith Grady, Robert Gibson, Heather Humphreys, Patrice LeBlanc, Rodney Northey, Husain Sadar and William Ross for their assistance with the second edition. I wish to record my thanks to Roger Creasey, Robert Gibson, Keith Grady, Ray Lamoureux and William Ross who were kind enough to comment on a draft of the Canadian material.

I reiterate my thanks to those Australians whose assistance I acknowledged in the first edition of this book. I would like to thank Jeff Angel, John Ashe, John Bailey, Lex Brown, Michael Coffey, Donna Craig, Frank Downing, Paul Garrett, Nick Gascoigne, Alistair Gilmour, Nick Harvey, Mary McCabe, Gerry Marvell, Helen Ross, Yolande Stone and Peter Waterman for their comments, responses and advice in preparing the second edition. Special thanks are due to John Bailey, Nick Harvey and, especially, Nick Gascoigne for commenting on a draft of the version presented here.

In addition to those whom I acknowledged in writing the account of New Zealand EIA in the first edition, I would like to thank Denise Church, Jenny Dixon, John Gallen, Sarah Giles, Paddy Gresham, Lindsay Gow, John Hutchings, Alisdair Hutchison, Bob McClymont, Richard Morgan, Ray Salter, Martin Ward and Barry Weeber for their help, comments and advice. Jenny Dixon, Lindsay Gow and Richard Morgan were kind enough to comment on an early draft of the New Zealand account and to provide further information and comment.

I am very grateful to all those in South Africa who provided me with information, often on many occasions: Keith Balchin, Janet Barker, Neal Carter, David Dewar, Richard Fuggle, Jan Glazewski, Stephen Granger, Richard Hill, Dennis Laidler, Marlene Laros, Farieda Khan, Darryll Kilian, Lynn O'Neill, John Raimondo, Merle Sowman, Alex Weaver and Keith Wiseman. Additional thanks are due to Richard Fuggle, Richad Hill, Darryll Kilian, Andries van der Walt and Alex Weaver, who commented on a draft of the South African EIA sections.

I owe a considerable debt to all these practitioners, environmentalists, consultants, central and local government officers, researchers and academics. Needless to say, any errors of fact or judgement are my, and not the respondents', responsibility.

Last, but by no means least, I wish to thank Josephine Wood for her patience and unwavering support.

We are grateful to the following for permission to reproduce copyright material:

Figure 5.2 from *Annual Report for 2000,* Environmental Impact Assessment Commission (EIAC, 2001); Table 5.1 from *Environmental Assessment in Canada – Frameworks, Procedures and Attributes of Effectiveness,* Canadian Environmental Assessment Agency (Doyle, D. and Sadler, B., 1996); Table 13.1 from *The Canadian Environmental Assessment Act – Responsible Authority's*

Guide, Canadian Environmental Assessment Agency (CEAA, 1994*); Figure 2.1 and Table 2.1 from *The NEPA Book: a Step-by-Step Guide on How to Comply with the National Environmental Policy Act*, Solano Press (Bass, R.E. *et al.*, 2001); Table 2.1 from *Environmental Law Series: NEPA Law and Litigation*, West Group (Mandelker, D.R., 2000); Figure 2.3 from the website of the US Environmental Protection Agency http://www.epa.gov/; Figure 4.3 from *Directory of Environmental Impact Statements*, Oxford Brookes University (Wood, G. and Bellanger, C., 1998); Figure 11.1 from *Farnborough Aerodrome Environmental Statement*, Terence O'Rourke plc (O'Rourke, T., 1999); Figure 19.4 from *Strategic Environmental Assessment in South Africa*, Department of Environmental Affairs and Tourism (DEAT, 2000).

Blackwell Publishers for an extract from 'Agriculture and the EC Environmental Assessment Directive: lessons for community policy making' by W.R. Sheate and R.B. Macrory published in *Journal of Common Market Studies 28* (1989); Butterworths for an extract from 'Environmental impact assessment' by A. Weaver, P. Morant, P. Ashton and F. Kruger published in *The Guide to Environmental Auditing in South Africa* ed. I.R. Sampson (Durban, 2000); Commission for Environmental Impact Assessment for extracts from *Information for Members of Working Groups of the Commission for Environmental Impact Assessment in the Netherlands* published by the EIAC (Utrecht, 1998), 'Success of EIA in Preventing, Mitigating and Compensating Negative Environmental Impacts of Projects' by S.A.A. Morel, published in *Environmental Impact Assessment in the Netherlands: Experiences and Views Presented by and to the Commission for EIA* by EIAC (Utrecht, 1996), 'Integrating Environmental Objectives in the Planning of Natural Gas Exploration Drillings in Sensitive Areas in the Netherlands: the North Sea coastal zone and the Wadden Sea' by S.A.A. Morel published in *New Experiences on EIA in the Netherlands: Process, Methodology, Case Studies* by EIAC (Utrecht, 1998), 'Reviewing EISs – the Netherlands Experience' by J.J. Scholten, published in *Environmental Assessment* 5(1) (1997), and 'Strategic Environmental Assessment: One Concept, Multiple Forms' by R. Verheem and J. Tonk, published in *Impact Assessment and Project Appraisal* 18 (2000); Council on Environmental Quality for an extract from *Environmental Quality 1996: Along the American River* (Washington, 1996); Department of the Environment for extracts from *Preparation of Environmental Statements for Planning Projects that Require Environmental Assessment: A Good Practice Guide* by HMSO (1995); Federal Environmental Assessment Review Office (FEARO) for extracts from *A Guide to the Canadian Environmental Assessment Act* by the Canadian Environmental Assessment Agency (1993), *Environmental Assessment in Policy and Program Planning: A Source Book* by FEARO (1992), and *EA Effectiveness Study* by FEARO (Quebec, 1994); Legislative Services Australia for extracts from *Environment Protection and Biodiversity Conservation Act Fact Sheet 3* (Canberra, 2000), and *Environmental Impact Statement Guidelines: Proposed National Low Level Radioactive Waste Repository* 2001/151 (Canberra, 2001) published by the

Department of the Environment and Heritage: Environment Australia; the author, Dr R. Morgan, for an extract from *A Structured Approach to Reviewing AEEs in New Zealand* published by the Centre for Impact Assessment and Research and Training (Dunedin, 2000); Pearson Education Limited for an extract from 'Scoping and public participation' by A. McNab published in *Planning and Environmental Impact Assessment in Practice*, ed. J. Weston (1997); Public Works and Government Services Canada for extracts from *Performance Report for the Period Ending March 31 1999*, and *Strategic Environmental Assessment: The 1999 Cabinet Directive on the Environmental Assessment of Policy, Plan and Program Proposals: Guidelines for Implementing the Cabinet Directive* by the Canadian Environmental Assessment Agency (Quebec, 1999); and Solano Press Books for extracts from *CEQA Deskbook: A Step-by-Step Guide on How to Comply with the California Environmental Quality Act* (1999) and *NEPA Book: A Step-by-Step Guide on How to Comply with the National Environmental Policy Act* (2001) both by R.E. Bass, A.I. Herson and K.M. Bogdan.

In some instances we have been unable to trace the owners of copyright material, and we would appreciate any information that would enable us to do so.

Abbreviations and national EIA terms

AEE	assessment of environmental effects (New Zealand)
ANZECC	Australian and New Zealand Environment and Conservation Council
	assessment report (Australia)
CEAA	Canadian Environmental Assessment Agency
CEARC	Canadian Environmental Assessment Research Council
CEC	Commission of the European Communities
CEPA	Commonwealth Environment Protection Agency (Australia)
CEQ	Council on Environmental Quality (United States)
CESD	Commissioner of the Environment and Sustainable Development (Canada)
COAG	Council of Australian Governments
CONNEPP	Consultative National Environmental Policy Process (South Africa)
	categorical exclusion (United States)
	class screening (Canada)
	comprehensive study (Canada)
DEA	Department of Environmental Affairs (South Africa)
DEAT	Department of Environmental Affairs and Tourism (South Africa)
DETR	Department of the Environment, Transport and the Regions (UK)
DoE	Department of the Environment (UK)
	direction (UK)
EA	environmental assessment (document: United States)(process: Canada, UK)
EARP	Environmental Assessment and Review Process (Canada)
EC	European Commission
ECW	Evaluation Committee on EIA (The Netherlands)
EIA	environmental impact assessment
EIAC	Environmental Impact Assessment Commission (The Netherlands)
EIR	environmental impact report (South Africa)
EIS	environmental impact statement (Australia, Canada, The Netherlands, United States)

EPA	Environmental Protection Agency (United States)
EPBC Act	Environment Protection and Biodiversity Conservation Act 1999 (Australia)
EPEP	Environmental Protection and Enhancement Procedures (New Zealand)
ES	environmental statement (UK)
ESD	ecologically sustainable development (Australia)
EU	European Union
	environmental information (UK)
FEARO	Federal Environmental Assessment Review Office (Canada)
FONSI	finding of no significant impact (United States)
	follow-up program (Canada)
IEM	integrated environmental management (South Africa)
LPA	local planning authority (UK)
MoE	Minister of the Environment (Canada)
MER	EIA (milieu-effectrapportage) (The Netherlands)
MfE	Ministry for the Environment (New Zealand)
	mediation (Canada)
NAPA	National Academy of Public Administration (United States)
NEMA	National Environmental Management Act 1998 (South Africa)
NEPA	National Environmental Policy Act 1969 (United States)
NOI	notice of intent (United States)
	notice of intention (Australia)
	notification of intent (The Netherlands)
OECD	Organisation for Economic Cooperation and Development
	opinion (UK)
PCE	Parliamentary Commissioner for the Environment (New Zealand)
PER	public environment report (Australia)
	plan of study (South Africa)
PPP	policy, plan and programme
	panel review (Canada)
	public registry (Canada)
	public review (Canada)
RA	responsible authority (Canada)
	relevant authority (South Africa)
RMA	Resource Management Act 1991 (New Zealand)
ROD	record of decision (South Africa, United States)
RSA	Republic of South Africa
SEA	strategic environmental assessment
	scoping report
	screening (Canada)
	self-directed assessment (Canada)
	specified information (UK)
UNECE	United Nations Economic Commission for Europe
UNEP	United Nations Environment Programme
VROM	Ministry of Housing, Spatial Planning and the Environment (The Netherlands)

Chapter 1

Introduction

Nature of environmental impact assessment

Environmental impact assessment (EIA) refers to the evaluation of the effects likely to arise from a major project (or other action) significantly affecting the natural and man-made environment. Consultation and participation are integral to this evaluation. EIA is a systematic and integrative process, first developed in the United States as a result of the National Environmental Policy Act 1969 (NEPA), for considering possible impacts prior to a decision being taken on whether or not a proposal should be given approval to proceed. NEPA requires, *inter alia*, the publication of an environmental impact statement (EIS) describing in detail the environmental impacts likely to arise from an action.

The EIA process should supply decision-makers with an indication of the likely environmental consequences of their actions. Properly used, EIA should lead to informed decisions about potentially significant actions, and to positive benefits to both proponents and to the population at large. As the UK Department of the Environment, Transport and the Regions put it (DETR, 1999b, paras 9, 14) formal EIA:

> is a means of drawing together, in a systematic way, an assessment of a project's likely environmental effects. This helps to ensure that the importance of the predicted effects, and the scope for reducing them, are properly understood by the public and the relevant competent body before it makes its decision. ... EIA can help to identify the likely effects of a particular project at an early stage. This can produce improvements in the planning and design of the development; [and] in decision-making.

As Glasson *et al.* (1999, p. 9) have noted:

> Underlying such immediate purposes is of course the central and ultimate role of EIA as one of the instruments to achieve sustainable development: development that does not cost the Earth!

Sadler (1996) also asserted that EIA is a key technique for incorporating concepts such as the precautionary principle and the avoidance of net loss of natural capital, central to the achievement of sustainable development, into decision making.

In principle, EIA should lead to the abandonment of environmentally unacceptable actions and to the mitigation to the point of acceptability of the environmental effects of proposals that are approved. EIA is thus an anticipatory, participatory environmental management tool, of which the EIA report is only one part. The objectives of the Californian EIA system make this very clear (Bass *et al.*, 1999, p. 1):

1. To disclose to decision-makers and the public the significant environmental effects of proposed activities.
2. To identify ways to avoid or reduce environmental damage.
3. To prevent environmental damage by requiring implementation of feasible alternatives or mitigation measures.
4. To disclose to the public reasons for agency approvals of projects with significant environmental effects.
5. To foster interagency coordination in the review of projects.
6. To enhance public participation in the planning process.

Appropriately employed, EIA is a key integrative element in environmental protection policy, but only one element in that policy (Lawrence, 1994). Because EIA is part of a wider approach to environmental protection it is influenced by the system of which it is an element. Generally, the more committed a jurisdiction is to environmental policy, the more influence EIA will have over decision making within that jurisdiction.

EIA is not just a procedure, or for that matter just a science. Its nature is dichotomous, rather like the duality of matter. As Kennedy (1988, p. 257) has put it, EIA is both science and art, hard and soft:

EIA as 'science' or a planning tool has to do with the methodologies and techniques for identifying, predicting, and evaluating the environmental impacts associated with particular development actions.
EIA as 'art' or procedure for decision-making has to do with those mechanisms for ensuring an environmental analysis of such actions and influencing the decision-making process.

Caldwell (1989, p. 9) has summarised the significance of EIA as follows:

1. Beyond preparation of technical reports, EIA is a means to a larger end – the protection and improvement of the environmental quality of life.
2. It is a procedure to discover and evaluate the effects of activities (chiefly human) on the environment – natural and social. It is not a single specific analytic method or technique, but uses many approaches as appropriate to a problem.
3. It is not a science, but uses many sciences (and engineering) in an integrated interdisciplinary manner, evaluating relationships as they occur in the real world.
4. It should not be treated as an appendage, or add-on, to a project, but regarded as an integral part of project planning. Its costs should be calculated as a part of adequate planning and not regarded as something extra.
5. EIA does not 'make' decisions, but its findings should be considered in policy- and decision-making and should be reflected in final choices. Thus it should be part of decision-making processes.
6. The findings of EIA should focus on the important or critical issues, explaining why they are important and estimating probabilities in language that affords a basis for policy decisions.

It is not clear precisely how the EIA process works. Bartlett and Kurian (1999, p. 415) suggested that:

> Writing about EIA has been guided by assumptions and models that have been implicitly assumed rather than explicitly and systematically explored, formulated, or articulated.

They advanced six categories of implicit models used in the EIA literature. One of these, the 'information processing model' assumed that the key to better decision making was the availability of high-quality information. This model, which was the most commonly used in the contemporary EIA literature, tended to underplay, or even disregard, the influence of politics in the decision-making process of which EIA forms part. Mostert (1996, p. 191) has highlighted the subjective nature of the supposedly rational EIA process assumed in this model: 'Subjectiveness occurs whenever the results of EIA are influenced by the subjective norms, values and interests of one or more of the parties involved.' (Weston, 2000, p. 190) emphasised this point: 'there are within the [EIA] process itself many key decisions to be made which will almost certainly not be based upon the rational principles of value free objectivity'.

The political nature of the decision-making context of EIA is inescapable. It cannot be assumed that the provision of high-quality environmental information, of itself, will lead to decisions that are consistently 'environmentally friendly'. It is increasingly acknowledged that the information generated by the EIA process is considered within a political decision-making arena, and is therefore influenced by its norms and values, as well as by its procedures. Any changes to the decision-making process that result from EIA will be changes made as a consequence of the evolution of the values and perspectives held by elected decision-makers and by their advisers or as a result of successful public intervention.

It should be emphasised that EIA is not a procedure for preventing actions with significant environmental impacts from being implemented, although in certain circumstances this could be the appropriate outcome of the process. Rather the intention is that actions are authorised in the full knowledge of their environmental consequences. Because EIA takes place in a political context, it is therefore inevitable that economic, social or political factors will outweigh environmental factors in many instances. This is why the mitigation of environmental impacts is so central to EIA: decisions on proposals in which the environmental effects have palpably been ameliorated are much easier to make and justify than those in which mitigation has not been achieved.

This chapter briefly describes the evolution and diffusion of EIA from its origins in the US National Environmental Policy Act 1969. It goes on to discuss the elements of the EIA process and the effectiveness of EIA systems, and to suggest a number of criteria against which EIA systems can be evaluated. The purpose of the comparative review of the selected EIA systems presented in this book is then explained. Finally, an overview and explanation of the structure of the book is presented.

Evolution and diffusion of EIA

California was the first of the American states to introduce an effective 'little NEPA', in 1970 (Bass *et al.*, 1999). (The majority of US states have still not done so.) International attention was soon being directed to EIA as a result of several celebrated legal cases in the United States, which clarified NEPA's significance. The ramifications of NEPA were beginning to be accepted at a time of unprecedented interest in the environment occasioned by the United Nations conference on the environment in Stockholm in 1972. The problems of burgeoning development, pollution and destruction of the natural environment that NEPA was intended to address were perceived as universal. The rigorous project-by-project evaluation of significant impacts inherent in EIA was seized upon as a solution to many of these environmental problems by many other jurisdictions, and elements of the US EIA process were adopted by them. Most were, however, cautious about importing NEPA-style litigation with EIA and made strenuous efforts to avoid doing so.

The methods of adoption varied; cabinet resolutions, advisory procedures, regulations and laws were employed. Probably the first overseas jurisdiction to declare an 'extremely rudimentary environmental impact policy' (Fowler, 1982, p. 8) was the Australian state of New South Wales in January 1972. The Commonwealth of Australia announced an EIA policy in May 1972 and passed the Environment Protection (Impact of Proposals) Act in December 1974. Canada preceded Australia, approving a federal cabinet directive on EIA in 1973. New Zealand instituted EIA procedures by cabinet minute in 1974. Columbia (Verocai Moreira, 1988) and Thailand (Nay Htun, 1988) established EIA systems through specific legislation in 1974 and 1975, respectively, followed by France in 1976. Ireland passed legislation that permitted, but did not require, EIA in 1976, and the cabinet of the West German government approved an EIA procedure by minute in the same year. The Netherlands governmental standpoint on EIA followed in 1979. There was also considerable EIA activity in numerous developing countries (Biswas and Agarwala, 1992; Lee and George, 2000). The diffusion of EIA was gathering pace and has continued unabated.

Several international agencies have involved themselves with EIA. In 1974 the Organisation for Economic Cooperation and Development (OECD) recommended that member governments adopt EIA procedures and methods and more recently, that they use EIA in the process of granting aid to developing countries (OECD, 1992). In addition, in 1985, the Council of the European Communities adopted a directive that required member states to implement formal EIA procedures by 1988. These procedures were strengthened by a further directive that came into effect in 1999.

In 1989 the World Bank ruled that EIA for major projects should normally be undertaken by the borrower country under the Bank's supervision (Council on Environmental Quality – CEQ, 1990, p. 45). The World Bank (1999) has recently updated its guidance on EIA. The United Nations Environment Programme (UNEP) also made recommendations to member states regarding the establishment of EIA procedures and established goals and principles for EIA. It subsequently issued guidance on EIA in developing countries (UNEP, 1988). The 1992 Earth Summit provided additional momentum to these developments. Principle 17 of the Rio Declaration (in Sadler, 1996, p. 24) stated that:

> Environmental Impact Assessment, as a national instrument, shall be undertaken for proposed activities that are likely to have a significant adverse impact on the environment and are subject to a decision of a competent national authority.

EIA is now practised in more than one hundred countries (Donnelly *et al.*, 1998).

This diffusion of EIA has resulted in a diverse vocabulary. In the Netherlands, EIA is known as MER (milieu-effectrapportage) and in Canada as environmental assessment. The EIS (EIA report) becomes an environmental statement in Britain, and an environmental impact report in South Africa and under the original New Zealand provisions. The Commonwealth of Australia has both an EIS and a public environment report. New Zealand requires an AEE (assessment of environmental effects) document and Canada utilises both the EIS and the 'comprehensive study'.

As EIA has spread, so has its nature been elaborated and clarified. There have been, perhaps, seven main themes as EIA has evolved over the years:

1. An early concern with the methodology of impact forecasting and decision making gave way first to an emphasis on administrative procedures for EIA, second to a recognition of the crucial relationship of EIA to its broader decision making and environmental management context, and then, more recently, to an acknowledgement of the subjective and political nature of the EIA process.
2. A tendency to codification and away from discretion. This is evident in CEQ graduating from the use of guidelines to regulations in the United States and in the enactment of federal Canadian EIA legislation after almost two decades of experience with administrative EIA procedures. New Zealand and South Africa provide additional examples of the codification of discretionary procedures and Australia has further refined, in considerable detail, procedures that were sketchily legislated 25 years previously.
3. The refinement of EIA systems by the adoption of additional elements as experience has been gained. These include detailed procedures for determining which projects should be subject to EIA (screening) (in the UK as a result of amendments to the European directive) and for determining the coverage of EIAs (scoping) (first in the United States and then in, for example, South Africa).
4. A concern to increase the quality of EIA by, for example, improving EIA reports, providing more opportunities for consultation and participation, emphasising linkages with sustainable development and increasing the weight given to EIA in decision making.
5. A concern to increase the effectiveness of EIA in reducing adverse

environmental impacts and to ensure efficiency in terms of its costs in time, money and manpower.

6. The linkage of EIA with ongoing environmental management systems by, for example, insistence on the use of environmental management plans to implement the mitigation measures contained in EIA reports.

7. The recognition that many variables are already resolved by the time the EIA of projects takes place and thus that some form of EIA of policies, plans and programmes (strategic environmental assessment) is necessary.

Elements of the EIA process

While not all EIA systems contain every element, the EIA process emanating from NEPA and subsequently diffused around the world can be represented as a series of iterative steps:

1. Consideration of alternative means of achieving objectives.
2. Designing the selected proposal.
3. Determining whether an EIA is necessary in a particular case (screening).
4. Deciding on the topics to be covered in the EIA (scoping).
5. Preparing the EIA report (i.e., *inter alia*, describing the proposal and the environment affected by it and assessing the magnitude and significance of impacts).
6. Reviewing the EIA report to check its adequacy.
7. Making a decision on the proposal, using the EIA report and opinions expressed about it.
8. Monitoring the impacts of the proposal if it is implemented.

Most of these steps (for example, screening) require a decision to be taken, quite apart from the proposal approval decision. As indicated in Figure 1.1, which summarises these steps, the EIA process is cyclical. Thus, the consideration of the environmental effects of alternative means of achieving the proponent's aims and the detailed design of the action are inextricably linked. Again, the results of consultation at the scoping stage or later may require the proponent to return to the design stage to increase the mitigation of impacts. Consultation and public participation should be important inputs at each stage in the EIA process, though the people and bodies invited, or enabled, to comment on the proposal may vary. Equally, the mitigation of environmental impacts should take place at each step in the process. Not every step in the EIA process shown in Figure 1.1 takes place overtly (or indeed, at all) in every EIA system. As mentioned above, scoping and project monitoring were not part of the original conception of EIA in NEPA and are still not required in many EIA systems. Indeed, there is a considerable diversity of views about the essential elements of an effective EIA system which should, in any event, be tailored to individual national circumstances.

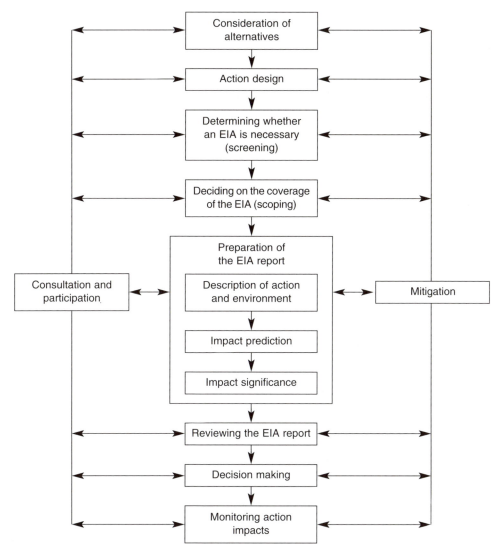

Figure 1.1 The environmental impact assessment process

EIA system effectiveness

Much of the debate about the effectiveness of EIA systems emanates from North America. It centres not so much on whether EIA can be viewed as effective, but on the factors that can be advanced to explain why an EIA system is effective, on which evaluation criteria are appropriate in judging the effectiveness of an EIA system and on how EIA can be improved.

While the view is not unanimous (see, for example, Fairfax, 1978), it is accepted very widely that:

At the US federal level, impact assessment works. We know how it works to influence project selection and design and to mitigate environmental impacts. (Wandesforde-Smith and Kerbavaz, 1988, p. 162)

CEQ (1990, p. 16) concurred with this view of the effectiveness of EIA:

The act unquestionably has had a profound effect on attitudes within the federal government, and its influence outside the federal government is almost as impressive.

Taylor (1984) believed that the US EIA system worked because it was an administrative reform in tune with its times: supportive forces both inside and outside government worked together to ensure the effective implementation of EIA, and the changes in organisational behaviour associated with it. Caldwell (1989, p. 10) has accepted this view:

To the question: has EIA reformed administration, my reply is yes – but as an instrument of a public opinion demanding administrative and policy reform.

Wandesforde-Smith and Kerbavaz (1988) and Wandesforde-Smith (1989) have gone further and emphasised the role of federal and state elections and of personnel ('public sector entrepreneurs') in implementing changes to both the US federal and the Californian EIA systems. They also emphasise the importance of financial and manpower resources and skill in bargaining. Perhaps more important is Caldwell's (1989, p. 12) point about environmental values:

EIA will be most effective where environmental values (1) are implicit and consensual in the national culture and (2) are explicit in public law and policy.

It is apparent that the success of EIA depends upon a large number of factors in addition to the precise nature of the procedures in force. As Hollick (1986, p. 159) has stated:

outside the USA it is commonly assumed not only that introducing procedural changes will change decision making, but also that agency procedures will change in accordance with promulgated procedures without some measure of coercion.

This assumption is clearly over-optimistic.

This is not to say that EIA procedure is not important in determining effectiveness. It is clearly of crucial importance but, while it is a necessary condition for EIA success, it is not a sufficient condition. Every EIA procedure operates within a political, legal, administrative and policy context peculiar to the jurisdiction concerned. To be successful in achieving a real shift in the weight given to the environment in decisions, the EIA procedure needs to interact positively with its jurisdictional context. As in the United States, this may not happen immediately. Wandesforde-Smith (1989, p. 165) has summarised these points as follows:

EIA effectiveness is associated with changing political regimes and with the changing level of support for the EIA process among courts, chief executives, and senior agency managers that this implies. The way an EIA process is formally structured and the way structure taps informal incentives for administrative behaviour are, equally clearly, important variables.

Ortolano *et al.* (1987, p. 286) felt that, to be effective, an EIA system needed to have a number of characteristics, including:

> Utilisation of proper methods in assessing impacts.
> Influence of environmental information on various aspects of planning and decision making, including formulation of alternative plans, selection of a proposed plan, and mitigation of adverse impacts.
> Placement of appropriate weight on environmental impacts relative to economic and technical factors.

Ortolano (1993) has subsequently emphasised the need to include procedural compliance and the completeness of EIA documents among the dimensions of EIA effectiveness.

Ortolano *et al.* (1987) believed that the effectiveness of EIA systems could be explained by reference to 'control mechanisms': intraorganisational and interorganisational processes and structures to ensure that the procedures actually worked. They advanced six types of control as causative: judicial, procedural, evaluative, instrumental, professional, and direct public and outside agency. They suggested that two or more of these mechanisms usually operated simultaneously and that opportunities for public involvement played a key role in the exercise of each.

This emphasis on the role of public involvement in the success of NEPA is widespread. Fairfax and Ingram (1981, p. 43) felt that:

> considerable public discussion and support has come less because of its uniquely cogent approach to the problems of fragmented decision-making, than because of the vocal and powerful environmental constituency which came to support the legislation after a spate of expansive judicial readings of its requirements.

CEQ (1990, p. 42) has also emphasised the crucial role of external review (which it takes to mean external agency review, public participation and judicial review) in the success of EIA.

Kennedy (1988, p. 262) reached the following conclusion to the question of which EIA procedures work:

> Generally speaking, however, it would appear that EIA works best when it is instituted in a formal-explicit way. That is to say, it works when there is a specific legal requirement for its application, where an environmental impact statement is prepared, and where authorities are accountable for taking its results into consideration in decision-making.
> In addition, for EIA to be successfully integrated in the project planning process it would appear that procedures for screening, scoping, external review and public participation need to be a part of it.

Evaluation of EIA system effectiveness

Sadler (1996) felt that concern over effectiveness was an overarching theme of EIA theory and practice. Although the theoretical underpinnings of EIA are very insecure (Bartlett and Kurian, 1999) the evaluation of EIA system effectiveness is necessary in order to advance understanding of the process, and ultimately to improve its performance. As Sadler (1998, p. 37) noted:

> Above all, there is an evident requirement to use effectiveness reviews as an integral strategy for building quality control and assurance throughout the EA process.

Sadler (1996) believed that the 'litmus test' of EIA effectiveness was the influence that the process had on decision making.

There has been, as yet, no reliable quantification of the effectiveness of EIA. It may be that this is not possible. As CEQ (1990, p. 15) has stated:

> Because NEPA was not designed to control specific kinds or sources of pollution, its benefit to society is difficult to quantify. The act was designed primarily to institutionalize in the federal government an anticipatory concern for the quality of the human environment, that is, an attitude, a heightened state of environmental awareness that, unlike pollution abatement, is measurable only subjectively and qualitatively.

Bartlett and Baber (1989, pp. 148, 149) endorsed the difficulty of empirically examining the effects of EIA on decision making within organisations:

> For that reason, it may be more desirable to judge the impact of impact assessment on bureaucratic decision making by examining the attitudes and opinions of those immediately responsible.

While the difficulties of reaching an objective overall judgement about any EIA system are apparent, there is a need for an evaluative framework for comparing the formal legal procedures, the arrangements for their application, and practice in their implementation in EIA systems. Sadler (1996, p. 39) has suggested that there should be three different components of an effectiveness review of the EIA process:

> *procedural:*– does the EA process conform to established provisions and principles?
> *substantive:*– does the EIA process achieve the objectives set, e.g. support well-informed decision making and result in environmental protection?
> *transactive:*– does the EA process deliver these outcome[s] at least cost in the minimum time possible, i.e. is it effective and efficient?

This evaluative framework could be constructed by analysing the extent to which various principles are met by EIA systems. Such principles might, for example, be based upon NEPA provisions, the requirements of the amended European EIA Directive, or upon the more detailed EIA principles for assessing authorities, for proponents, for the public and for government put forward by the Australian and New Zealand Environment and Conservation Council (1991). Perhaps the most rigorous example of the use of this type of evaluative framework is Gibson's (1993) analysis of the Canadian federal and Ontario EIA systems on the basis of eight 'interdependent principles for the design of effective environmental assessment processes' (Box 1.1).

Various alternative approaches for evaluating EIA systems have been advanced (for example, by Hollick (1986) and the Canadian Environmental

Box 1.1   Eight basic principles for evaluating EIA processes

1. An effective environmental assessment process must encourage an integrated approach to the broad range of environmental considerations and be dedicated to achieving and maintaining local, national and global sustainability.
2. Assessment requirements must apply clearly and automatically to planning and decision making on all undertakings that may have environmentally significant effects and implications for sustainability within or outside the legislating jurisdiction.
3. Environmental assessment decision making must be aimed at identifying best options, rather than merely acceptable proposals. It must therefore require critical examination of purposes and comparative evaluation of alternatives.
4. Assessment requirements must be established in law and must be specific, mandatory and enforceable.
5. Assessment work and decision making must be open, participative and fair.
6. Terms and conditions of approvals must be enforceable, and approvals must be followed by monitoring of effects and enforcement of compliance in implementation.
7. The environmental assessment process must be designed to facilitate efficient implementation.
8. The process must include provisions for linking assessment work into a larger regime including the setting of overall biophysical and socio-economic objectives and the management and regulation of existing as well as proposed new activities.

Source: Gibson, 1993.

Assessment Research Council (CEARC, 1988)). More recently, Sadler (1996, p. 22) advanced a number of interdependent principles governing the design and development of effective EIA processes.

- Clear mandate and provisions.
- Explicit goals and objectives.
- Uniform, consistent application.
- Appropriate level of assessment.
- Relevant scope of consideration.
- Flexible, problem solving approach.
- Open, facilitative procedures.
- Necessary support and guidance.
- 'Best practice' standards.
- Efficient, predictable implementation.
- Decision oriented.

- Related to condition setting.
- Follow-up and feedback inbuilt mechanisms.
- Cost-effective outcomes.

Evaluation criteria are, in effect, shorthand versions of principles for EIA and, carefully articulated, have considerable advantages in terms of brevity and clarity. Box 1.2 presents a set of evaluation criteria that are based upon the representation of the stages in the EIA process shown in Figure 1.1, the aims of EIA, and the various evaluation frameworks discussed above. The focus of the criteria is on the requirements and operation of the EIA process, i.e. mainly on procedural effectiveness, though they also encompass efficiency and equity considerations (Sadler, 1998). Assessing the substantive dimension, the effectiveness of EIA in delivering its desired outcome, the enhancement of environmental protection, is a different, and ultimately more difficult, task. Consequently, only the penultimate criterion involves an overall evaluation of

---

Box 1.2   EIA system evaluation criteria

1. Is the EIA system based on clear and specific legal provisions?
2. Must the relevant environmental impacts of all significant actions be assessed?
3. Must evidence of the consideration, by the proponent, of the environmental impacts of reasonable alternative actions be demonstrated in the EIA process?
4. Must screening of actions for environmental significance take place?
5. Must scoping of the environmental impacts of actions take place and specific guidelines be produced?
6. Must EIA reports meet prescribed content requirements and do checks to prevent the release of inadequate EIA reports exist?
7. Must EIA reports be publicly reviewed and the proponent respond to the points raised?
8. Must the findings of the EIA report and the review be a central determinant of the decision on the action?
9. Must monitoring of action impacts be undertaken and is it linked to the earlier stages of the EIA process?
10. Must the mitigation of action impacts be considered at the various stages of the EIA process?
11. Must consultation and participation take place prior to, and following, EIA report publication?
12. Must the EIA system be monitored and, if necessary, be amended to incorporate feedback from experience?
13. Are the discernible environmental benefits of the EIA system believed to outweigh its financial costs and time requirements?
14. Does the EIA system apply to significant programmes, plans and policies, as well as to projects?

the EIA system. For the reasons outlined above, this relies mainly on the opinions of those involved in the EIA process. These criteria can be employed to judge the effectiveness of any EIA system and to enable an international comparison to be made between systems. Such a comparative review provides the basis for suggesting how the effectiveness of EIA can be improved.

Comparative review of EIA systems

Because every EIA system is unique and each is the product of a particular set of legal, administrative and political circumstances, the examination of several EIA systems comparatively by analysing each element in the EIA process should achieve three objectives. The first is explanatory. By placing the EIA process and the stages in EIA procedures in their international context, it should be possible to explain their nature much more clearly than by studying the system in a single jurisdiction. Second, analysis across EIA systems provides a means of better understanding practice in any particular jurisdiction. It is known that some EIA systems work better than others, and step-by-step comparative analysis may help to throw more light on the factors which are essential to the success of EIA processes. The third objective stems from the first two. As Lundquist (1978) has stated: 'comparative studies of national approaches to solving environmental problems have often led to valuable and practical suggestions to improve the effectiveness of the national processes examined.' If this comparative review leads to one such suggestion it will have been successful.

Seven different EIA systems are compared in this book: those in the United States, UK, the Netherlands, Canada, Commonwealth of Australia, New Zealand and South Africa. The US and UK systems chose themselves. The United States possesses the original EIA system and, as with so much else in the environmental policy field, examination of American experience is often a pointer to the future elsewhere. Many of the problems currently facing, for example, the UK in improving the quality of EIA have been apparent in the United States over the years since 1970, and attempts have been made to resolve them that are relevant to experience elsewhere.

The UK is the only one of the seven jurisdictions to have introduced a formal EIA system with initial reluctance. Comparison of the UK's system with more mature EIA systems should provide a valuable insight into the remedies for the problems which have become apparent, many of which have been experienced elsewhere. However, such comparisons need not only indicate improvements in the UK system: there are some respects in which the British system may provide pointers to others.

It is important to see EIA in the United States and in the UK in their international context. The Netherlands is generally acknowledged as having a sophisticated system of environmental controls, including an EIA system regarded by many observers as the most effective in Europe. The Netherlands, following numerous studies, had almost put its EIA system in place when the European Directive on EIA was adopted. The origin of its system is, therefore, in marked contrast to that in the UK, which was instituted as a direct response to the Directive.

The Canadian federal environmental assessment (EA) system was established on an informal basis in 1973 and has been refined substantially over the years. It provided the model for the Netherlands' EIA system and, in particular, for the Dutch use of panels to review EIA reports. It also had an influence on the design of the EIA system in New Zealand. The substantial Canadian programme of research on EIA topics won worldwide admiration. The introduction of formal Environmental Assessment and Review Process (EARP) guidelines in 1984 was followed, in 1995, by the proclamation of the Canadian Environmental Assessment Act. The provisions of this Act establish a formal and tightly prescribed second generation EIA system and are supported by considerable financial and manpower resources. It is likely that aspects of the Canadian federal EA system will continue to provide models for other jurisdictions.

Formal provisions for EIA in the Commonwealth of Australia date from 1974, 14 years before the UK system was introduced. The original requirements were derived, with amendment to avoid frequent recourse to the courts, from NEPA. The Commonwealth procedure evolved over the years and, again following the US system, was reinforced by the addition of scoping. In 1999, the original EIA legislation was repealed and, unlike in the United States, a new (Mark II) EIA system was introduced. This new EIA process is designed to complement, but not to duplicate, the legislative and/or administrative procedures which the six states and two self-governing territories of Australia have put in place to extend EIA to their own activities. The federal EIA system of the Commonwealth of Australia therefore provides a valuable contrast to those in the United States and Canada (Holland *et al.*, 1996) and to that in South Africa.

New Zealand first introduced EIA procedures by means of a cabinet minute in 1974, the same year as Australia. From the outset it employed formal audit (review) procedures to provide an independent check on the EIA reports prepared. After considerable debate, environmental management in New Zealand generally, and EIA in particular, were revolutionised in 1991. One of the aims of this far-reaching reform was sustainable management. EIA is now inextricably interwoven into local (regional and territorial) authority procedures for determining various types of applications. In principle at least, EIA is now comprehensive in that it applies, at the appropriate level of detail (as determined by the regional and territorial authorities), to virtually all projects. EIA in New Zealand is thus largely locally administered (as in the UK) and has become an almost infinitely flexible approach: reason enough for its inclusion in a comparative study.

For some of its population, South Africa is a developed federal country but, for many of its people, conditions are more typical of a developing country. It has a long history of academic (for example, Shopley and Fuggle, 1984) and professional interest in EIA and, in particular, in the integration of EIA into a broader system of integrated environmental management. This system, designed in part to overcome the environmental effects of apartheid, was practised on a discretionary basis for many years as legislated EIA powers lay dormant. Shortly after regulations activating these powers were brought into effect in 1997, the National Environmental Management Act 1998 became law but without the necessary EIA regulations. This somewhat contradictory position is further complicated by the enactment by various provinces of their own EIA requirements. These provinces, which are largely responsible for the implementation of the national EIA requirements, frequently lack the necessary staff resources.

Each country was revisited, and researched anew (along with South Africa) for this edition of the book. Apart from reviewing the documentary and electronic literature (including many documents available only as 'grey' literature), the interview was the main research method employed in this comparative study for the reasons advanced by Bartlett and Baber (1989 – see above). Interviews were conducted with government and agency officials, with researchers in universities and research establishments, with representatives of industry, with lawyers, with consultants and with pressure group campaigners in each of the jurisdictions analysed. A structured approach was employed, interviews being conducted on the basis of a set of questions derived from the criteria set down in Box 1.2.

Wherever possible, an attempt was made to overcome potential inaccuracies by cross-checking participants' accounts with those of other participants in the EIA process and with documentary evidence. Drafts of parts of earlier versions of much of the material in this edition were reviewed by many of those interviewed (see Acknowledgements). Generally, the approach adopted was in close accord with the principles for conducting EIA evaluations enunciated by Sadler (1998) (Box 1.3).

Structure of the book

The first chapters of the book provide the background to, and an overview of, the seven EIA systems. Chapter 2 describes the first of the EIA systems: it explains the evolution of the NEPA provisions, recounts the main features of the US EIA system and discusses its implementation.

Chapter 3 deals with the European Directive on EIA, using the same format as the American chapter. This provides the context for analysis of the UK and Dutch EIA systems. The following chapter covers EIA in the UK, again using the same format, but commencing with an account of pre-European-Directive British activity in EIA. Chapter 5 presents an overview of the EIA systems in the Netherlands, Canada, the Commonwealth of Australia, New Zealand and South Africa.

The next fourteen chapters each follow the same pattern. Chapter 6 deals with the legal basis of EIA systems and Chapter 7 with their coverage of proposals and impacts. The subsequent nine chapters review, in turn, the various steps in the EIA process shown in Figure 1.1. They cover: (Chapter 8) the consideration of the environmental impacts of alternative actions in the design

---

Box 1.3   Principles for undertaking EIA effectiveness evaluation

- Take a systematic approach, placing EA in the overall context of the decision-making process and the forces and factors bearing on practice and performance.
- Specify performance criteria, measures, and indicators for evaluating the overall effectiveness of EA and its operational characteristics.
- Adopt a multiple-perspective approach, canvassing views of participants to gain a full appreciation of process effectiveness.
- Recognise that participant judgements of success are relative and vary with role, affiliation, values and experience.
- As far as possible, corroborate and cross-reference these views with data and information from project files, inspection reports, effects monitoring and environmental auditing.
- Qualify the issues and challenges by comparison to accepted standards of good practice (e.g. complex problem relatively poorly/well handled in the circumstances).
- When drawing conclusions, focus on the 'art of the possible', contrasting what was accomplished with what could be achieved realistically.
- Identify cost-effective improvements that can be implemented immediately, as well as longer-term structural changes that appear necessary (e.g. to law, procedure and methods).

Source: Sadler, 1998, p. 35.

---

process; (Chapter 9) screening; (Chapter 10) scoping; (Chapter 11) the preparation and content of the EIA report; (Chapter 12) reviewing the EIA report; (Chapter 13) the consideration of EIA in decision making; (Chapter 14) monitoring the impacts of projects; (Chapter 15) the mitigation of environmental impacts; (Chapter 16) consultation and participation. There follow chapters on EIA system monitoring (Chapter 17), the benefits and costs of EIA systems (Chapter 18) and, in Chapter 19, the assessment of the environmental impacts of programmes, plans and policies (strategic environmental assessment). In each case a discussion of the relevant aspect of the EIA process is presented. This is followed by a description of how this aspect of the EIA process is dealt with in the United States, UK, the Netherlands, Canada, Commonwealth of Australia, New Zealand and South Africa, and the extent to which the appropriate evaluation criterion is met (fully, partially or not at all). Finally, a comparative summary table is presented.

Chapter 20 draws the main threads of the earlier chapters together by summarising the performance of each of the seven EIA systems against the evaluation criteria, and discussing their shortcomings. Finally, the chapter puts forward a number of suggestions, based upon the comparative review, for improving the various EIA systems and EIA generally.

Chapter 2

EIA in the United States

The United States has a total area of 9.4 million square kilometres and a population of over 280 million people. With a population density of only 29 people per square kilometre (almost one-ninth of the UK's) it is hardly surprising that a frontier ethic developed in which land was seen as a disposable asset and in which controls over land use were regarded as a curtailment of the individual liberty which was one of the principal goals of the original settlers. Partly as a result of this frontier ethic there is an historical distrust of government institutions in the United States and a consequent desire for decision making which is open to inspection and intervention by the public.

The history of environmental concern in the United States is remarkable (Smythe, 1997). That of environmental control is brief but typically vigorous. Prior to 1970 there was little effective federal control over the environment. While federal control over land use remains very weak (and while much state and local control over the use of land is not much stronger), the United States now has an imposing array of detailed and complex controls over air pollution, water pollution, hazardous wastes, etc. (Wood, 1989; Ortolano, 1997). The Environmental Protection Agency (EPA), formed in 1970, has become the largest federal regulatory agency with more that 18,000 employees and an annual budget of nearly $8 billion. There are estimated to be some 20,000 lawyers specialising in environmental matters in the United States (a country with more lawyers per capita than any other). Mandelker (1993) cautioned that the complexity, duplication, jurisdictional fragmentation and expense of US environmental regulation were not viable indefinitely.

In its fifth report, the Council on Environmental Quality (CEQ, 1974, p. 54), reflecting the difficulties of using zoning to control US land-use change effectively, stated:

> There is an increasing recognition that development proposals must be examined on an individual basis under a system of review that has both clearly defined standards and the flexibility to take into account changing community values and the special characteristics of each project.

Environmental impact assessment is perhaps the best-known technique for individual project appraisal. The EIA system was introduced in the United States on 1 January 1970, under the provisions of broad enabling legislation, the National Environmental Policy Act 1969 (NEPA). In retrospect, NEPA can be seen as the first step in an environmental revolution in the United States. One of its authors said at the time that it was 'the most far reaching environmental and conservation measure ever enacted by the Congress' (Jackson, quoted in Fogleman, 1990, p. 1).

It is remarkable that what has frequently, but not completely accurately, been described as a 'standard administrative reform measure' should have received so much attention and been so widely imitated around the world. This chapter describes the evolution of the NEPA provisions and their subsequent refinement through the use of regulations. It provides an overview of the US EIA system at the federal level and comments on the implementation of NEPA in practice.

Evolution of the NEPA provisions

During the 1960s it became apparent to many in the US Congress that pollution and other environmental problems were both complex and interrelated.[1] It was clear to some that a comprehensive approach to the environment was needed, one that was capable of anticipating environmentally disruptive activities and avoiding them, rather than merely reacting to episodes of pollution by passing specific abatement laws. Because of activities like international airport and interstate highway construction, the federal government was perceived to be a major cause of environmental degradation. However, environmental responsibilities were divided and enforcement powers were lacking. An advisory council to coordinate the prevention of environmental degradation, reporting annually, was proposed in a bill introduced by Representative Dingell in 1969. This became the Council on Environmental Quality, one important outcome of NEPA.

A second important element was the national environmental policy introduced by Senator Jackson. This 'motherhood and apple-pie' policy (Box 2.1) can now be seen to have anticipated the world's concern about sustainable development, intergenerational equity, resource usage and the integration of environmental considerations into decision making generally (World Commission on Environment and Development, 1987). The inclusion of the policy was intended to provide guidance in making decisions where environmental values were in conflict with other values. However, the policy by itself, though laudable, was seen to be insufficient if environmental degradation was to be reduced.

An 'action-forcing' mechanism was needed to ensure implementation of the policy. The belatedly introduced and justifiably famous Section 102(2)(C) of NEPA required a detailed statement by federal agencies evaluating the effect of their proposals on the state of the environment (Box 2.1). Caldwell, a consultant to the responsible Senate Committee who is acknowledged as one of the principal architects of NEPA, has stated that:

> The impact statement was required to force the agencies to take the substantive provisions of the Act seriously, and to consider the environmental policy directives of the Congress in the formulation of agency plans and procedures. (Caldwell, 1976, quoted in CEQ, 1990, p. 21)

Box 2.1   The US National Environmental Policy Act 1969: ends and means

Sec. 101. (a) The Congress, recognizing the profound impact of man's activity on the interrelations of all components of the natural environment, particularly the profound influences of population growth, high-density urbanization, industrial expansion, resource exploitation, and new and expanding technological advances and recognizing further the critical importance of restoring and maintaining environmental quality to the overall welfare and development of man, declares that it is the continuing policy of the Federal Government in co-operation with State and local governments, and other concerned public and private organizations, to use all practicable means and measures, including financial and technical assistance, in a manner calculated to foster and promote the general welfare, to create and maintain conditions under which man and nature can exist in productive harmony, and fulfil the social, economic, and other requirements of present and future generations of Americans.

(b) In order to carry out the policy set forth in this Act, it is the continuing responsibility of the Federal Government to use all practicable means, ... to the end that the Nation may:

(1) fulfil the responsibilities of each generation as trustee of the environment for succeeding generations; ...

Sec. 102. The Congress authorizes and directs that, to the fullest extent possible:

(1) the policies, regulations, and public laws of the United States shall be interpreted and administered in accordance with the policies set forth in this Act, and

(2) all agencies of the Federal Government shall ...

(c) include in every recommendation or report on proposals for legislation and other major Federal actions significantly affecting the quality of the human environment, a detailed statement by the responsible official on:

  (i)   The environmental impact of the proposed action,
  (ii)   Any adverse environmental effects which cannot be avoided should the proposal be implemented,
  (iii) Alternatives to the proposed action,
  (iv)   The relationship between local short-term uses of man's environment and the maintenance and enhancement of long-term productivity, and
  (v)   Any irreversible and irretrievable commitments of resources which would be involved in the proposed action should it be implemented.

This detailed statement, the 'environmental impact statement' (EIS), was the third important element in the Act and became the central document in the EIA process.

Interestingly, while most environmental legislation in the United States has become increasingly prescriptive, detailed and complex, NEPA was short, simple, comprehensive and couched in language reminiscent of the US constitution. It was, perhaps, the last of the New Deal legislation. NEPA remains substantially as drafted, 30 years later. However, Senator Muskie, who guided the Clean Air Act through Congress, insisted on far more detailed directives and left less scope for agency discretion. He forced the newly created Environmental Protection Agency (EPA) to become the environmental evaluator of all other agencies' actions by requiring it to review and comment on the impact of those projects for which EISs were prepared (Clean Air Act 1970, section 309). The Office of Federal Activities within EPA undertakes this function. It therefore provides important support to CEQ's lead role in overseeing the operation of NEPA (Environmental Law Institute, 1995a).

Operationalising the Act was no simple matter, however. While CEQ, located in the Executive Office of the President, issued guidelines dealing with the preparation of EISs, federal agencies reacted to the new requirements in ways varying from avoidance to amateurism. Although Congress had not anticipated that the courts would have a major role in implementing NEPA, a celebrated series of cases clarified the substantive nature of the EIS requirement. Most importantly, it was determined that the 'action-forcing' procedural provisions were not ends in themselves but were designed to ensure that the environmental policy, from which the Act takes its name, was implemented. However, it was believed that application of the essentially procedural requirements of NEPA was almost certain to affect the decisions made by agencies and thus achieve the Act's aims.

One effect of this litigation (apart from making foreign observers quake at the prospect of importing it) was to lead to the writing of voluminous documents designed to resist legal challenge rather than to meet the policy objectives of NEPA. The EIS for the Trans-Alaska Pipeline was reputed to be more than two metres thick. There were two separate sets of problems in relation to EISs:

1. The usefulness of EISs reviewed was being impaired by several common failings: inadequate discussion of the identified environmental impacts, inadequate treatment of the reviewing agencies' comments on environmental impacts, and inadequate consideration of alternatives and their environmental impacts.
2. Too many statements were boring, voluminous and obscure and lacked the necessary analysis and synthesis. They were often inordinately long, with too much space devoted to unnecessary description rather than to analysis of impacts and alternatives.

As a result of the confusion caused by the various court decisions, President Carter instructed CEQ to prepare regulations to make the EIA process more relevant and to reduce the length of EISs.

The CEQ legal team consulted widely and issued several drafts during the process of framing the regulations which were eventually put forward in 1978. They were given effect by an executive order (CEQ, 1978) and have remained almost unchanged since they were drafted. As Yost (1990) has noted, there are two important differences between the CEQ Regulations and the previous guidelines. First, they are mandatory requirements binding on all agencies. Second, the Regulations cover the whole of the EIA process, whereas the guidelines dealt only with EISs.

The Regulations standardised basic NEPA compliance practice throughout the federal government. Agencies responsible for the preparation of EISs adapted, supplemented and revised the CEQ Regulations to meet their own needs. Twenty years on, CEQ was entitled to be self-congratulatory:

> Through mechanisms such as scoping, classification of actions, incorporation by reference, tiering, and other procedures, the Council's regulations have been instrumental in keeping the NEPA process focused and useful to both decision-makers and the public alike, while providing agencies substantial discretion to adapt those procedures to their programs. (CEQ, 1990, p. 26)

CEQ was subjected to savage staff cuts by the Reagan Administration in the early 1980s (Vig and Kraft, 1984). However, despite the hostility of the Administration, the breadth of support (including that of the US Chamber of Commerce) for NEPA and the CEQ Regulations ensured that both emerged unscathed. CEQ was then restaffed under President Bush only, ironically, to be threatened anew by the environmentally supportive Clinton regime. CEQ survived, but with a much reduced budget, which was maintained by the G.W. Bush Administration. Figure 2.1 shows the fluctuations in CEQ staff levels over the years. That NEPA and CEQ have endured is testament to the widespread perception of NEPA as a standard-bearer of US environmental protection policy. As Train, a past chairman of CEQ who was instrumental in its implementation, has stated:

> I can think of no other initiative in our history that had such a broad outreach, that cut across so many functions of government, and that had such a fundamental impact on the way government does business. . . . I believe I had a unique familiarity with the whole EIS process from inception to implementation and am qualified to characterise that process as truly a revolution in government policy and decision-making. (Train, quoted in Bartlett, 1989, p. 2)

The US federal EIA system

The main steps in the US federal EIA system are shown in Figure 2.2.[2] A 'lead' federal agency is designated to implement the various steps in the EIA process. This agency is usually involved in actually constructing a project, in funding it, in granting a permit for it, or in proposing a programme, plan or legislation. However, it relies heavily on the developer (if funding or permit-granting activities are involved) for information and upon other agencies and the public for comment.

Environmental impact assessment

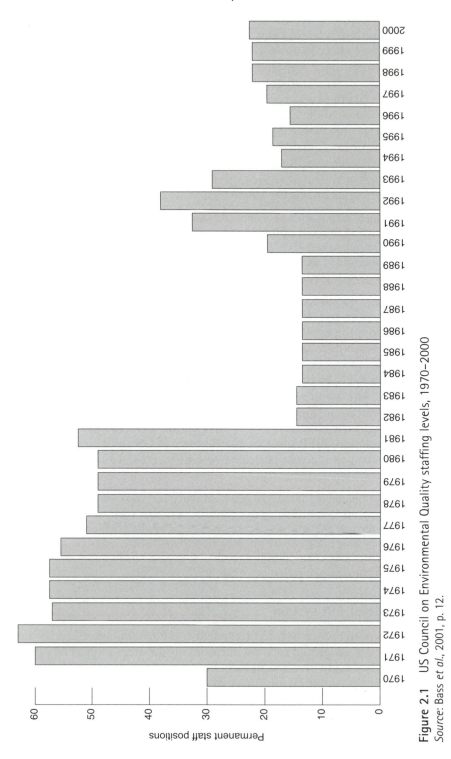

Figure 2.1 US Council on Environmental Quality staffing levels, 1970–2000
Source: Bass *et al.*, 2001, p. 12.

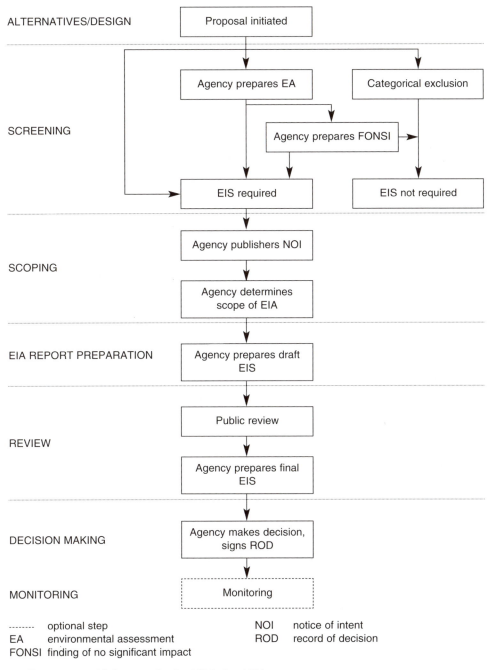

ALTERNATIVES/DESIGN

SCREENING

SCOPING

EIA REPORT PREPARATION

REVIEW

DECISION MAKING

MONITORING

-------- optional step
EA   environmental assessment
FONSI   finding of no significant impact

NOI   notice of intent
ROD   record of decision

Figure 2.2   Main steps in the US federal EIA process

The first step in the EIA process is the identification of the proposal leading to the action by the agency (i.e. construction of, funding, or permit-granting for a project; or proposed programmes, plans or regulations). The agency will then undertake a preliminary environmental analysis to determine whether there is a need for an environmental impact statement (preparation of which can commence forthwith), whether the environmental impacts are clearly so insignificant as to permit a categorical exclusion from the EIA process (for which documentation is optional) or whether an environmental assessment (EA) should be prepared so that the significance of impacts can be more clearly identified.[3] In effect, this EA, again prepared by the agency, is an abbreviated EIS as it covers many of the topics required in an EIS. However, it is not subject to the same rigorous consultation and participation provisions. Depending on the findings of the EA, an EIS may be required or, as in almost every case (Chapter 9), the agency may decide that none is necessary. In this case, a finding of no significant impact (FONSI) must be written, summarising the reasons for this decision.

When an EIS is required, a notice of intent (NOI) has to be published by the agency in the Federal Register and scoping commences. Scoping is a procedure intended to bring those with different interests in the proposal (including members of the public) to an agreement about which of the environmental impacts associated with it are significant and thus require investigation. Agency regulations may require analysis of some issues but other significant matters are agreed by consultation, frequently at a meeting (or a series of meetings) at which the various interested parties are represented. These issues are then addressed in the draft EIS. This is written by, or on behalf of, the agency, though the developer provides a great deal of the relevant information upon which it is based if funding or permitting is involved. In practice, most EISs are drafted by environmental consultancies.

The draft EIS normally follows a set pattern dictated by the relevant agency's regulations or guidelines: it will describe the existing environment, explain what the proposed project is and analyse the effects of the project on the environment. It is these effects which constitute the substance of the draft EIS, which should supposedly not normally be more than 150 pages in length. They are generally discussed at some length and mitigation measures are usually proposed. In accordance with the requirements of NEPA (Box 2.1), it is usual to provide: a summary of probable adverse environmental effects which cannot be avoided; a discussion of alternatives to the action; a discussion of the relationship between local, short-term uses of the environment and maintenance and enhancement of long-term productivity; and a discussion of irreversible and irretrievable commitments of resources. Most agencies follow the tighter structure specified in their own regulations in organising this material in their EISs.

The draft EIS is sent to EPA for critical review and filing and is forwarded to all the relevant federal, state, tribal and local organisations likely to wish to comment. This review process involves reading the draft EIS and commenting both on the way the reviewing agency's interests are affected and on the content of the EIS generally (though this latter type of comment is less common than the former). There are arrangements for local groups and for the public to participate and there has to be a minimum period for deposit of the documents of 45 days to allow this participation to take place. Once the lead agency has received the comments of the various consulted agencies and bodies it is in a position to prepare the final EIS.

The final EIS describes the modified form of the proposed action, including any changes that have been made since the draft EIS was published, and responds to the comments received from the various bodies consulted. This document usually contains quite extensive proposals for mitigation of impacts. A record of decision must also be prepared, indicating the decision that has been made and the reasons for it. This is sometimes circulated for a period of time and agencies with an administrative appeals process can adopt a procedure whereby they release the final EIS and the record of decision simultaneously. Generally, however, the record of decision is issued after a 30-day waiting period following the filing of the final EIS with EPA.

There are somewhat inadequate provisions for monitoring the environmental impacts arising from an action and for ensuring that the various conditions or mitigation measures that have been included in the final EIS are implemented. This may be done in the form of conditions appended to permits that have to be obtained from the lead agency or in the form of conditions attached to grants that are made by the agency. If the agency itself is carrying through the measures, there is usually a system of inspection to ensure that the project is constructed as described in the final EIS (unless, of course, there are overwhelming and unforeseen reasons for change, in which case a supplementary EIS may have to be prepared).

There are provisions for 'tiering' EISs: the preparation of broad-programme EISs followed by site-specific EISs cross-referenced to the overall document. The use of these is increasing but still limited. There are also provisions for mediation by CEQ if EPA and other agencies such as the Department of the Interior are unable to agree that the impacts of the action are acceptable.

Implementation of NEPA

There is no doubt that the EIA process is biting.[4] Most federal agencies have updated their regulations to ensure that all the requirements of NEPA, as they have become clearer, are met. NEPA nowhere provides for the termination of a major federal action because of its environmental consequences, but actions in the courts have stalled or stopped such projects if their consequences have not been properly documented. The buying-out of mineral rights at the New World Mine, Montana (Box 2.2), provides a celebrated example.

Box 2.2   The New World Mine EIA, Montana

The EIA of a controversial proposal by Crown Butte Mines to exploit their mineral rights by developing a gold, copper and silver mining complex less than three miles from the north-east border of Yellowstone National Park began in 1993.

A preliminary draft of the EIS showed that there could be major adverse impacts on a federally designated wild and scenic river, on grizzly bear habitat, and on Yellowstone National Park. Interagency review showed a need, *inter alia*, for additional studies on groundwater conditions at the mine site and for a risk assessment of the proposed tailings impoundment.

It was also abundantly clear that there would be years of contentious litigation over the mine, regardless of whether the federal government approved or denied the company's application. Yellowstone was the first national park in the world and the proposal accordingly attracted opponents who were able to wield great national influence. In the face of this apparent stalemate, environmental groups and the company began discussing creative ways to resolve the conflicts.

Following a number of influential interventions and innovative negotiations, President Clinton (in person) and the parties announced an agreement in 1996. The essential details were that Crown Butte would agree to drop plans to develop the site, that the federal government would agree to transfer $65 million in federal assets in exchange for title to all the lands essential to development of the mine, that the company would place $22.5 million in a trust fund to remediate historical environmental contamination in the area; and that the parties would agree to settle the existing litigation by the environmental groups and potential environmental claims by the federal government.

Source: adapted from CEQ, 1998, pp. 99–101.

Because no special tribunal or board was established to enforce NEPA, litigation has emerged as an important means of public participation in the EIA process (Holland, 1996). There has been substantial EIA litigation, predominantly by environmental or citizen groups. The volume of litigation declined from that during the early years as issues were clarified, generally running at less than 100 lawsuits a year from the mid-1980s to the mid-1990s. The Supreme Court has always ruled that NEPA is essentially procedural (Holland, 1996; Mandelker, 2000). The most common causes of legal action have been the absence or inadequacy of EISs or EAs. Of 102 cases in 1997, two resulted in injunctions (CEQ, 1999, p. 355).[5]

Several projects have been aborted as a result of the adverse impacts revealed in preparing an EIS and it appears that a majority of projects is modified as a result of the assessed impacts. This mitigation of impacts appears to be 'where the action is' and is widely cited as one of the main justifications of the process. To a large extent, EIA has been assimilated into federal decision-making processes and is meeting many (but not all) of the objectives of its proponents. It has been argued that the role of NEPA has declined and become largely symbolic as more specific environmental legislation has been enacted, shifting the focus of project impact evaluation to, for example, the air pollution control process. It seems more likely that NEPA's current lack of notoriety may well be a measure of its success in internalising the consideration of environmental quality in federal agencies.[6] However, this process is not complete, as CEQ (1997d, p. iii) has admitted:

> agencies may sometimes confuse the purpose of NEPA. Some act as if the detailed statement called for in the statute is an end in itself, rather than a tool to enhance and improve decision-making.

It is normal for the EIS to address the procedural requirements of NEPA, as refined in the agency guidelines, and to rely on scoping for the identification of issues, rather than to use any 'comprehensive EIA methodology'. Widespread use is, however, made of specialised technical methods for assessing particular impacts (for example, air pollution modelling: Canter, 1996). The trend is to make greater use of the information generated for other purposes (for example, the granting of an air pollution permit) in preparing the EIS and to combine the granting of permits to reduce the number of hurdles that applicants must negotiate in realising their proposals.

The number of EISs has fallen (Figure 2.3).[7] Well over 1,000 draft, final and other EISs were produced each year in the first decade, but the number has dropped steadily since to around 500 per annum (CEQ, 1997c). On the other hand, around 50,000 EAs are being produced each year (Blaug, 1993). The emphasis in EIA has moved away from EISs toward mitigated FONSIs in which negotiation takes place very early in the process, on the basis of the EA, and no EIS is produced.

Despite the generally accepted improvement in the quality of EISs over recent years, there is scope for further amelioration in their analytical content and for closer adherence to the spirit of NEPA rather than its letter, which has tended to be overemphasised as a result of litigation (Bear, 1989; Caldwell, 1998). There have been several unsuccessful attempts by Congress to strengthen NEPA's monitoring provisions.

EIA has been mainly (but not exclusively) confined to projects and probably owes much of its success to the general weakness of the US land-use planning system, especially at the federal level. Another factor in NEPA's success, which has been empirically demonstrated by Caldwell *et al.* (1983), is that it was directed at government agencies, particularly those responsible for the undertaking of development activities of potential environmental significance, rather than at private developers (Fowler, 1982; von Moltke, 1984). To implement NEPA, these agencies recruited and developed interdisciplinary environmental review staffs which grew increasingly influential over the years (Environmental Law Institute, 1981; Taylor, 1984). More recently, however, senior staff numbers have declined, leading to a loss of corporate EIA memory.

Fifteen states, Puerto Rico and the District of Columbia have enacted their own EIA legislation (Welles, 1997). Table 2.1 (Mandelker, 2000; Bass *et al.*, 2001) summarises these. Some states require EISs only for projects proposed

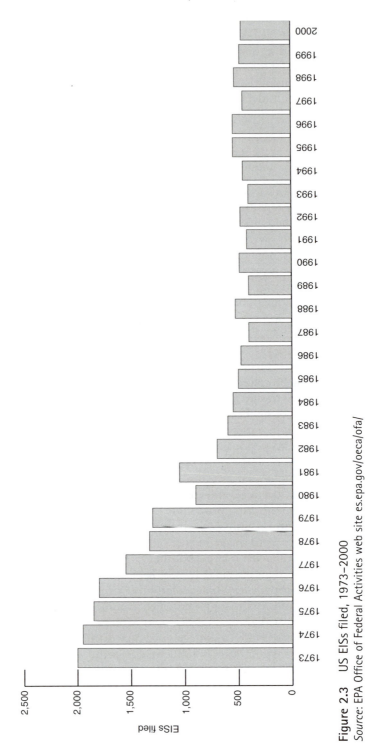

Figure 2.3   US EISs filed, 1973–2000

Source: EPA Office of Federal Activities web site es.epa.gov/oeca/ofa/

Table 2.1 American state EIA legislation

State	Legal reference	Contact office	Comments
California	California Environmental Quality Act (Cal. Pub. Res. 21000–21177)	State Clearinghouse Office of Planning and Research	Requires environmental impact report similar to federal statement and including mitigation measures and growth-inducing effects. Applies to state agencies and local governments. Detailed provisions governing preparation of impact report and judicial review. State agency to prepare guidelines. Statutory terms defined
Connecticut	(Conn. Gen. Stat. 22a–1 to 22a–1h)	Environmental Protection Department	State agencies to prepare environmental impact evaluations similar to federal impact statement and including mitigation measures and social and economic effects. Actions affecting environment defined
District of Columbia	District of Columbia Environmental Policy Act (D.C. Code Ann. 6–981 to 6–990)	Environmental Regulation Administration	Mayor, district agencies and officials to prepare impact statements on projects or activities undertaken or permitted by district. Impact statement to include mitigation and cumulative impact discussion. Action to be disapproved unless mitigation measures proposed or reasonable alternative substituted to avoid danger
Georgia	Georgia Environmental Policy Act (Ga. Code Ann. 12–16 to 12–16–8)	Environmental Protection Division	Applies to projects proposed by state agencies for which it is possible to expect significant effect on the natural environment. Limited primarily to land-distributing activities and sale of state land. Decision on project not to create cause of action
Hawaii	Hawaii Environmental Impact Statement Law (Hawaii Rev. Stat. 343–1 to 343–8)	Land and Natural Resources Department, Conservation and Resource Enhancement Division	State agencies and local governments to prepare impact statements on use of public land or funds and land uses in designated areas. Statements must be 'accepted' by appropriate official. Judicial review procedures specified
Indiana	(Ind. Code Ann. 13–1–10–1 to 13–1–10–8)	Department of Environmental Management	Similar to NEPA. Applies to state agencies

State	Legal reference	Contact office	Comments
Maryland	Maryland Environmental Policy Act (Md. Nat. Res. Code Ann. 1–301 to 1–305)		State agencies to prepare environmental effects reports covering environmental effects of proposed appropriations and legislation, including mitigation measures and alternatives
Massachusetts	Massachusetts Environmental Policy Act (Mass. Gen. Laws Ann. Ch. 30, ss 61, 62–62H)	Environmental Affairs Executive Office, Environmental Impact Review Section	State agencies and local authorities to prepare environmental impact reports covering environmental effects of actions, mitigation measures and alternatives
Minnesota	Minnesota Environmental Policy Act (Minn. Stat. Ann. 116D.01 to 116D.06)	Minnesota Environmental Quality Board	State agencies and local governments to prepare environmental impact statements covering environmental effects of actions, mitigation measures and economic, employment and sociological effects. Procedures for preparation of statements and for judicial review specified. State environmental quality board may reverse or modify state actions inconsistent with policy or standards of statute
Montana	Montana Environmental Policy Act (Mont. Code Ann. 75–1–101 to 75–1–105; 75–1–201 to 75–1–207)	Legislative Environmental Analyst	Similar to NEPA. Applies to state agencies
New York	State Environmental Quality Review Act (NY Envtl. Conserv. Law 8–0101 to 8–0117)	Environmental Conservation Department	State agencies and local governments to prepare impact statements similar to federal impact statement and including mitigation measures and growth-inducing and energy impacts. Statutory terms defined. Procedures for preparing statement specified. State agency to adopt regulations on designated topics
North Carolina	North Carolina Environmental Policy Act (NC Gen. Stat. 113A–1 to 113A–13)	Department of Environment, Health and Natural Resources	Similar to NEPA. Applies to state agencies. Local governments may also require special-purpose governments and private developers of major development projects to submit impact statement on major developments. Certain permits and public facility lines exempted

State	Legal reference	Contact office	Comments
Puerto Rico	(P.R. Laws Ann. tit. 12, ss 1121–1127)		Similar to NEPA. Applies to Commonwealth agencies and political subdivisions
South Dakota	(S.D. Codified Laws Ann. 34A–9–1 to 34A–9–13)	Environment and Natural Resources Department	State agenices 'may' prepare environmental impact statements similar to federal impact statement and adding mitigation measures and growth-inducing 'aspects'. Statutory terms defined. Ministerial and environmental regulatory measures exempt
Virginia	(Va. Code 3.1–18.8, 10.1–1200 to 10.1–1212)	Environmental Quality Department, Enforcement and Policy Division	Similar to NEPA. Applies to state agencies for major state projects. Impact statements also to consider mitigation measures and impact on farmlands
Washington	State Environmental Policy Act (Wash. Rev. Code 43.21C.010 to 43.21C.910)	Department of Ecology, Environmental Review Section	State agencies and local governments to prepare impact statements identical to federal impact statement but limited to 'natural' and 'built' environment. Proposal may be denied if it has significant impacts or mitigation measures insufficient. Judicial review procedures specified. State agency to adopt regulations on designated topics
Wisconsin	Wisconsin Environmental Policy Act (Wis. Stat. Ann. 1.11)	Department of Natural Resources, Environmental Analysis and Liaison	Similar to NEPA. Applies to state agencies. Statements also to consider beneficial aspects and economic advantages and disadvantages of proposals

Sources: adapted from NEPA Law and Litigation, © West Group, (Mandelker, 2000), pp. 12.4–12.7; Bass *et al.*, 2001, p.136.

within specific areas, others demand them for actions undertaken by state agencies or using state funds, and yet others require them for these categories together with actions which need state permits. The requirements of other states apply to all these types of actions plus a number of actions taken by local agencies. Three states – California, New York and Washington – have a comprehensive system covering local government and private activities as well as those of the state itself (Hart and Enk, 1980; Pendall, 1998). The various state legal requirements differ from the federal system and most have proved to be less effective, the comprehensive systems being among the exceptions. Some counties and cities in other states have also introduced their own EIA requirements.

To summarise, the essential elements of NEPA and of the US EIA process which have emerged over the years are:

1. Statement of national environmental policy that anticipated the concept of sustainability.
2. Central body responsible for EIA policy, system monitoring and annual reporting.
3. Formal set of procedural EIA steps which includes the consideration of alternatives, screening, an environmental assessment, scoping, a draft EIS and a final EIS (but not impact monitoring) for most environmentally significant actions.
4. Enforceability through the courts.
5. Requirement of participation and consultation at several stages in the EIA process.
6. Completion of the EIA process prior to the decision on the action being taken.
7. Application to actions other than projects.

Despite the impression that NEPA may no longer be as honoured in its own land as was once the case, it was undoubtedly prophetic. NEPA has been imitated more than any other US legislation. More than 100 countries and international organisations have introduced EIA systems (see Chapter 1; Donnelly *et al.*, 1998). As Clark (1997, p. 22) has stated:

> NEPA has generated an international, interdisciplinary profession dedicated to EIA, which now includes professional organisations, peer-reviewed journals, academic programmes and other training courses, and a broad range of statutes and other formal requirements for EIA at the international, national, state and local levels of government.

Notes

1. This description owes much to the account of 20 years of NEPA (CEQ, 1990, pp. 15–51); see also Blumm, 1990; CEQ, 1997c; Caldwell, 1998).
2. This description is derived from the text of the relevant regulations first promulgated in 1978 and now codified in the Code of Federal Regulations at *40 Code of Federal Regulations 1500–1508* Regulations for Implementing the Procedural Provisions of the National

Environmental Policy Act. The regulations are reproduced on CEQ's NEPA web site (ceq.eh.doe.gov/nepa/nepanet.htm) together with the text of NEPA and the last CEQ annual reports. For thorough descriptions of federal EIA procedures (which also reproduce both NEPA and the regulations), see Fogleman (1990), Bass *et al.* (2001), Kreske (1996), Mandelker (2000) and Eccleston (1999). See also Environmental Law Institute (1995a), Canter (1996), Marriott (1997), Ortolano (1997) and, for a UK perspective, Glasson *et al.* (1999) and Sheate (1996).

3. Most agencies (which are frequently organised by regions or divisions) have promulgated rules or guidelines to help identify which projects can be categorically excluded or should move straight to an EIS.

4. This account is partially derived from Wood, 1989, pp. 63–6.

5. More recent statistics are not available because *Environmental Quality* (CEQ's annual report) is no longer published.

6. Various major gatherings have been held and publications launched to mark NEPA's maturity and to suggest improvements. See *Environmental Law* **20**: (1990) 447–810 (summarised by Blumm, 1990), *The Environmental Professional* **15** (1993), Hildebrand and Cannon (1993) and Clark and Canter (1997).

7. Numbers of EISs are presented on the EPA Office of Federal Activities web site: es.epa.gov/oeca/ofa/

Chapter 3

The European Directive on EIA

The Treaty of Rome, dating from 1957, contained no reference to environmental policy but this oversight was addressed in the Paris Declaration on the Environment in 1972 (the year the UK, Ireland and Denmark joined the original six member states of the European Union (EU). The Commission of the European Communities (CEC) published its first *Action Programme on the Environment* in 1973, justifying this on harmonisation and competition grounds. From that time, a trickle of environmental legislation has become a stream, covering water and air pollution, waste disposal, control of chemicals, noise, wildlife protection, recycling and packaging, and environmental impact assessment (Bell and McGillivray, 2000). EU legislation is directly applicable in national courts (regulations) or binding on government as to the ends to be achieved (directives) without the need for ratification. Furthermore, the Commission has a duty to enforce EU legislation, eventually bringing matters to the European Court of Justice if necessary (Haigh, 1991; Kramer, 2000).

Partly as a result of proportional representation, the environment has been higher on the political agenda of many influential European countries than it has in the UK. This concern is driving EC policy, so that the stream of environmental legislation is swelling to become a river. This legislation is impinging more and more on British practice, as is its enforcement in the European Court of Justice (Kramer, 2000). For example, decisions have been brought against the UK by the Court for failure to implement European legislation on time (Bell and McGillivray, 2000).

The Single European Act 1986 introduced a clear environmental policy in the form of articles which established principles of environmental protection. This policy developed further with the Treaties of Maastricht (1992) and Amsterdam (1997), and now includes the prevention principle by which pollution nuisances are best avoided at source rather than their effects being subsequently counteracted (Kramer, 2000). The Directive on EIA exemplifies the way in which European influence has led to an increase in planning (and other) controls over the environment, affecting large numbers of people, by applying this principle, first articulated in 1977 (CEC, 1977).

The original EIA Directive represented the first EU intrusion into the planning domain, and had major repercussions on member state decision making and practice. This was undoubtedly the reason it took so long to progress from the Commission's original proposal to adoption. However, this environmental directive (as amended) is, though far-reaching, but one of the many which are being promulgated with the changing nature of the EU. Some of these directives are likely to prove just as significant in altering member state regulation as that on EIA. The implementation of the Single European Act, which provided that 'environmental protection requirements shall be a component of the Community's other policies' (Article 174(2)), is unlikely to reduce the flow of the river of environmental regulation. Rather, it is likely to raise the importance of the environment in many fields (Kramer, 2000).

This chapter describes the evolution of the original European EIA Directive and of the subsequent amending Directive. In particular, the changes which took place between the published draft directives and the versions eventually adopted are discussed. The chapter then presents an overview of the provisions of the amended Directive and briefly mentions its implementation by the 15 member states of the EU.

Evolution of the Directive

The Commission has stated that 'too much economic activity has taken place in the wrong place, using environmentally unsuitable technologies' (CEC, 1979, p. 49), and that 'effects on the environment should be taken into account at the earliest possible stage in all the technical planning and decision-making processes' (CEC, 1977). Because of this concern to anticipate environmental problems, and hence to prevent or mitigate them, the Commission became interested in EIA in the early 1970s, like many other bodies.

The Commission instigated research investigations on EIA in 1975. Lee and Wood (1976, 1978b) reported that many aspects of EIA procedure already existed within member states. As a consequence, it was felt that the requirements of a European EIA system could be integrated into member state decision-making processes without the disruption or the litigation which characterised early American experience. It was also suggested that project EIA should be the first stage of a European EIA system which would eventually encompass policies and plans, once more than rudimentary experience of project assessment had been gained.

Following this research programme, the Commission decided that an EIA system should meet two objectives:

1. To ensure that distortion of competition and misallocation of resources within the EU was avoided by harmonising controls.
2. To ensure that a common environmental policy was applied throughout the EU.

The Commission issued its first preliminary draft directive in 1977. After some 20 drafts, not all of which were released, and substantial consultation (this is reliably reported to have been the most discussed European draft directive to date (Wathern, 1988, 1989; Sheate and Macrory, 1989; Sheate, 1996)), the Commission put forward a draft to the Council of Ministers in June 1980 (CEC, 1980).

The draft directive specified that projects likely to have a significant effect on the environment were to be subject to EIA. Such an assessment was obligatory for some 35 types of project listed in Annex 1. EIA was also required for certain projects in other specified categories listed in Annex 2 subject to criteria and thresholds to be established by member states. There were provisions for Commission coordination of criteria and thresholds, for a simplified form of assessment in certain cases, and for the impacts upon the environment within another member state affected by the proposal to be assessed. The information to be supplied by the developer was to include a justification of the rejection of reasonable alternatives to the proposed project where these were expected to have less significant adverse effects on the environment.

Annex 3 specified the required content of the assessment in detail. The impacts to be considered included those arising from the physical presence of the project, the resources it used, the wastes it created, and its likely accident record. The competent authority had to release its assessment, a summary of the main comments received, the reasons for granting or refusing permission and the conditions, if any, to be attached to the granting of the permission. In the event of the project being authorised, the competent authority was expected to check periodically whether any conditions attached to the approval were being satisfied, and whether the project was having any unexpected environmental effects that might necessitate further measures to protect the environment.

The Council of Ministers did not approve the draft directive in June 1980. The British government, for reasons elaborated in Chapter 4, was reluctant to accept the imposition of a mandatory system of EIA, at least in the form set out in the published draft directive. Numerous representations were made to the Commission as a result of consultations and a number of alterations were made to the draft directive. These were mostly minor in nature and, on balance, strengthened rather than weakened the EIA provisions (CEC, 1982).

During subsequent negotiations between the Commission and the British government, several of the more controversial aspects of the draft directive, including the provision of reasons for granting consent, were deleted to meet the British position. In the course of the negotiations many types of industry were shifted from Annex 1, where their assessment would have been compulsory, to Annex 2, where their assessment was to be much more discretionary (Sheate and Macrory, 1989; Wathern, 1989; Sheate, 1996). The British government withdrew its objections to the amended draft directive late in 1983, only for the Danish government to continue to express serious reservations about the undermining of the sovereign power of the Danish parliament to approve development projects. A provision exempting projects approved by specific acts of national legislation (Wathern, 1988) paved the way to adoption of a much modified version of the draft directive in June 1985.

The adopted version of the directive (CEC, 1985)[1] limited the Annex I projects to oil refineries, large coal gasification and liquefaction plants, large power stations, radioactive waste disposal sites, integrated steelworks, asbestos plants, integrated chemical plants, motorways, railways and large airports,

ports, canals, and toxic waste disposal facilities. The list of Annex II projects grew substantially but the requirement for Commission coordination of criteria and thresholds was dropped.

The main text of the Directive did not mention the discussion of alternatives or the consideration of impacts upon neighbouring member states. The requirements for publication of the authority's own assessment, and of its synthesis of public comments, were also substantially weakened. The monitoring provision was deleted. There was no mention of the EIA of plans or programmes in the adopted Directive. Annex III, specifying the desirable (rather than the required) content of the information supplied by the developer, reflected the reduced sweep of the requirements.

The net effect of these changes was the emasculation of the provisions in the earlier drafts of the Directive. These early versions were themselves criticised as being over-cautious and for not containing provisions for the Commission to monitor and oversee the EIA system effectively, let alone use it to make substantive and constructive inputs to problem solving (Wandesford-Smith, 1979). As Brouwer (quoted in Wathern, 1988, p. 201) stated, from a Dutch perspective:

> This EC-directive, like so many others, is a very weak compromise. It is more the result of the cumulative resistance from the development promoters and bureaucracies in the member countries than a synthesis of the best ideas for the protection of the environment.

The Directive in its adopted form provided a flexible framework of basic EIA principles to be implemented in each member state through national legislation. While many of the original provisions were removed, there was nothing to prevent member states from instituting EIA systems which were more comprehensive and rigorous than the provisions put forward by the Commission. Several countries, including the Netherlands, considerably exceeded the requirements of the original directive in their national EIA systems (Coenen, 1993). As the Council on Environmental Quality (CEQ, 1990, p. 46) stated, the European Directive represented a first step 'towards establishing effective, efficient EIA processes to help reconcile economic growth and development with maintenance and enhancement of environmental quality'.

In many ways, the process leading to the notification of the amended EIA Directive mirrored that leading to the original Directive, which contained a provision for 'a report on its application and effectiveness' to be prepared five years after notification (Article 11(3)). This five-year review, which was completed in 1993 (CEC, 1993) found that much of the transposition necessary to enshrine the provisions of the Directive into member state legislation was still incomplete. The review revealed weaknesses in the coverage of certain projects, in the consideration of alternatives, in screening, in scoping, in consultation and participation, and in monitoring.

As a result of these findings, the Commission proposed the restoration of many of the elements which had been included in drafts of the original Directive but which were whittled away during negotiations. Following lengthy negotiations between the Commission and the member states, a draft of the new directive was published in 1994 (CEC, 1994). This included an increase in the number of Annex I projects, a new set of screening criteria, provisions relating to the consideration of alternatives and of transboundary impacts and to the publication of the reasons for the competent authority's decision. In addition, provision for scoping involving the developer, the competent authority and relevant environmental authorities (but not the public) was included. There was no provision for impact monitoring, though this had been included in a preliminary version of the new directive (Sheate, 1996, 1997; Glasson *et al.*, 1999).

Following a series of drafts and considerable further debate between the Commission and the member states, and some weakening of the provisions, the amended Directive was notified in March 1997 (European Commission, 1997d). The main casualty of the negotiations was scoping, which became discretionary, at the developer's request. The member states were to implement the changes introduced by the amended directive (shown in Table 3.1) by 1999. A further modification to the Directive was proposed in 2001 to strengthen its public participation provisions to accord with the Aarhus Convention on access to information by specifying requirements in greater detail (European Commission, 2001e). Table 3.2 shows how one aspect of the Directive, the selection of projects for assessment, evolved between 1976 and 1997.

Table 3.1   Main changes brought about by the amended European Directive on EIA

Screening
- Greater number of Annex I projects (incorporating former Annex II projects): now 21 categories.
- New Annex II projects: 12 categories.
- New Annex III selection criteria for use in Annex II screening decisions.
- Screening decisions made in relation to Annex II projects must be made public.

Scoping
- A scoping opinion (detailing the information to be supplied) may be obtained by the developer, at his/her request, from the competent authority.

Alternatives
- An outline must be given of the main alternatives to the proposal studied, and the main reasons for the choice made must be indicated.

Transboundary effects
- Consultation and public participation must take place with any other member state likely to be affected by a proposed project.

Mitigation
- Main mitigation measures proposed must be made public.

Decision making
- Main reasons for the decision made on the proposal must be made public.

Table 3.2 Selection of projects for assessment in the amended European Directive and its progenitors

1976	1977	1979	1979	1980	1982	1985	1994	1997
ENV/197/76	EIE/OU/10	EIE/OU/14	EIE/OU/18	COM (80) 313	COM (82) 158	85/337/EEC	COM (93) 575	97/11/EC
No list system. Projects subject to EIS determined by 'applicability' guidelines	List of projects subject to mandatory assessment. Criteria for selection of other projects	List of projects subject to mandatory assessment. Criteria for selection of other projects	List of projects subject to mandatory assessment (Annex 1). List of projects (and modifications to Annex 1 projects) subject to assessment when so required according to criteria set by competent authority (Annex 2). Provision for simplified form of assessment. Screening criteria for selection of other projects	List of projects subject to mandatory assessment (Annex 1). Provision for exemption and simplified assessment where appropriate. List of projects (and modifications to Annex 1 projects) subject to assessment when so required (Annex 2). Competent authority(ies) establish criteria and thresholds. Provisions for determining other projects	As previous draft except more detailed provision for exemption	As before except detailed exemption clause (paragraph). Commission to report annually to Council on the application of the paragraph. Provisions for simplified assessment and determining other projects deleted. Transfer of some projects from mandatory Annex I to more discretionary Annex II	As before but member state criteria and thresholds must accord with new Annex IIa selection criteria. Clarification of definitions of some mandatory Annex I projects; clarification of, and additions to, Annex II projects	As before but Annex IIa becomes Annex III. Transfer to some projects from Annex II to mandatory Annex I. Minor changes to Annex II project descriptions

Source: data for 1976 to 1982 and part of 1985 based on Sheate and Macrory, 1989.

The European Directive EIA system

The *legal basis* of the EIA system, a European directive, is clear. It is left to member states to implement the requirements of the EIA Directive in whatever legislation they consider to be appropriate. The Directive, as mentioned above, provides a skeletal framework and leaves a great deal of detail to be determined by member states (Coenen, 1993; Lee, 1995).

The Directive, as amended, consists of 14 articles and 4 annexes. The main steps in the EIA process are shown in Figure 3.1. The Directive places a general obligation on each member state 'to ensure that, before consent is given, projects likely to have significant effects on the environment by virtue, inter alia, of their nature, size or location, are made subject to ... an assessment' (Article 2(1)). This assessment may be integrated into existing project consent procedures or into other procedures (Article 2(2)). Acts of national legislation (Article 1(5)) are excluded and specific projects may be exempted in exceptional cases, after making relevant information available to the public and to the Commission (Article 2(3)).

The *coverage* of the Directive is confined to projects. The Directive, as amended, applies to a considerably longer list of projects than did the original Directive, and now includes a number of environmentally sensitive projects previously excluded from coverage (for example water treatment plants).

The word 'environment' is used to mean the physical environment. The social and economic environments are not overtly included in this definition, as they are in many other jurisdictions, for example in the United States (Mandelker, 2000). Article 3 of the Directive requires that:

> The environmental impact assessment shall identify, describe and assess, in an appropriate manner ... the direct and indirect effects of a project on ...:
> – human beings, fauna and flora;
> – soil, water, air, climate and the landscape;
> – material assets and the cultural heritage;
> – the interaction between the[se] factors ....

Other types of effect (below) are consigned to Annex IV where their use is largely discretionary.

The amended Directive introduces a requirement for any *alternatives* to the proposal studied to be documented. Thus, the information to be provided by the developer must include:

> an outline of the main alternatives studied by the developer and an indication of the main reasons for his choice, taking into account the environmental effects. (Article 5(3))

The wording of this requirement is identical to that in the original Directive but, by including it in an article, rather than in an annex, it has now become mandatory rather than optional. However, this requirement still falls short of demanding the study of alternatives.

The European Directive requirements regarding *screening* (i.e. choosing which projects should be subject to EIA) by virtue 'of their nature, size or

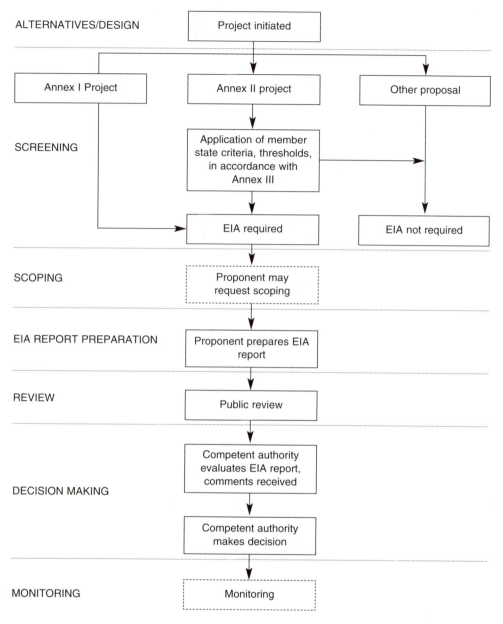

ALTERNATIVES/DESIGN

Project initiated

Annex I Project

Annex II project

Other proposal

SCREENING

Application of member state criteria, thresholds, in accordance with Annex III

EIA required

EIA not required

SCOPING

Proponent may request scoping

EIA REPORT PREPARATION

Proponent prepares EIA report

REVIEW

Public review

DECISION MAKING

Competent authority evaluates EIA report, comments received

Competent authority makes decision

MONITORING

Monitoring

-------- optional step

Figure 3.1   Main steps in the European Directive EIA process

location' are summarised in Figure 3.1. The strengthening of these arrangements was probably the most significant outcome of the amendments to the Directive. All projects listed in Annex I are subject to assessment (Article 4(1)). Annex I has been considerably expanded by the amended Directive, to incorporate many projects originally placed in Annex II. The list includes 21 types of project, such as power stations, major industrial installations, major communication infrastructure, waste disposal installations, groundwater abstraction schemes, quarries and overhead electrical power lines. The annex lays down thresholds for most of these categories.

Annex II covers projects for which member states are to determine whether or not EIA is required, on the basis of either a case-by-case examination or the application of thresholds or criteria (Article 4(2)). The amended Directive introduces a new Annex III which sets down selection criteria, indicating the likelihood of significant environmental effects, which are to be used in Annex II screening decisions. The list of Annex II projects is wide-ranging, and includes categories not specifically referred to in the original Directive, such as deforestation, wind farms, asbestos production, certain coastal works, ski runs and theme parks. The amended Directive also requires screening decisions made in relation to Annex II projects to be made available to the public (Article 4(4)).

There is no requirement in the Directive for the provision of preliminary information equivalent to the US environmental assessment to assist in screening. Nor is there any provision for different levels of EIA in the Directive, though proposals for a simplified type of environmental assessment were advanced at one stage in the discussions leading to the adoption of the original Directive (see above).

There is no formal requirement for *scoping* (i.e. for determining the topics to be studied in an EIA for a particular project) in the Directive. However, the amended Directive does require the competent authority to give an opinion on the scope of the information to be supplied, should the developer request this (Article 5(2)). Residual evidence of the negotiations on the amendments to the Directive is provided by the admonition that member states 'may require the competent authorities to give such an opinion, irrespective of whether the developer so requests' (Article 5(2)). There is no provision in the Directive that the commencement of work on an EIA be announced.

The general nature of the content of the EIA information to be utilised by the proponent in the *preparation of the EIA report* is, however, specified. Article 5(3) of the Directive sets down the minimum information which must be provided by the developer:

> a description of the project comprising information on the site, design and size of the project,
> a description of the measures envisaged in order to avoid, reduce and, if possible, remedy significant adverse effects,
> the data required to identify and assess the main effects which the project is likely to have on the environment,
> an outline of the main alternatives studied by the developer...,
> a non-technical summary.

Article 5 also indicates that the developer should furnish all the information listed in Annex IV where member states consider that it is relevant and reasonable to do so. This includes: descriptions of the project, of alternatives to the project, of baseline environmental conditions, of the likely significant environmental effects of the project, and of mitigating measures; a non-technical summary; and an indication of difficulties encountered in compiling the information. Annex IV makes it clear that the description of the likely significant effects of the project should cover the direct effects and any indirect, secondary, cumulative, short-, medium- and long-term, permanent and temporary, positive and negative effects, of the project. This information corresponds closely to that specified in most other EIA systems.

In order to facilitate the assessment, Article 5(4) provides that member states 'shall, if necessary, ensure that any authorities holding relevant information' make this available to the developer. However, there is no requirement that liaison between the developer and relevant authorities takes place while the assessment is being undertaken. Having carried out the assessment, the developer is obliged to supply the competent authority responsible for the authorisation of the project with the resulting information. The form in which this information is submitted is not specified in the Directive. (It is referred to hereafter as the EIA report.)

The Directive does not provide for a formal *review* of the EIA report by the competent authority (or any other body) or for the preparation of draft and final EIA reports. However, Article 6 of the Directive provides for the EIA report to be made widely available as a basis for consultation and public participation. Article 7 of the amended Directive, on transboundary effects, also requires consultation and public participation on the EIA report to take place within any other member state affected by a proposal. There is no provision for the developer to respond to the points raised by the public or consultees on the content of the EIA report, or for these comments to be made public.

Article 8 requires that the results of this exercise, together with the developer's EIA report, must be taken into consideration in taking the *decision* on the project. The Directive requires, when the competent authority has reached a decision on the consent application, that the public (and any member state that was consulted under Article 7) be informed and that any conditions attached to that decision be made public. Article 9 of the amended Directive also requires that the main reasons upon which the decision has been based should be provided along with a description of any mitigating measures.

The amended Directive is still silent on the question of the *monitoring* of project impacts.

The *mitigation* of project impacts is one of the main aims of the European Directive. As mentioned above, it is a requirement of Article 5(3) that mitigation measures be specified in the proponent's EIA report and of Article 9 that those incorporated into the consent be communicated to the public. These mitigation measure requirements are also listed in Annex IV. These are, however, the only points in the EIA process where mitigation measures must be shown to be considered, although the precautionary principle underlies the whole Directive.

Consultation and participation are limited, under the provisions of the Directive, to commenting upon the EIA report. Member states are required to designate the environmental authorities which should receive copies of the environmental information and who must be consulted about their opinions on the consent application (Article 6(1)). Similarly, member states must ensure that both the consent application and the environmental information are made available to the public and that the public concerned is given an opportunity to comment before the project is initiated (Article 6(2)). In addition, to accord with the Espoo Convention on Environmental Impact Assessment in a Transboundary Context (United Nations Economic Commission for Europe – UNECE, 1994), the amended Directive strengthened the provision requiring member states to supply the above information, as a basis for consultation and public participation, to other member states if the project is likely to have significant transboundary effects on their environments (Article 7). As in other EIA systems, there are provisions for the protection of industrial and commercial secrecy (Article 10). There is no provision for third-party appeals against decisions involving EIA. Consultation and public participation provisions are likely to be strengthened further by the proposed amendments to implement the Aarhus Convention on Access to Public Information (UNECE, 1998).

There is provision in the Directive for EIA *system monitoring*. Article 11 of the Directive provides for the exchange of information between member states and the Commission on experience in applying the Directive and for member states to inform the Commission about the criteria and thresholds they have used in the selection of Annex II projects. There is also a requirement for the preparation of a five-year review of the amended Directive's application and effectiveness, just as there was for the original Directive. Article 2 of the amending directive commits the Commission, should it be necessary, to submit additional proposals, to ensure 'further co-ordination in the application of this Directive'.

Member states were required to take the necessary measures to comply with the Directive within two years of its notification (i.e. by 14 March 1999) and to inform the Commission accordingly (Article 3 of the amending Directive). The European Commission has not issued any published guidance as to how the amended Directive is to be implemented, nor has it reinstated the substantial EIA training programme it funded to improve the implementation of the original Directive. However, the expert meetings and information exchange it arranged in relation to the original Directive are continuing.

The *costs and benefits* of the EIA system are not referred to in either the original or the amended directives but the benefits are obviously intended to exceed the costs.

The European Directive on EIA contains no provisions relating to *strategic environmental assessment* (SEA). However, conscious that European EIA, by its nature, occurs late in the planning process (i.e. at the project level), the Commission has pursued the issue of SEA since the mid-1970s, with drafts of a proposed directive first appearing in the early 1990s. These culminated in a separate directive on the assessment of plans and programmes (Feldmann *et al.*,

2001) that was formally published on 21 July 2001 (European Commission, 2001a). In some ways (for example, scoping quality control and monitoring) it is, surprisingly, stronger than the amended EIA Directive. Member states must implement the SEA Directive in national law by July 2004.

The compromises made in the gestation of the original Directive were very evident in its final 'minimax' form. At its minimum, it required that a limited list of projects be subjected to a limited form of EIA. At its maximum, it recommended that a much longer list of projects be subjected to a more universally recognised form of EIA. The Commission no doubt hoped that practice in member states would prove to be well above the minimum required. More realistically, it may also have hoped that, once the benefits of the flexible EIA system had become as apparent as they have in the United States (CEQ, 1990), it would become possible to strengthen it, as indeed has happened. Unfortunately, further compromises mean that weaknesses relating to scoping, consultation and public participation, and impact monitoring remain (Barker and Wood, 2001). Table 3.3, in which the European Directive EIA system is evaluated against the criteria advanced in Chapter 1, demonstrates these shortcomings.

Implementation of the Directive

A directive is binding in that it specifies ends which must be achieved, while leaving member states the choice of means. Member states must not only introduce the necessary *legal provisions* but also ensure that they work, i.e. that the ends specified in the Directive are achieved in practice (Haigh, 1991). Several countries (for example Belgium, France, Ireland, Luxembourg and the Netherlands) had already introduced some legal requirements for EIA by 1985 but these were insufficient to implement the Directive fully.

Several studies have indicated how the original European Directive was implemented in member states, including a five-year review of the implementation of the original Directive (CEC, 1993; European Commission, 1996a,b, 1997c; Bond and Wathern, 1999; Bjarnadóttir, 2001). By July 1991, all member states had incorporated some EIA provisions within their own legislation. However, years after it was required, transposition of the Directive into national legislation was still not complete (CEC, 1993; Kramer, 2000). The European Court of Justice has taken an active role in ensuring member state compliance with the directive, in particular with the definition of projects considered within its remit (Kramer, 2000).

By 1996, all member states had transposed the original Directive (European Commission, 1997c, p. 34). It was possible to observe considerable achievements in carrying into practice both the letter and the spirit of the Directive, but these varied considerably between member states. A study of eight member states showed that prior to the introduction of the amended Directive they had all taken steps, to a greater or lesser extent, to strengthen procedures in different areas of the EIA process (European Commission, 1996b; Barker and Wood, 1999). Not all member state arrangements for implementing the revised Directive had been implemented by mid-2001, when formal warnings were issued to several countries, including the UK and the Netherlands.

Table 3.3 Performance of the amended European Directive EIA system

Criterion	Criterion met	Comment
1. Is the EIA system based on clear and specific legal provisions?	Yes	Directive must be transposed into member state legislation to have direct effect
2. Must the relevant environmental impacts of all significant actions be assessed?	Yes	Comprehensive coverage of significant projects. Marginal discretion in impact coverage
3. Must evidence of the consideration, by the proponent, of the environmental impacts of reasonable alternative actions be demonstrated in the EIA process?	Partially	Directive now requires previously discretionary discussion of alternatives (where studied)
4. Must screening of actions for environmental significance take place?	Yes	Annex I lists projects for which EIA is compulsory. Annex II lists projects for which EIA is necessary if member state criteria and thresholds, framed to meet Annex III requirements, are met
5. Must scoping of the environmental impacts of actions take place and specific guidelines be produced?	Partially	Competent authority and environmental authorities must now participate in scoping if developer so requests. Member states encouraged to make scoping mandatory
6. Must EIA reports meet prescribed content requirements and do checks to prevent the release of inadequate EIA reports exist?	Content: Yes Checks: No	Content prescribed in Article 5(3) and (where reasonable and relevant) Annex IV. No pre-release checks on content required
7. Must EIA reports be publicly reviewed and the proponent respond to the points raised?	Partially	Strengthened requirements for public review. No provision for proponent to respond to points raised
8. Must the findings of the EIA report and the review be a central determinant of the decision on the action?	Partially	EIA report and comments upon it must be taken into consideration but not necessarily a central determinant. Reasons for decision must be provided
9. Must monitoring of action impacts be undertaken and is it linked to the earlier stages of the EIA process?	No	No provision for monitoring
10. Must the mitigation of action impacts be a central consideration at the various stages of the EIA process?	Yes	EIA report must cover mitigation (Article 5(3)) Consent conditions and mitigation measures must now be made public
11. Must consultation and participation take place prior to, and following, EIA report publication?	Partially	Consultation of environmental authorities (not public) in developer-requested scoping. Consultation and public participation must take place following EIA report release

Table 3.3   *continued*

Criterion	Criterion met	Comment
12. Must the EIA system be monitored and, if necessary, be amended to incorporate feedback from experience?	Partially	Member states must notify Commission of screening criteria and thresholds and of any exemptions. Commission must prepare five-year report on operation of amended Directive and propose necessary amendments
13. Are the discernible environmental benefits of the EIA system believed to outweigh its financial costs and time requirements?	Yes	General consensus (but not unanimity) as to increasing utility of EIA in improving project design and mitigation measures
14. Does the EIA system apply to significant programmes, plans and policies, as well as to projects?	Not applicable	Separate directive on SEA of plans and programmes (not policies), notified in 2001 for implementation by 2004, involves separate procedures

The Commission has published a number of guidance documents on different aspects of EIA, including the assessment of cumulative impacts (European Commission, 1999a). Originally published in the mid-1990s, guides to screening, scoping and EIS review (European Commission, 2001c,d,b, respectively) have subsequently been updated. There is, however, no comprehensive EC EIA guidance. Several individual member states have also issued procedural guidance and been active in the provision of training to support the implementation of the Directive. However, the widely perceived need, in many member states, for further guidance on different aspects of EIA has been reinforced by the changed requirements necessitated by the amendments to the Directive.

The arrangements for dealing with *alternatives* vary between member states (Bond and Wathern, 1999) but many countries have not introduced mandatory requirements. Practice has been reported to have frequently been unsatisfactory (CEC, 1993; Dresner and Gilbert, 1999). It remains to be seen to what extent the strengthened provisions of the amended Directive lead to improvements in practice.

While some member states had transposed the requirements of Article 5 and Annex III (now IV) regarding the content of EIA reports into law, others had confined EIA report *coverage* to the minimum requirements specified in Article 5(2) and made inclusion of Annex III requirements fully discretionary.

There were significant differences in the interpretation of the *screening* provisions of the original Directive between member states (European Commission, 1997c, p. 32). These were reflected in the total number of projects subjected to EIA, and their composition, which differed greatly between member states, even after differences in population had been taken into

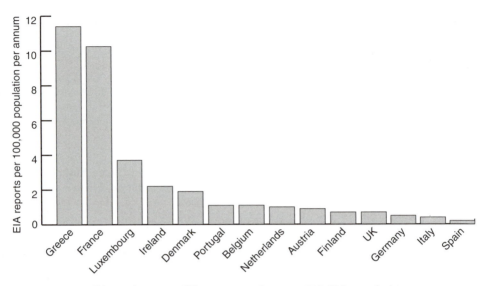

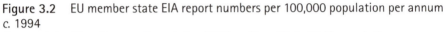

Figure 3.2   EU member state EIA report numbers per 100,000 population per annum c. 1994

Sources: derived from European Commission, 1997b, p. 15, and United Nations web site www.un.org

account – due to differences in size, legislation and economic development (European Commission, 1997c, p. 34). Figure 3.2 shows numbers of EIA reports prepared annually in the early to mid-1990s per 100,000 member state population.[2] The Commission originally issued guidance (European Commission, 2001c) designed to demystify the screening process in 1995.

Although the original directive was silent on *scoping*, a number of member states (including Germany) had made provision for scoping in their legislation. Other member states (for example, Ireland) either had some non-mandatory arrangements for scoping or encouraged developers to use this practice. Generally, developers were allowed to access environmental data in member states, but instances of lack of cooperation and, more seriously, of lack of data, existed. The Commission originally issued guidance on scoping (European Commission, 2001d) to help with this process in 1996. It is expected that the discretionary scoping provisions in the amending Directive will improve practice.

Although it commissioned research on EIA methods in the 1970s, the Commission has limited its guidance on the *preparation of EIA reports* to the treatment of cumulative impacts in EIA reports (European Commission, 1999a).

A *review* of 112 EIA reports covering earlier and later years during the period 1990 to 1996 showed an increase in the number of 'satisfactory' reports during this time, from 50 per cent to 71 per cent (European Commission, 1996b; Barker and Wood, 1999). The most important single factor in bringing about improvement was found to be the increasing experience of the participants in the EIA process. Changes in EIA legislation also appear to have been influential in some member states, while the type of project was a factor in others. Other research showed that the quality of EISs varied according to the size of the projects and the degree of controversy associated with them (European Commission, 1996b). Few member states have followed the example of the well-known Netherlands EIA Commission and set up independent review bodies (European Commission, 1997c, p. 34). The Commission's guidance on the reviewing of EIA reports (European Commission, 2001b), originally issued in 1994, was not as widely used as it could have been.

It is clear that, for some projects, both the *decisions* themselves and the form in which the projects have been approved have been influenced by the EIA process. The Commission (CEC, 1993; European Commission, 1996a, paras 3.99–3.109) found that there was evidence to suggest that EIA was helping to strengthen and streamline the decision-making process. However, there were also significant numbers of cases where the EIA was considered to have had little or no effect. On the other hand, Dresner and Gilbert (1999), on the basis of case studies in several member states, reported widespread criticism that EIA was taking place too late in the decision-making process and was sometimes taking too long.

Although the Directive contains no requirements for the *monitoring* of project impacts, many member states have made specific legal provision for this (European Commission, 1997c, p. 11). While other member states have relied on existing monitoring procedures and practice, there appear to be serious deficiencies in these arrangements in many member states.

While *mitigation* practice varied, it was apparent that amelioration has not always taken place, and certainly not at all the various stages of the EIA process. Despite the briefness of the implementation period examined, CEC (1993, p. 61) reported that:

> there is clear evidence that project modifications have and are taking place, due to the influence of the EIA process. However, there is also evidence that, as yet, its impact is not as widespread as intended and that modifications are mainly confined to those of a minor or non-radical nature.

More recent research showed that, although modifications were taking place to projects as a result of EIA, there appeared to be no overall trend over time in the number or significance of the modifications (Barker and Wood, 1999).

Access by the public to copies of EIA reports varied from the provision of complimentary copies, through purchase of copies and rights of reference, to instances where considerable persistence was necessary to consult even the non-technical summary. Generally, though practice relating to participation and consultation has been reported as having improved (European Commission, 1997c, p. 34), it has varied from satisfactory to unsatisfactory (Dresner and Gilbert, 1999).

One of the most impressive aspects of the development of EIA in Europe during the 1990s has been the commitment by the European Commission to

EIA system monitoring. In essence, this has taken the form of the commissioning of a series of staggered comparative research studies aimed at gauging the manner in which member states have responded to the EIA directive (CEC, 1993; European Commission, 1996a,b, 1997a,c, 1998). This has resulted in the derivation of an EU-wide collective picture from the different member state experiences (Barker and Wood, 1999, 2001).

While it was hard to generalise from the varied experience of EIA both within and between member states, CEC (1993, p. 59, emphasis added) found that:

> The financial *costs* of carrying out an assessment for an EIS are typically a small fraction of one percent of the capital cost of the project. ... The overall timescale of implementing projects does not appear to be significantly affected by EIA.

A study of 18 EIAs in four member states concluded that, typically, EIAs have cost between €10,000 and €1.8 million and the EIA process has been accomplished well within 24 months (European Commission, 1996a, paras 3.25, 3.51). These costs, together with the administrative costs of involvement in the EIA process, appeared to be broadly acceptable to the principal participants (paras 3.99–3.108). The Commission (CEC, 1993, p. 63, emphasis added) believed that the 'planning, design and authorisation of projects are beginning to be influenced by the EIA process and that environmental *benefits* are resulting'. Again, in the later study it was found that the costs of EIA were outweighed by the benefits of reducing adverse environmental impacts (European Commission, 1996a, para. 3.118). While there appear to have been considerable differences in the ways in which EIA has been operated in the member states, there has been general agreement that EIA has been beneficial in reducing adverse environmental impacts (European Commission, 1996a, paras 3.96, 3.112).

The European Commission has initiated several studies on *strategic environmental assessment* procedures, methodologies and practice (see, for example, European Commission, 1994, 1995, 1996b, 1997a, 1999b; Feldman *et al.*, 2001). It has been demonstrated (European Commission, 1997b, 2001f) that there has been considerable SEA practice within member states, a finding endorsed by Fischer *et al.* (2002).

Despite considerable progress, it was apparent that, many years after the original Directive finally came into effect, EIA practice frequently left much to be desired. A number of measures needed to be taken before the full realisation of the benefits obtainable from the implementation of the original Directive could be achieved. These included the requirements for better screening of projects, partial scoping, enhanced consultation and participation (including better international consultation), improved treatment of alternatives, and increased importance of EIA in decision making which characterised the amended Directive (European Commission, 1997d). Ironically, as mentioned above, these amendments restored many of the provisions originally contained within the draft directive published in 1980. It is already clear that many member states failed to implement the provisions of the amended Directive on time. It remains to be seen to what extent the remaining weaknesses relating to the treatment of alternatives, scoping, the quality of EIA reports, monitoring and consultation and participation influence the effectiveness of the amended Directive.

Notes

1. The texts of the original Directive, of the amended Directive and of many research reports and guidance documents, are reproduced on the web site of the European Commission Environment Directorate-General at www.europa.eu.int/comm/environment/eia/home/htm
2. No EIA report figures were available for Sweden.

Chapter 4

EIA in the UK

The UK has possessed a land-use planning system since 1948 which allows substantial discretion in the consideration of the environmental implications of new development. It is possible for local planning authorities (LPAs) to prepare plans in which environmental policies are emphasised and to refuse development, or to impose conditions to planning permissions, for environmental reasons. LPAs can thus make a powerful contribution to environmental protection by determining the nature and location of new development and redevelopment (Wood, 1999a). They are now being encouraged to use their long-standing powers actively to assist in environmental protection as an integral part of planning for sustainable development (Department of the Environment, Transport and the Regions – DETR, 1999a).

The flexibility of the town and country planning system and the ability of LPAs, and central government in call-in and appeal cases, to take environmental factors into account in decision making is the principal reason for the belated acceptance that EIA could provide a valuable additional safeguard in the UK. The UK's initial reluctance to accept the imposition of EIA by Brussels and its later implementation of the original Directive almost to the letter by integration into existing decision-making procedures can be seen as a reflection of the larger British relationship with Europe, and especially of its attitude to environmental regulation (Bond and Wathern, 1999). The UK has traditionally tended to take an insular view of environmental policy (accentuated under the Thatcher government) assuming that nature (wind and an encircling sea) would provide solutions to environmental problems without the need for rigid regulation. Once European directives have been adopted, however, British governments have prided themselves on their prompt and effective transposition. (Critics, for example Sheate (1996), would dispute whether implementation has always followed the spirit as well as the letter of the various directives.)

This chapter describes the early interest in EIA in the UK and the government's response to draft versions of the original and amended European directives on EIA. It then explains the evolution of the various UK regulations made to implement these directives. The chapter goes on to provide an overview of the UK EIA system and describes briefly how the UK regulations have been implemented in practice.

Evolution of the UK regulations

There has been a largely bipartisan attitude to EIA in the UK since the early 1970s.[1] The development of onshore oil facilities in Scotland, where planning authorities were not accustomed to handling complex projects, was one of the factors leading the Scottish Development Department, with the support of the Department of the Environment (DoE), to commission the preparation of guidance. Aberdeen University developed a systematic procedure for planning authorities to make a balanced appraisal of the environmental, economic and social impacts of major industrial developments within the existing statutory planning framework. The authors recommended a procedure relying on a discretionary, flexible and cooperative approach by developers and local planning authorities (Clark *et al.*, 1976). Their report was distributed to every local planning authority in the country, free of charge, and 'commended for use by planning authorities, government agencies and developers'.

Concurrently, a further study was commissioned by DoE to consider the administrative, rather than the methodological, aspects of EIA. Catlow and Thirwall (1976) recommended legislative changes to ensure the formal integration of EIA within the existing UK planning system. DoE delayed the publication of this report by a year, eventually distributed it to a much smaller audience than the previous study, and expressed grave reservations about the introduction of any new procedures.

Central government was careful not to commit itself to EIA, despite its positive reception of the Aberdeen report. For example, it made no response to the Royal Commission on Environmental Pollution's (1976) endorsement of the need for EIA for certain major developments. Eventually, after much deliberation, the government guardedly announced that it favoured the limited use of EIA. There was no hint, however, of implementing any legislative changes to encourage its use.

As the various drafts of the European Commission's original Directive on EIA (see Chapter 3) began to appear, the government continued to endorse the principle that EIA was a useful element in the planning process for considering large and significant proposals while rejecting the idea of any mandatory EIA system. The government's cautious reception of the updated Aberdeen assessment manual, which it had commissioned, contrasted with that of the original document: 'it is important that the approach suggested in the report should be used selectively to fit the circumstances of the proposed development and with due economy' (Clark *et al.*, 1981, p. v). Local authorities were left to purchase the document themselves.

The European Commission's draft directive was deliberated by a select committee of the House of Lords which heard a great deal of evidence from both supporters and antagonists of the proposed directive. The committee expressed some reservations on points of detail but, in an important report (House of Lords, 1981) which summarised the various views for and against mandatory EIA, came down firmly in its favour. The Royal Town Planning Institute, among other bodies, responded positively to the select committee's report in its submission to members prior to its being debated in the House of Lords. The report received widespread support from those peers who spoke.

The lone voice of dissent was that of the Minister for the Environment. The government, notwithstanding the recommendations of the well-informed select committee, decided that the proposed European requirements for EIA might duplicate or complicate the existing town and country planning procedures which it was striving to simplify. Meanwhile, considerable experience of the use of non-mandatory EIA was being gained in the UK.

As a result of the period of intensive negotiation in Brussels which ensued, major concessions limiting both the coverage of an EIA report and the range of projects to be subject to mandatory EIA (see Chapter 3) were felt to reduce the ramifications of EIA to the point where the British environment minister could accept the Directive. It seems probable that the House of Lords committee report was one factor influencing this decision. Another factor seems to have been the view that the British government need not extend the coverage of EIA to Annex II projects (Wood, 1988b; Sheate and Macrory, 1989). Box 4.1 summarises the history of EIA in the UK (see also Bond, 1997).

The Department of the Environment set up a working party in 1984 with members drawn from industry, local government, the planning profession, environmental groups and other government departments to consider the implementation of the forthcoming Directive, now that Britain had withdrawn its objections. The working party made a number of recommendations to implement the Directive largely within the existing planning system with minimal procedural complexity. A consultation paper reflecting these recommendations was circulated in 1986, accompanied by a draft advisory booklet which the working party had drawn up.

The consultation paper excluded Annex II projects, except on the direction of the minister, from mandatory EIA and stated that LPAs should not demand an EIA for such developments. However, the paper sanctioned, and even encouraged, the use of voluntary EIA by authorities and developers. It thus reflected government policy since the early 1970s, save for the expectation that there would now be perhaps half a dozen mandatory Annex I project EIAs per annum.

The draft advisory booklet indicated, perhaps for the first time, a positive attitude by government towards EIA. Certainly, the EIA report was not to be a minimalist document; it was expected to provide as objective a statement of the environmental effects of the project as possible.

Following the publication of the consultation paper, legal advisers at DoE and at the Commission of the European Communities (CEC) independently indicated that the exclusion of Annex II projects could not be justified in law. The British acceptance of the Directive may therefore have been based on a misconception and DoE officials had to return to the drawing-board. They produced a further consultation paper in early 1988. This draft reflected a much less reluctant approach to implementing the Directive and advanced indicative criteria to help determine which Annex II projects were to be subject

Box 4.1   Brief history of EIA in the UK

1974   Aberdeen University, Catlow and Thirwall research commissioned

1976   Research reports published favouring EIA
Bipartisan reluctance to formalise EIA
British Gas, BP, Shell and others practise EIA

1981   House of Lords report favours EIA
Aberdeen University revised manual published

1984   Government withdraws objections to EIA on misunderstanding

1988   Government implements most original regulations on time

1989   Government publishes procedural guidance

1990   Three hundred environmental statements (ESs) per annum prepared
EIA consultancies grow

1991   Government publishes monitoring report

1994   Government amends regulations, publishes research report and good
practice guide on evaluating environmental information

1995   Government publishes good practice guide for preparing environmental statements

1996   Government argues against scoping in forthcoming amended
Directive
Government publishes research report on ES quality

1997   Government publishes research report on mitigation measures in EIA

1999   Government implements many new regulations on time

2000   Government publishes new procedural guidance
Six hundred ESs per annum prepared

to EIA. The expectation was that the number of these would be no more than a few dozen a year. The emphasis on voluntary EIA disappeared.

The DoE officials publicised the proposals energetically and made a number of changes in the ensuing Planning Regulations and Circular published on 15 July 1988 as a result of the representations received (Wood and McDonic, 1989). It is significant that DoE adopted the term 'environmental assessment' rather than the US 'environmental impact assessment', given its earlier opposition to a formal EIA system. Whether it took the term from the name of the US preliminary (screening) document (see Chapter 2) or from the Canadian name for EIA (see Chapter 5) is a matter of conjecture.

Further regulations relating to projects authorised under separate consent systems (i.e. not under town and country planning legislation) and in Scotland and Northern Ireland were promulgated subsequently. Over the years, various minor amendments to the Planning Regulations were made to extend the range of projects subject to EIA and to rectify anomalies in the implementation of the original Directive.

Once it became apparent that the amended Directive was going to be adopted, the renamed Department of the Environment, Transport and the Regions[2] commissioned research on alternative approaches to meeting the anticipated screening requirements. A consultation paper was issued in July 1997 seeking responses to DETR's proposals relating to screening, to the publicising of screening decisions, to dealing with scoping requests by developers, to the publicising of consent decisions, and to various other issues. This paper contained the most unequivocal endorsement of EIA to date:

> The Government wholly endorses the use of the EA process to ensure that the likely significant effects of development projects are fully assessed and taken into account in deciding whether the projects may proceed. (DETR, 1997b, para. 18)

A further consultation paper devoted to detailed screening proposals was issued in December 1997 (DETR, 1997a) and the proposed planning regulations were issued as a draft in July 1998. Various comments on the consultation documents were received, many of which urged that scoping be made mandatory. Minor modifications to the proposals were made (scoping remained discretionary) and the Town and Country Planning (Environmental Impact Assessment) (England and Wales) Regulations 1999 implementing the amended Directive came into force on 14 March 1999, accompanied by a circular (DETR, 1999b). Interestingly, the new Planning Regulations and Circular refer throughout to EIA, rather than to environmental assessment, bringing British usage closer to European and international terminology.

The UK EIA system

For projects requiring planning permission, the amended Directive was given legal effect in England and Wales through the Town and Country Planning (Environmental Impact Assessment) (England and Wales) Regulations 1999, and in Northern Ireland through the Planning (Environmental Impact Assessment) Regulations (Northern Ireland) 1999. The first legislation approved by the Scottish Parliament for over two centuries was the Environmental Impact Assessment (Scotland) Regulations 1999. The 1999 Regulations implement the provisions of the amended European Directive almost to the letter for those projects subject to planning control. The four annexes to the Directive become Schedules 1 to 4, with some modifications. The EIA Regulations apply to two separate lists of projects, based on Annexes I and II of the amended Directive. These lists now cover a wider range of projects than the previous regulations. In particular, Schedule 2 contains screening thresholds and criteria not specified in the Directive for each category of development.

The Planning and Compensation Act 1991 contains a section which enables the Secretary of State for the Environment to require environmental assessment of planning projects other than those listed in the original European Directive. The 1988 Planning Regulations were amended to cover the EIA of private motorways, motorway service areas, wind generators and coast protection works. The amended Directive now covers some of these projects but the 1999 Regulations include motorway service stations, sports stadia, leisure centres and golf courses, which are not listed in Annex II.

The 1999 Regulations contain provisions for LPAs to give a formal 'screening opinion' that EIA is required when they are requested to do so by developers. They must also notify developers that EIA is required when a planning application is submitted without an environmental statement (ES). In either event, the Regulations permit the developer to appeal to the relevant Secretary of State for a 'screening direction' that EIA is, or is not, required.

As a result of the amendments to the original Directive, a developer may request a formal 'scoping opinion' from the LPA or, where the LPA fails to provide one, a scoping direction from the Secretary of State, regarding the information to be included in an ES. Another important requirement introduced to implement the amended Directive relates to the treatment of alternatives in the ES.

Certain statutory consultees (including the Environment Agency) are required to provide the developer with information should it be requested. The Regulations also set down the nature of prescribed consultation and publication arrangements and extend the amount of time available to LPAs to reach a decision on planning applications involving EIA from 8 to 16 weeks.

Part II of Schedule 4 contains a list of the mandatory information required by Article 5 of the amended Directive and Part I replicates Annex IV of the amended Directive. An environmental statement is defined by reference to Schedule 4. It must include the information referred to in Part II and

> such of the information referred to in Part I ... as is reasonably required to assess the environmental effects of the development and which the applicant can, having regard in particular to current knowledge and methods of assessment, reasonably be required to compile. (Article 2 (1))

This is a formulation which follows closely the wording of the amended Directive and which should avoid many of the disagreements about whether the content requirements for ESs specified in the 1988 Planning Regulations accurately reflected those of the original Directive.

The 'environmental information' which must be considered in reaching a decision consists of the environmental statement, together with any further information and the representations of consultees and members of the public about the impacts of the development. A further new requirement of the 1999 Planning Regulations is that the reasons for the decision to grant or to refuse planning permission in cases involving EIA must be stated.

The Regulations have been modified by the Town and Country Planning (Environmental Impact Assessment) (England and Wales) (Amendment) Regulations 2000 to require EIA to be undertaken in certain cases when conditions on old minerals planning permissions are reviewed. An outline of the main steps in the EIA process for planning decisions is shown in Figure 4.1. There are special provisions relating to developments in enterprise zones and simplified planning zones (DETR, 1999b).

Environmental impact assessment

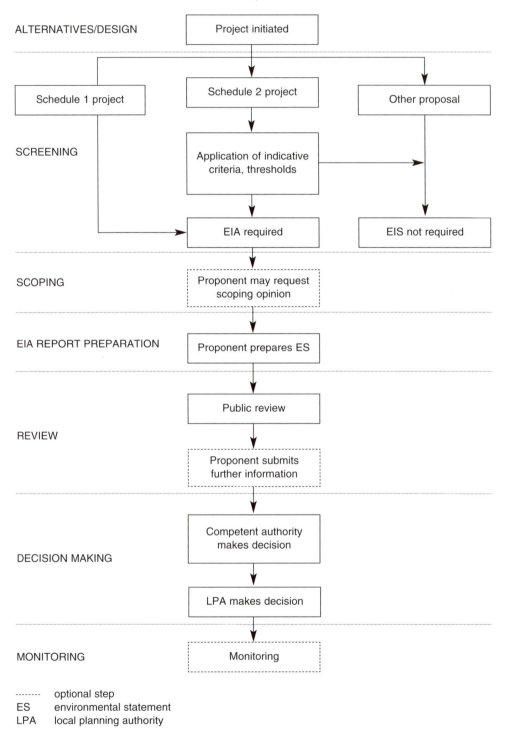

ALTERNATIVES/DESIGN

Project initiated

Schedule 1 project

Schedule 2 project

Other proposal

SCREENING

Application of indicative criteria, thresholds

EIA required

EIS not required

SCOPING

Proponent may request scoping opinion

EIA REPORT PREPARATION

Proponent prepares ES

Public review

REVIEW

Proponent submits further information

Competent authority makes decision

DECISION MAKING

LPA makes decision

MONITORING

Monitoring

-------- optional step
ES   environmental statement
LPA   local planning authority

Figure 4.1   Main steps in the EIA process for UK planning decisions

Advice on procedures and on the implementation of the EIA Planning Regulations in England and Wales is presented in DETR Circular 2/99 (DETR, 1999b). Equivalent circulars apply in Wales (Welsh Office 11/99), in Scotland (Scottish Executive Development Department 15/99) and in Northern Ireland (Development Control Advice Note 10). These circulars set out indicative criteria and thresholds to help determine whether certain projects (Annex II projects) should be subject to EIA. Further guidance on EIA procedures is provided for England and Wales (DETR, 2000a)[3] and for Scotland (Scottish Executive, 1999).

The Circular (DETR, 1999b) provides guidance on the operation of the procedures. The detailed indicative criteria and thresholds to be used by LPAs in reaching a judgement about whether EIA is to be required for Schedule 2 projects are contained in the advisory Circular. The criteria and thresholds (and the other advice contained in the Circular) can be changed easily and, in any event, do not have regulatory force. However, the Regulations provide a right of appeal against an LPA determination that EIA is required. Normal town planning appeal provisions also apply and the Secretaries of State can call applications in for determination by central government. There is, therefore, relatively little discretion left to LPAs in determining whether or not the Regulations apply to particular applications. Further details about the various stages of the UK EIA system for planning decisions are contained in Chapters 6–20.

Certain types of project listed in Annexes I and II of the European Directive (see Chapter 3) are authorised outside the British planning system. Accordingly, it was necessary to adopt additional regulations to extend EIA to these projects. The arrangements relating to these other regulations are broadly similar to those for planning projects (Glasson et al., 1999; DETR, 2000a).

The Highways (Assessment of Environmental Effects) Regulations 1999 require the Secretary of State for the Environment, Transport and the Regions (in England) to publish an ES for the preferred route at the time when draft orders are published. Regulations requiring EIA for new railways and other transport projects (Transport and Works (Applications and Objections Procedure) (England and Wales) Rules 2000) have come into force. Arrangements for the provision of ESs with any private and hybrid parliamentary bills to initiate projects not covered by these rules are also in place in the UK.

The Electricity Works (Environmental Impact Assessment) Regulations 2000 require an ES to be submitted to the Secretary of State for Energy (in England) for the construction of nuclear power stations, for some other power stations and for certain overhead power lines. The Nuclear Reactors (Environmental Impact Assessment for Decommissioning) Regulations were promulgated in 1999 to ensure that EIA was carried out where decommissioning took place.

The EIA of major onshore oil pipelines is covered by the Pipe-line Works (Environmental Impact Assessment) Regulations 2000 and of gas pipelines by the Public Gas Transporter Pipe-line Works (Environmental Impact

Assessment) Regulations 1999. EIA for offshore oil development is required by the Offshore Petroleum Production and Pipe-lines (Assessment of Environmental Effects) Regulations 1999.

The Environmental Impact Assessment (Forestry) Regulations 1999 require the EIA of afforestation projects in any case where, in the opinion of the Forestry Commission, the project is likely to have significant environmental effects. These significantly strengthen the 1988 Afforestation Regulations which were triggered only by grant applications. The UK did not include the Directive's Annex II rural landholdings restructuring projects in any of the original regulations, because the government agreed with the European Commission that they were unlikely to occur in the UK in a form that would have significant environmental effects and hence require EIA (Sheate and Macrory, 1989). Projects for the use of uncultivated land were also excluded. However, EIA was extended to this latter type of project through the promulgation of uncultivated land regulations and the publication of guidelines in early 2002.

Improvements to existing land drainage works undertaken by drainage bodies and the Environment Agency do not require an express grant of planning permission but fall under the Environmental Impact Assessment (Land Drainage Improvement Works) Regulations 1999. These require the drainage body to consider whether or not the proposed works would be likely to have significant environmental effects and ought therefore to be subject to an EIA.

There are regulations relating to ports and harbours, to reflect the existing authorisation procedures. The Harbour Works (Environmental Impact Assessment) Regulations 1999 empowered the (then) Minister of Agriculture, Fisheries and Food (for fishery harbours) or the (then) Secretary of State for the Environment, Transport and the Regions (for English harbours more generally) to decide whether EIA is needed.

Offshore salmon farming facilities within two kilometres of the coast require a lease from the Crown Estate Commissioners. The Environmental Impact Assessment (Fish Farming in Marine Waters) Regulations 1999 require the Commissioners to consider an ES provided by the developer before granting a lease in circumstances where the development may have significant effects on the environment. Dredging for minerals offshore requires a dredging licence from the Crown Estate Commissioners. In cases where dredging is likely to have significant environmental effects, the Environmental Impact Assessment and Habitats (Extraction of Minerals by Marine Dredging) Regulations, which require the applicant to provide an ES, were due to be made in 2002. Regulations relating to the EIA of certain water resource projects were also expected in 2002.

The arrangements for non-planning projects in Wales are similar to those in England, save that the National Assembly for Wales is the responsible body. The arrangements for Scotland are broadly similar to those for England and Wales, though there are some differences due to the different legal and administrative arrangements which apply in Scotland. Separate provision for EIA has been made in Northern Ireland, which has its own legal and administrative procedures. However, the general principles of the EIA system covering the rest of the UK apply (DETR, 2000a).

The Commission of the European Communities pursued the compatibility of UK regulations with the requirements of the original Directive with some vigour and publicly queried whether seven previously approved projects, including the Channel Tunnel Rail Link, ought to have proceeded without EIA. This European intervention caused a considerable political storm, since the UK has prided itself on its implementation of European directives (see above). This misunderstanding was the only occasion to date that the subject of EIA and its integration within British decision-making procedures achieved prominence in the UK media. It is, however, not uncommon for the media to refer to EIA in relation to controversial proposals. Further European enforcement action in the form of a formal request for the EIA of a leisure complex at Crystal Palace and for the promulgation of the uncultivated land regulations was made in 2001.

Implementation of the regulations

Of about 4,000 ESs prepared between 1988 and 2000, over 2,700 were produced under the Planning Regulations for England, Wales and Scotland (see Figure 4.2). Less than 10 per cent of ESs related to projects falling within Annex I of the European Directive (see Chapter 3). Most ESs related to Annex II infrastructure, waste disposal, energy and extractive industry projects (Figure 4.3) (Wood and Bellanger, 1998). Most LPAs have received at least one ES but the county councils (in England) have received many more ESs than the districts and unitary authorities.

It should be observed that, in comparison to the United States (Mandelker, 2000), there have been few court cases relating to EIA. On this criterion, the introduction of EIA into the UK land-use planning system may be judged to have been implemented successfully. Nevertheless, the number of legal actions involving EIA has increased and several important court cases relating to EIA have revealed weaknesses in local implementation – LPAs have been found to be wrong in not requiring EIA, and planning decisions have been quashed as a result. These decisions reflect considerable variations in practice in implementing the various requirements of the Planning Regulations (DoE, 1991a, 1996; Leu *et al.*, 1996; Jones *et al.*, 1998; Glasson *et al.*, 1999). These variations are described in Chapters 6 to 20. There is less evidence about the implementation of the other types of EIA regulation but these variations in practice are believed to apply equally (see, for example, Weston and Prenton-Jones, 1997).

In relation to the planning system, DoE published advice, additional to the circular and the procedural guide, on the preparation of ESs (DoE, 1995) and on reviewing and dealing with ESs (DoE, 1994a). Other guidance has been published by the Department of Transport (1993) and DETR (1998a) (on roads), by various other national bodies, and by Essex County Council (2000). Commentaries have been published by, *inter alia*, Sheate (1996) and Glasson

Environmental impact assessment

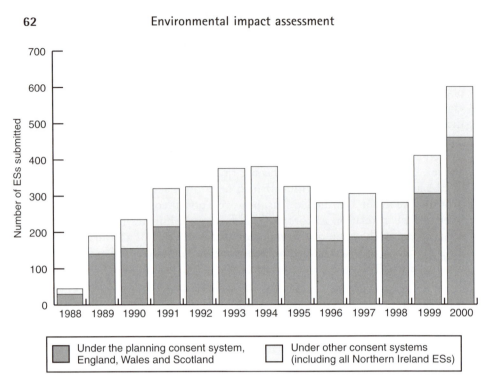

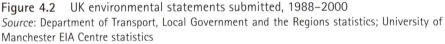

Figure 4.2 UK environmental statements submitted, 1988–2000
Source: Department of Transport, Local Government and the Regions statistics; University of Manchester EIA Centre statistics

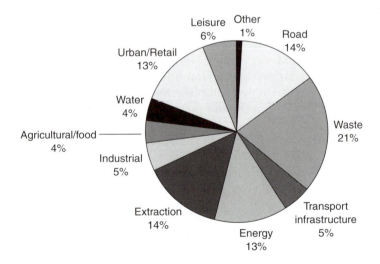

Figure 4.3 Types of project subject to UK EIA, July 1988 to April 1998
Source: Wood and Bellanger, 1998

et al. (1999). Numerous EIA training activities have also been arranged by professional bodies and by universities, several of which have been supported by speakers from DETR. The UK has several successful established masters programmes in EIA.

There has been considerable research interest in the implementation of EIA in the UK. Apart from work funded by research councils and undertaken by university researchers, both DETR and the European Commission have initiated substantial studies. DoE commissioned an early monitoring study (DoE, 1991a), work on the evaluation of ESs (DoE, 1994b), on the preparation of ESs (DoE, 1995), on the quality of ESs (DoE, 1996) and on mitigation in EIA (DETR, 1997d). The European Commission's five-year review of the operation of the EIA Directive (CEC, 1993), study of the costs and benefits of EIA (European Commission, 1996a) and evaluation of the effect of experience on the performance of the EIA process (European Commission, 1996b) all included substantive treatment of the UK EIA system.

There is no formal requirement for the strategic environmental assessment of policies, plans and programmes in the UK. In accordance with the government's avowed intention to put 'sustainable development at the heart of every Government Department's work' (Her Majesty's Government, 1999) a brief guide to the incorporation of environmental considerations into policy appraisal (environmental appraisal) was published in 1998 (DETR, 1998b). In addition the government has commissioned research into methods of environmental appraisal (DETR, 1998c). Therivel (1998) reported that, notwithstanding the absence of regulatory requirement, numerous environmental appraisals of development plans and a number of SEAs of European Structural Fund applications, and of water and other sectoral programmes, had been undertaken.

Notes

1. Part of this account is derived from Wood and McDonic (1989). See also Sheate (1996) and the first edition of this book.
2. DETR became the Department of Transport, Local Government and the Regions in 2001, losing the environment portfolio to another department.
3. This guidance, published to clarify and place the 1999 Regulations in context, replaced a similar, widely read, earlier guide which explained the 1988 regulations (DoE, 1989).

Chapter 5

EIA in the Netherlands, Canada, the Commonwealth of Australia, New Zealand and South Africa

This chapter provides an overview of the EIA systems in the Netherlands, Canada, the Commonwealth of Australia, New Zealand and South Africa. In each case a brief account of the origins and development of the EIA system is presented. This is followed by a summary of the legislative provisions for EIA and an outline of the main procedural steps in the EIA process. The principal organisations responsible for EIA are also mentioned. The accounts of the EIA systems in this chapter are intended to provide an introduction to the discussion of the different stages or aspects of the EIA processes presented in Chapters 6–20.

The Dutch EIA system

The Netherlands has a population of 16 million people, not much smaller than that of Australia, and an area of 41,000 square kilometres (about 0.5 per cent of the size of Australia or the contiguous United States of America). The country is not only densely populated but low lying: a significant proportion of its land area has been reclaimed from the sea and is protected by dikes. It is not surprising, therefore, that there should be a deep national concern for the environment (Mostert, 1995). This has manifested itself in a strong land-use planning system and in a high and long-standing level of public interest in environmental protection. The existence of a proportional representation system has led to the translation of this public concern about the environment into political interest and consequent legislative and budgetary action.

The Dutch EIA (or m.e.r. – milieu-effectrapportage) system, which owes its parentage more to Canada than to the United States, was very carefully considered before implementation. North American EIA specialists were employed by the government to assist in this process (Jones, 1984). The Netherlands commissioned a series of research studies in the late 1970s which resulted in proposals for the introduction of a formal EIA system. These proposals were largely independent of the European Commission's initiative on EIA, despite taking a similar time to come to fruition. Nine 'trial run' EIAs were undertaken to ensure that the procedures envisaged were practicable. A governmental standpoint was submitted to the Dutch parliament in 1979 and a bill was introduced in 1981. Twelve EIAs were initiated using the prospective legislation and an 'interim' EIA Commission was set up (Mostert, 1995;

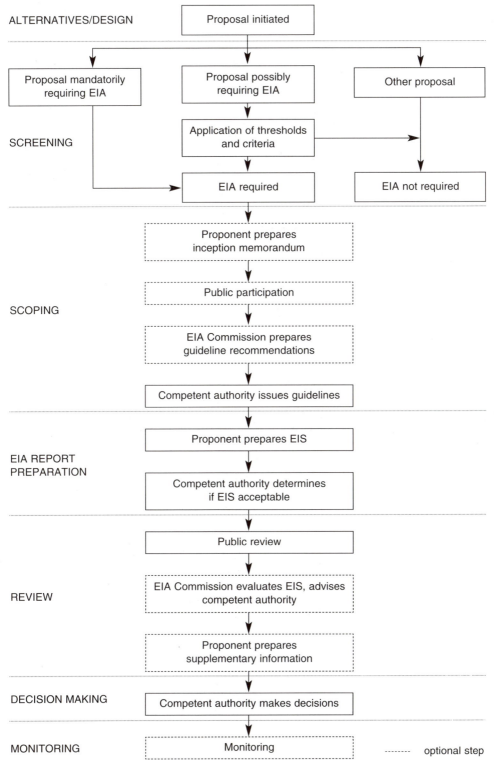

ALTERNATIVES/DESIGN

Proposal initiated

Proposal mandatorily requiring EIA

Proposal possibly requiring EIA

Other proposal

SCREENING

Application of thresholds and criteria

EIA required

EIA not required

SCOPING

Proponent prepares inception memorandum

Public participation

EIA Commission prepares guideline recommendations

Competent authority issues guidelines

EIA REPORT PREPARATION

Proponent prepares EIS

Competent authority determines if EIS acceptable

REVIEW

Public review

EIA Commission evaluates EIS, advises competent authority

Proponent prepares supplementary information

DECISION MAKING

Competent authority makes decisions

MONITORING

Monitoring

-------- optional step

Figure 5.1 Main steps in the Dutch EIA process

Arts, 1998). By 1987, when the EIA provisions incorporated into the Environmental Protection (General Provisions) Act came into effect, considerable experience of EIA had been gained. These provisions were consolidated in the Environmental Management Act 1994, to which subsequent amendments have been made. They are supported by the Environmental Impact Assessment Decree 1994 (No. 224),[1] as amended, and by the Environmental Impact Assessment (Inception Memorandum) Regulations 1993. Five of the twelve provinces had introduced supplementary requirements to extend the coverage of EIA to additional activities by the end of 2001.

The main features of the Dutch EIA system are shown in simplified form in Figure 5.1 (see also Arts, 1998; Glasson *et al.*, 1999). Several features should be noted:

1. The EIA process is integrated into existing decision-making procedures.
2. The EIA process is not confined to projects.
3. There are statutory requirements relating to the treatment of alternatives, to scoping (including the preparation of project-specific guidelines), to the review of EIA reports and to the monitoring and evaluation of the impacts of implemented projects.
4. There are provisions for public participation at both the scoping and EIA report review stage and there is a third-party right of appeal against decisions.
5. The Dutch EIA Commission (see below) plays a central and very influential role in the EIA process generally and at the scoping and EIA report review stages in particular.

The EIA Commission (EIAC or Commissie voor de milieu-effectrapportage) is an independent body able to call upon about 200 appointed members (experts) to provide guidance (Box 5.1).[2] The involvement of this body is required by law and EIAC's opinions are made public (Mostert, 1995; Arts 1998). EIAC furnishes part-time chairpersons, from its presidium of a chairperson and several deputies, to guide small EIA working parties of two to seven Commission members with expertise relevant to the activities assessed (EIAC, 1998). Other experts from a further pool of 500 persons are co-opted where necessary. The working parties advise on the guideline recommendations prepared and assess the accuracy of the scientific content of the EIA report and its completeness in relation to the statutory requirements and the guidelines. EIAC, which has staff of about 35, provides a technical secretary for each of these working parties (EIAC, 1998). EIAC also provides scoping and review advice on projects and programmes supported by Dutch development cooperation funding (EIAC, 2000).

The Dutch government is committed by law to submit a report to parliament on the functioning of the EIA system every five years. The first evaluation report was published on schedule in 1990 and, while giving the EIA system a generally favourable review, it suggested a number of amendments to the system, including (at the request of the government) several to implement

Box 5.1   The Dutch EIA Commission

The Environmental Impact Assessment Commission fulfils an important role as an *independent adviser* to the competent authority. The EIA procedure includes fixed moments when the Commission is to provide advice, and these are governed by strict deadlines. The Commission's main task consists of providing advice on:

- scoping guidelines for the content of the EIS to be adopted by the competent authority (advisory guidelines);
- reviewing the EIS drawn up by the project proponent (advisory review).

In addition, the Commission advises on requests for exemption from the duty to carry out an EIA and, when requested by the environment and nature management ministries, reviews EISs prepared in neighbouring countries for environmental consequences in the Netherlands (under the EU directive on environmental assessment and the UNECE Espoo Convention concerning transboundary EIAs).

The Commission consists of about 200 experts (members) appointed by Royal Decree. These experts mainly hold posts at universities, research institutes, consultancies and government organisations. They are invited to take part in working groups on a project-by-project basis, depending on the expertise required. Commission members are appointed for a five-year period and are eligible for reappointment. If no Commission members are available, or if there is a lack of specific expertise required for a project within the Commission, other experts (referred to from now on as advisers) may be invited to take part. Informally, these advisers have the same status and authority as the appointed Commission members.

The Commission members and advisers are supported in their work by the Commission's secretariat. The secretariat comprises the general secretary, three deputy secretaries, ten working group secretaries and sixteen administrative, computer, documentation and accounts staff members. The Commission has a chairman and five deputy chairmen who also chair the individual working groups.

Source: EIAC, 1998, p. 2.

---

the European Directive more faithfully (Evaluation Committee on the Environmental Protection Act – ECW, 1990). A further report was released in 1996 (ECW, 1996).

Somewhat surprisingly, the original Dutch legal provisions did not fully implement the requirements of the European Directive on EIA (see Chapter 3). The European Commission formally noted deficiencies relating to the failure to ensure the EIA of certain Annex I projects and to the exclusion of some Annex II projects in 1990. Further, the Commission noted shortcomings in relation to exemption provisions and to the consultation of neighbouring countries. These deficiencies appear to have been due to a misunderstanding and the amendments made to the Act and the EIA Decree in 1994, especially the introduction of a screening procedure for Annex II projects, were designed to meet the comments both of the European Commission and of the Evaluation Committee. The 1999 amendments were intended to implement the requirements of the revised European EIA Directive. However, in 2001 the European Commission formally requested the Netherlands to tighten its legislative requirements for impact coverage, screening and EIS content.

A considerable body of literature on EIA has been established in the Netherlands, including substantial contributions in English (see, for example, Mostert, 1995; EIAC, 1996, 2001b; de Jong, 1997; Arts, 1998). A comprehensive handbook on environmental impact assessment, published by the Ministry of Housing, Spatial Planning and Environment (VROM) was completely rewritten in 1994. Various other guides, including three on scoping, one on alternatives and three on monitoring and auditing have been released. In addition, a series of guides (some in English) on the application of methods of impact forecasting, covering air, surface water, soil, plants/animals/ecosystems, landscape, noise, radiation and health impacts has been published. There are also guides on the EIA of certain types of project (for example motorways, published by the Ministry of Transport).

Several training courses have been run by the authorities involved in EIA and by universities. A substantial body of EIA research findings exists (see, for example, VROM, 1994a,b). There is now an influential and growing community of over 400 EIA professionals working for developers, competent authorities, ministries, agencies, consultancies and universities, apart from those associated with the EIA Commission, VROM and the Ministry of Transport. Many members of this community are members of the EIA section of the Dutch Society of Environmental Professionals. There is a Dutch journal (*KenMERken)* devoted to EIA. In addition, the Ministry of Transport publishes its own EIA journal. While there remain some difficulties in the operation of the Dutch EIA system (ECW, 1996; Sadler, 1996), national EIA information, knowledge and skill levels are now substantial.

The level of EIA activity has been considerable over the years (see Figure 5.2). The combined numbers of guideline recommendations and reviews of environmental impact statements increased year by year until 1993 after which the number of projects for which EIA procedures were initiated levelled off. It currently averages 60–80 per annum. Decisions had been made on 467 projects by the end of 2000, with suspensions or cancellations taking place in 199 cases (EIAC, 2001a). Non-environmental factors are generally the reason for those project cancellations which do occur.

The Canadian EIA system

Canada is a vast country, the second largest in the world, covering nearly 10 million square kilometres, with a population of 30 million, most of whom live in a band about 150 kilometres wide immediately north of the US border. The

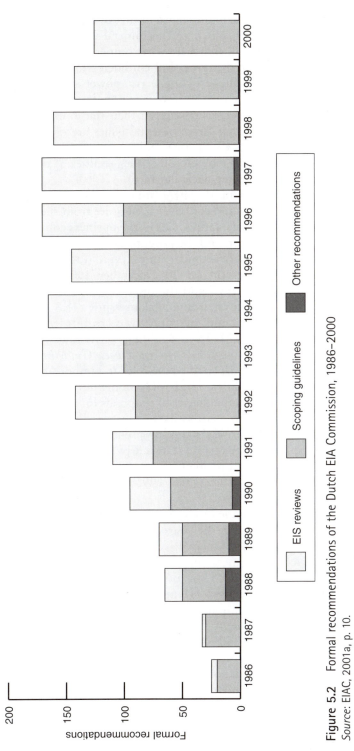

Figure 5.2　Formal recommendations of the Dutch EIA Commission, 1986–2000
Source: EIAC, 2001a, p. 10.

population density of Canada, less than one-tenth of that of the United States, is thus somewhat misleading, since much of the mineral-rich north is very sparsely populated. Canada's fragile northern environment has sometimes fallen victim to the frontier ethic in the form of ill-considered logging, mining or so-called mega-projects such as huge hydroelectric power stations. More recently, enough of the multi-ethnic urban immigrant population of this out-door country have joined the indigenous peoples in their traditional concern about the environment to ensure that environmental issues are often placed high on the political agenda (Rees, 1987). Inevitably, however, this environ-mental concern has been somewhat tempered by other realities as the econ-omic pendulum of recession has swung back and forth.

It was inevitable that interest in the EIA provisions in the National Environmental Policy Act 1969 should spill over the border from the United States to Canada. A group of officials, individuals in the business community, and citizens' groups saw the need for, and the potential of, EIA, and the non-mandatory federal Environmental Assessment and Review Process (EARP) was established by a cabinet decision on 20 December 1973. This was amended by a second decision in 1977 and the responsibility of the federal Minister of the Environment for the environmental assessment (EA, not EIA) of federal projects, programmes and activities was reaffirmed in the Government Organisation Act 1979. In 1984, these provisions were for-malised in the *Environmental Assessment and Review Process Guidelines Order* (Government of Canada, 1984) which clarified the roles and responsibilities of the participants in the EARP procedures. The EARP Guidelines were intended to ensure that the environmental consequences of proposals for which the fed-eral government had decision-making authority were assessed. EARP, which was principally (but not solely) a project planning process, was based on the principle of self-assessment (i.e. EA was undertaken by each federal depart-ment, as in the United States) but provided for review by an EA panel in a limited number of cases (Northey, 1994; Sadler, 1995; Sadar and Stolte, 1996; Hazell, 1999).

The EARP Guidelines were intended to be advisory but, following success-ful and highly publicised challenges in the courts by environmental groups in 1989 in the Rafferty/Almeda and Oldman River cases, the 1984 Order-in-Council was held to be a law of general applicability, binding on the federal government (Hazell, 1999). The need for EARP reform had been widely recognised for some time and the court cases emphasised the urgency of change and raised EA decision making to public prominence. Accordingly, in fulfil-ment of a 1984 election commitment (Fenge and Smith, 1986), the federal government introduced the Canadian Environmental Assessment Bill in 1990. After extensive consultation and substantial revision, the bill was given Royal Assent in June 1992. Four key regulations required for its operation came into effect when the Canadian Environmental Assessment Act was proclaimed in force early in 1995.

The Act is intended to entrench in law the federal government's obligation to integrate environmental considerations in all its decisions relating to projects (but not to policies, plans and programmes, to which EARP, in principle, applied). The Act was part of a package intended not only to reduce the uncertainties associated with EARP but to make the EA process more efficient, effective, fair and open. A second element of the package was a non-statutory requirement for strategic environmental assessment. A summary of the environmental assessment of every proposal submitted to cabinet had to be released at the time the cabinet decision was announced (see Chapter 19). The final element of the reform package was a participant funding programme to help individuals and organisations to involve themselves in public reviews of projects (see Chapter 16).

The EA process under the Act bears many similarities to EARP but is intended to be less discretionary. A new, more autonomous agency – the Canadian Environmental Assessment Agency (CEAA) – replaced the pre-existing Federal Environmental Assessment Review Office and was given additional powers over the EA process. The principal steps in the process are shown in Figure 5.3. The federal Canadian EA system consists of two separate, sometimes successive, procedures: the self-directed assessment and the public review. Each of these procedures contains two assessment tracks, each with its own steps. Most of the federal departments or agencies having decision-making authority for projects (the 'responsible authorities') have developed their own environmental assessment procedural guidance. Federal EA coordination regulations have been promulgated to clarify the nature of the necessary cooperation between responsible authorities. Where projects subject to 'screening' have potentially significant adverse environmental impacts, or if the project is the subject of public concern because of its potential environmental effects, the responsible authority's minister is supposed to refer the proposal to the Minister of the Environment for public review (Northey, 1994; Hazell, 1999).

The system is almost comprehensive but allows for the vast majority of federally controlled project EAs to be dealt with as screenings. The number of screenings in 1999–2000 was reported to be nearly 5,700 (CEAA, 2000). A small number of projects (less than 10 per annum)[3] is subject to 'comprehensive study', and an even smaller number (about two a year) to panel review, requiring the preparation of an environmental impact statement. There have been no mediations to date. Notable positive features of the Act are the provisions relating to cumulative environmental effects, to the provision for mediation (Sadler, 1994) and to follow-up provisions. One of the four purposes of the Act, along with consideration of environmental effects in decision making, consideration of transboundary effects and the encouragement of public participation is:

> to encourage responsible authorities to take actions that promote sustainable development and thereby achieve or maintain a healthy environment and a healthy economy. (section 4(b))

Following an exhaustive review of the Act after five years of operation, the Minister of the Environment (2001) suggested a series of amendments which were taken forward in a bill (Government of Canada, 2001).

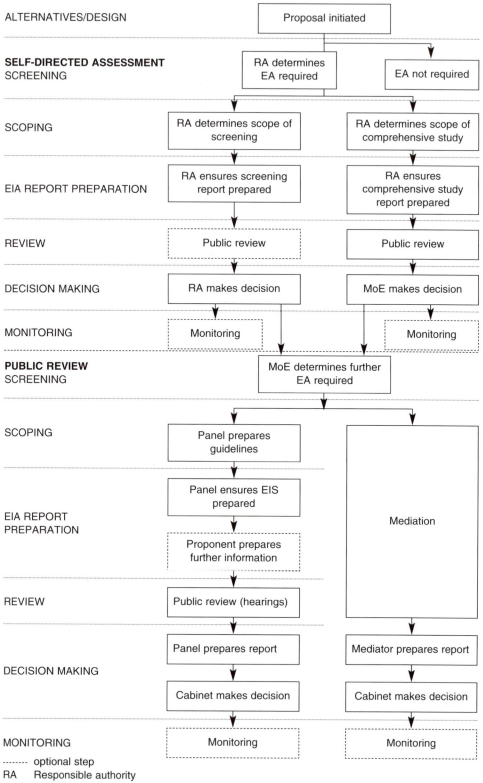

ALTERNATIVES/DESIGN — Proposal initiated

SELF-DIRECTED ASSESSMENT
SCREENING — RA determines EA required / EA not required

SCOPING — RA determines scope of screening / RA determines scope of comprehensive study

EIA REPORT PREPARATION — RA ensures screening report prepared / RA ensures comprehensive study report prepared

REVIEW — Public review / Public review

DECISION MAKING — RA makes decision / MoE makes decision

MONITORING — Monitoring / Monitoring

PUBLIC REVIEW
SCREENING — MoE determines further EA required

SCOPING — Panel prepares guidelines

EIA REPORT PREPARATION — Panel ensures EIS prepared / Proponent prepares further information / Mediation

REVIEW — Public review (hearings)

DECISION MAKING — Panel prepares report / Cabinet makes decision / Mediator prepares report / Cabinet makes decision

MONITORING — Monitoring / Monitoring

-------- optional step
RA   Responsible authority
MoE   Minister of the Environment

Figure 5.3   Main steps in the Canadian EA process

The president of the Canadian Environmental Assessment Agency reports directly to the federal Minister of the Environment. The Agency, which has a staff of about 100, a budget of over C$10 million, and six regional offices in addition to its headquarters, is responsible for providing policy direction and procedural information on the Act and for periodic reports on its implementation (CEAA, 2000). It also coordinates certain aspects of comprehensive studies and provides the secretariat for the reviews conducted by federal environmental assessment panels (often jointly with the provinces).

In addition, each of the ten provinces has its own EA legislation (Doyle and Sadler, 1996). About a dozen municipalities, including the City of Ottawa, have also instituted their own formal or informal EA requirements or are putting them into place (Lawrence, 1994). These are not normally directly complementary to either the federal or provincial processes. Most of the land north of 60 degrees latitude (the northern boundary of most of the provinces) is under federal jurisdiction but there are several land claims settlements with aboriginal peoples. Special EA provisions apply within these areas and within the James Bay and Northern Quebec Agreement area – the site of the Great Whale 10,000 MW hydroelectric mega-project. Responsibility for environmental and resource matters is split under the constitution between the federal government and the provinces, and many projects require EA by more than one government (Clark and Richards, 1999). EA has been the subject of significant jurisdictional conflicts between federal and provincial governments. Generic joint EA arrangements have now been put in place through a Canada-wide accord on environmental harmonisation and through a series of bilateral agreements on EA (for example, between Canada and Alberta). As a result of these, almost all federal panel reviews are conducted jointly with the provinces.

Many aspects of the environmental assessment provisions in the ten provinces of Canada are stronger than the equivalent parts of the federal EA system (Gibson, 1993; Smith, 1993, pp. 40–1). Unfortunately, a change of provincial government in the late 1990s led to the weakening of the widely admired Ontario EA system, which was originally enshrined in legislation in 1976. Table 5.1 compares the federal and provincial systems in 1994.

Between 1984 and 1992 much EA research was administered through a grant to the Canadian Environmental Assessment Research Council (CEARC). The objective was to promote research to improve the scientific, technical and procedural basis for EA and to advance the theory and practice of EA. This much-admired independent body sponsored research and commissioned reviews, studies and workshops. It produced about 150 reports, many of which have been disseminated overseas earning Canada an international reputation as a leader in the EA field (see, for example, Jacobs and Sadler, 1990). It was disbanded in 1992 largely as a result of public expenditure cuts. CEARC's legacy survives, however, as the Canadian Environmental Assessment Act 1992 requires CEAA to promote or conduct research on EIA in Canada (section 62(c)). The Agency now has a reduced research budget, much of which has, been directed to national studies focused on improving the practice of EA under the Act.

Table 5.1 Characteristics of Canadian federal and provincial EA procedures, 1994

JURISDICTION	Project screening	Scoping to key decision topics	Project terms of reference	Specific mitigation	EA or EIS document filed	Review of EA by governments and public	Terms and conditions of approval	Surveillance of construction or implementation	Monitoring of effects or post-construction evaluation	Periodic EA audits of approvals	Periodic evaluation EA process
British Columbia	●	●	●	●	●	●	●	X	●	●	●
Alberta	●	●	●	●	●	●	●	●	●	●	●
Saskatchewan	●	●	●	●	●	●	●	○	○	○	●
Manitoba	●	X	○	●	●	●	●	○	●	X	○
Ontario	●	X	X	●	●	●	●	X	○	X	●
Quebec	X	X	●	●	●	●	●	○	○	X	○
New Brunswick	●	●	●	●	●	●	●	○	○	X	○
Nova Scotia	●	X	●	●	●	●	●	X	○	X	●
Prince Edward Island	○	○	○	●	●	●	●	○	○	●	●
Newfoundland	●	●	●	●	●	●	○	●	○	X	○
Canada (CEAA)	●	●	●	●	●	●	○	X	○	X	●

● Yes   ○ Partially/Optionally   X No

Source: adapted from Doyle and Sadler, 1996, p. 15.

Environmental assessment is a high-profile process in Canada, partly because its application provides one of the most visible manifestations of the government's commitment to the environment and partly because it often provides the best available opportunity for public participation in environmental decision making. (Much of Canada possesses little in the way of enforceable land-use planning outside the more populated areas (Rees, 1987).) Despite major cuts to EA budgets (Hazell, 1999), the Canadian 'EA world' probably consists of over 2,000 people.

The Commonwealth of Australia EIA system

Australia, as well as taking many of its immigrants and an essentially British system of parliamentary government from the old world during the last two centuries, has, in more recent years, taken much from the older new world: the United States. Although the population of Australia (19 million) is much smaller than that of the contiguous United States, its land area is similar and its history of colonisation and its frontier ethic are, not surprisingly, also similar. By the 1960s and early 1970s, it had become apparent that Australia's state-controlled land-use planning systems, though stronger than those in the United States, lacked the power to prevent some of the worst excesses of development. The Commonwealth of Australia (Australia's federal government) recognised the need for stronger environmental control and looked with interest at the advantages of the US federal EIA process.

The Commonwealth and the six states and two main territories could not agree on the need for EIA, and the Commonwealth constitution, dating from 1901, made no mention of the environment. While many of the states and territories (each of which has its own government) resented Commonwealth interference in their right to control their own environment, others were more enthusiastic. (Some states proceeded to implement their own informal EIA provisions.) Despite several meetings of the environment ministers forming the (then) Australian Environment Council, the Commonwealth eventually recognised that uniformity was unattainable and passed its own legislation (Fowler, 1982). The Environment Protection (Impact of Proposals) Act became law in 1974, some 14 years before the European Commission's directive on EIA came into effect. Like the US National Environmental Policy Act 1969 (NEPA), this Act related only to federal activities (and not to those undertaken by state or local governments) and was independent of existing procedures. However, quite deliberately, far more discretion was built into the Australian Commonwealth system, and, consequently, there has been far less opportunity for resort to the courts than in the United States.

Since the Act was passed, there has been widespread recognition by all the states of the need for stronger environmental controls and each has passed legislation or set procedures in place to extend the scope of EIA to their own activities. There is no doubt that 'the main plank of the Commonwealth's powers with respect to environmental matters' (Fowler, 1996, p. 246; see also Brown and McDonald, 1995), the Impact of Proposals Act, established the framework within which all subsequent EIA processes in Australia have been developed. Some of these state EIA systems (between which the similarities are more apparent than the differences) became stronger, of wider application, and generally more effective than their Commonwealth precursor (Porter, 1985; Harvey, 1998; Thomas, 1998). There is much greater EIA activity at state than at Commonwealth level and most (but not all) of the EIA systems have been progressively strengthened over the years (Harvey, 1998; Thomas, 1998). Considerable harmonisation of EIA requirements has taken place over the 1990s in the interests of trade and competition in Australia (Australian and New Zealand Environment and Conservation Council – ANZECC, 1991, 1996, 1997; Commonwealth Environment Protection Agency – CEPA, 1992; Council of Australian Governments – COAG, 1997). The relevant state and territory EIA legislation, authorities and levels of assessment are shown in Table 5.2.

Despite numerous reviews by a large number of bodies, there was little fundamental change to the Commonwealth EIA system beyond the introduction of scoping, and of the less rigorous and detailed 'public environmental report' (PER) procedure, together with some encouragement for strategic environmental assessment, for 25 years. While responsibility for the administration of the Impact of Proposals Act always rested with the Environment Minister, both the name and organisational structure of the responsible department changed repeatedly (Harvey, 1998, p. 162). During the early 1990s, a semi-autonomous Commonwealth Environment Protection Agency (CEPA) oversaw an exhaustive review which formalised much detailed information about the EIA system, together with numerous reform proposals (CEPA, 1994; Carbon, 1998). The incoming Liberal government instigated a review of federal/state roles in environmental protection through the Council of Australian Governments. Agreement on a number of issues was reached (COAG, 1997) and this facilitated the publication of a ministerial consultation paper (Hill, 1998) which incorporated many of the earlier EIA reform proposals. This rapidly led to the passing of the Environment Protection and Biodiversity Conservation Act 1999 (EPBC Act) (Padgett and Kriwoken, 2001).

This legislation was enacted, following countless amendments, as a result of intense political manoeuvring by the powerful Environment Minister and Leader of the Government in the Senate, Robert Hill. The EPBC Act replaces not only the Impact of Proposals Act but several other laws and has its main objective:

> to provide for the protection of the environment, especially those aspects of the environment that are matters of national environmental significance. (section 3(1) (a))

Chapter 4 of the EPBC Act (which, in total, has 528 sections) details the procedures for environmental assessments and approvals and has 105 sections. (The Impact of Proposals Act had, by comparison, only 25 sections.)

The main steps in the Commonwealth EIA system, which came into effect on 16 July 2000, are summarised in Figure 5.4. The Act (which has been

State or territory	Legislation	Responsible authority	Levels of EIA			
			Lowest			→ Highest
Australian Capital Territory	Land (Planning and Environment) Act 1991	Environment Minister	PA	PER	EIS	Inquiry
New South Wales	Environmental Planning and Assessment Act 1979, and Amendment 1993	Department of Urban Affairs and Planning; Minister for Environment	SEE, REF		EIS	Inquiry
Northern Territory	Environmental Assessment Act 1982, and Amendment 1994	Minister for Environment		PER	EIS	Inquiry
Queensland	Integrated Planning Act 1997	Department of Environment and Heritage; local government			EIS or IAS	
South Australia	Development Act 1993, and Amendment 1996	Minister for Housing and Urban Development; Minister of Mines	DR	PER	EIS	Inquiry
Tasmania	Environmental Management and Pollution Control Act 1994	Local government; Board for Environmental Management; Sustainable Development Advisory Council	Level 1	Level 2 DPMP	Level 3 EIS, SECIS, or EMP	
Victoria	Environmental Effects Act 1978, and Amendment 1995	Minister for Planning and Local Government			EES (often includes Inquiry)	
Western Australia	Environmental Protection Act 1986	Minister for Environment	CER	PER	ERMP	Inquiry

CER   consultative environmental review
DPMP   development proposal and management plan
DR   development report
EES   environmental effects statement
EIS   environmental impact statement
EMP   environmental management plan
ERMP   environmental review and management program

IAS   impact assessment study
PA   preliminary assessment
PER   public environmental report (*Western Australia:* Public Environmental Review)
REF   review of environmental factors
SECIS   social, economic and community impact statement
SEE   statement of environmental effects

Source: derived from Harvey, 1998, especially p. 51.

Environmental impact assessment

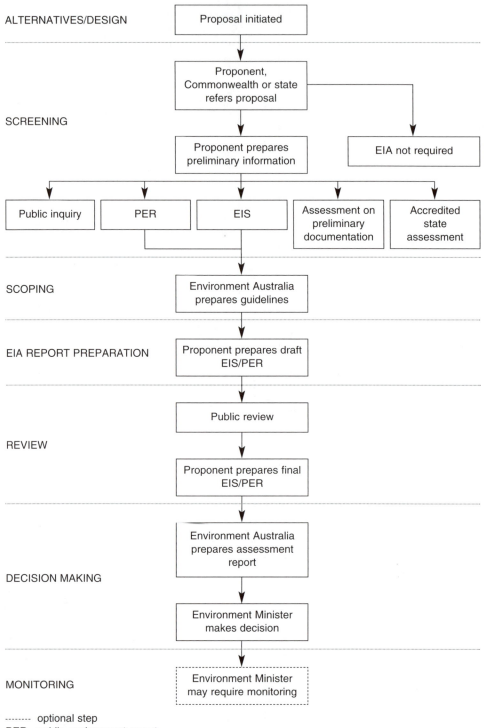

ALTERNATIVES/DESIGN

SCREENING

SCOPING

EIA REPORT PREPARATION

REVIEW

DECISION MAKING

MONITORING

-------- optional step
PER   public environment report

Figure 5.4   Main steps in the Commonwealth of Australia EIA process

amended) contains powers relating to screening, to scoping, to the preparation of draft and final environmental impact statements, to consultation and participation, to the preparation of an assessment report, to the taking of this assessment report into account in decision making, to strategic environmental assessment and to the holding of inquiries. There also exists a discretionary power to require the auditing of approved actions. The EPBC Act relies on significant federal/state cooperation to implement the accreditation provisions which are intended to avoid duplication of processes and to allow state EIA processes to be undertaken in place of those of the Commonwealth (Hughes, 1999; Ogle, 2000; Scanlon and Dyson, 2001).

Unlike NEPA and the Impact of Proposals Act, the EPBC Act provides a legal framework that contains considerable procedural detail about the various stages of the EIA process. In addition, it provides for the approval of regulations and administrative guidelines. Accordingly, the Australian government issued the detailed Environment Protection and Biodiversity Conservation Regulations 2000 (No. 181) in July 2000, together with the administrative guidelines on significance (Environment Australia, 2000a). The regulations and guidelines set out further details of how the Act is to be administered. Various explanatory memoranda (Environment Australia, 1999b, 2000c), leaflets (Environment Australia, 1999a), fact-sheets and other documents (for example, Environment Australia, 2000b), all of which can be accessed on the Environment Australia web site,[4] have been produced to explain the legislation. There has also been a number of commentaries on the EPBCA (see, for example, Hughes, 1999; Ogle, 2000; Padgett and Kriwoken, 2001; Scanlon and Dyson, 2001).

The EIA function of the Commonwealth is currently administered by the Environment Assessment and Approvals branch of Environment Australia, a part of the Department of the Environment and Heritage. The branch has a staff of about 35 devoted to the implementation of those EPBC Act assessment and approval requirements for which the Commonwealth has environmental assessment and approval responsibility. Between the Impact of Proposals Act's commencement in December 1974 and July 2000, when it was repealed, about 4,000 referrals were received (an average of about 160 per annum) and 178 EISs, 37 PERs and 6 public inquiry directions were made, an average of fewer than 9 EIA reports per annum. During 1999–2000, prior to the EPBC Act, Environment Australia received 314 proposals to assess, and directed that 5 EISs and 4 PERs be prepared and that one public inquiry be undertaken. Thus, only 10 particularly controversial projects (i.e. less than 5 per cent of proposals) resulted in EIA reports or an inquiry, with the remainder being otherwise assessed. The number of referrals during the first full year of operation of the EPBC Act was 303, resulting in 6 EISs and 3 PERs.[5]

The New Zealand EIA system

New Zealand underwent a revolution in environmental management in the late 1980s. Several government departments were abolished or restructured, privatisation prevailed, local government was reorganised and environmental law was utterly transformed. The Resource Management Act 1991, which replaced numerous previous Acts (including the Town and Country Planning Act 1977, the Clean Air Act 1972 and the Water and Soil Conservation Act 1967) introduced EIA as a central element in a decision-making process designed to promote the goal of sustainable management (Box 5.2). This is EIA Mark II in New Zealand, the Mark I system having been introduced in 1974. EIA is now, in principle at least, comprehensive and flexible in that it applies, at the appropriate level of detail, to virtually all projects and, in addition, may apply to policies and plans prepared under the Resource Management Act 1991 provisions.

New Zealand has a slightly larger land area than the UK, but, with 3.6 million people, its population density is only 5 per cent of the UK's. New Zealand local authorities, which are generally not as well staffed as those in the UK, serve average populations about twice as small (less than 50,000). Under the current arrangements, New Zealand local government (rather than central government as under the Mark I system) administers the EIA system. In particular, the planning departments of local authorities (especially of district councils) are responsible for dealing with proponents and making recommendations on the basis of the EIA or assessment of environmental effects (AEE) undertaken.

New Zealand introduced its Environmental Protection and Enhancement Procedures, partially modelled on the Canadian EIA system, in 1974 by means of a cabinet minute. These evolved over the years and in the latest version of the Procedures (Ministry for the Environment – MfE, 1987) public agencies were required to undertake screening, agree project-specific scoping guidelines with the Ministry for the Environment, consult with various statutory, local

---

Box 5.2   Purpose of the New Zealand Resource Management Act 1991

Section 5. Purpose: (1) The purpose of this Act is to promote the sustainable management of natural and physical resources.

(2) In this Act, 'sustainable management' means managing the use, development and protection of natural and physical resources in a way, or at a rate, which enables people and communities to provide for their social, economic, and cultural well-being and for their health and safety while:

(a) Sustaining the potential of natural and physical resources (excluding minerals) to meet the reasonable needs of future generations; and

(b) Safeguarding the life-supporting capacity of air, water, soil, and ecosystems; and

(c) Avoiding, remedying, or mitigating any adverse effects of activities on the environment.

and other authorities, publish an environmental impact report, submit it to a formal published 'audit' by the Parliamentary Commissioner for the Environment (PCE, an 'ombudsman' reporting not to the government but to parliament) and agree monitoring arrangements with the Ministry for the Environment. This EIA process was subject to considerable public oversight but was largely discretionary and, in practice, related to only a limited number of public sector projects, virtually all of which were approved. Wells and Fookes (1988) reported that an average of three formal environmental impact reports were formally audited (reviewed) each year between 1977 and 1989 but that hundreds of impact statements of variable size, scope and quality were being produced annually for local authorities and other bodies but not being subjected to the Procedures. In principle, but not in practice, the Procedures also applied to policies.

Considerable debate about the Procedures developed (see, for example, Morgan, 1988). Wells and Fookes (1988) summarised the need for reform to produce an EIA process which was 'simple but effective, comprehensive but not complex, flexible yet consistent, authoritative yet accessible'. At the same time, wide-ranging reviews of environmental administration, in which the Minister for the Environment (later Prime Minister) took a strong personal interest, and of local government, resulted in complete reorganisations of both (Bührs and Bartlett, 1993). The various objectives of these reorganisations included the decentralisation of decision making, increased consideration of the environment at all levels of decision making, enhanced public involvement, the elimination of administrative and legislative fragmentation, reducing the size and cost of central government and realising public assets.

New Zealand now has a system of 12 regional and 69 territorial (city and district) with 4 unitary authorities (replacing hundreds of local and special-purpose bodies) which are responsible for several duties formerly undertaken by central government (Bührs and Bartlett, 1993; Memon, 1993). Environmental impact assessment is one of these: EIA is now, in principle, inextricably integrated into local authority procedures for determining applications for land use and subdivision consents and for coastal, water and discharge permits under the provisions of the Resource Management Act. Central government must also comply with this system.

The Ministry for the Environment has issued a number of guides to the Resource Management Act over the past ten years. These include a guide to scoping (MfE, 1992b) and several guides mentioning the environmental assessment of regional policies and plans and of district plans (including MfE, 1992a, 1993). It also commissioned a more general guide on AEE (Morgan and Memon, 1993). Good practice guidance for applicants for resource consents on preparing a basic AEE report (MfE, 1999a), and for local authorities when reviewing AEE reports prepared by applicants (MfE, 1999b), has been published. The most recent guide is intended to increase public involvement in the resource management processes controlled by the Act (MfE, 2001d).[6] Several commentaries on the EIA system have been published (see, for example, Dixon, 1993; Montz and Dixon, 1993; Dixon and Fookes, 1995;

Morgan, 1995, 2000b; PCE, 1995; Smith, 1996; Bartlett, 1997; Williams, 1997; Fookes, 2000).

A simplified version of the main steps in the EIA system in New Zealand (derived from Montz and Dixon (1993) and PCE (1995)) is shown in Figure 5.5. The Act makes broad provisions in relation to the EIA system and has devolved almost all responsibility for the administration of EIA from central to local government. The Act provides the outline of the EIA process, but leaves much detail to be provided by individual regional authorities in their regional policy statements and regional plans and by territorial authorities in their district plans.

The Resource Management Act 1991 contains provisions which effectively provide for a two-phase screening process and encourage scoping. It indicates the content requirements for an AEE report (which include alternatives), provides for public participation and consultation and requires that the report be considered in the decision. It also contains provisions relating to monitoring. In addition, the Act provides for public council hearings into applications at which the Ministry for the Environment (and any other government agency) may submit its views. It also permits the call-in of requests for resource consents by the Minister for the Environment where issues of national significance are raised.

There is a universal right of appeal against local authority decisions to the Environment Court. This court, which supersedes the Planning Tribunal, has all the powers of a district court. It hears cases *de novo* and it considers both the merits and legal dimensions of cases. Its findings on the merits and related matters of fact are final but its decisions on matters of law can be appealed to higher courts. Apart from the value of the precedents created by the Environment Court's findings, the EIA system is also subject to scrutiny by the Parliamentary Commissioner for the Environment.

It is notable that EIA is not mentioned by name in the Resource Management Act but that the procedures and aims of EIA (in its New Zealand guise of AEE – used to describe both the EIA process and the EIA report) suffuse its policy and plan preparation provisions and its resource consent provisions (Dixon, 1993; Bartlett, 1997). An AEE, along with submissions on resource consents and council investigations, effectively provides the relevant environmental information for decision making. In this sense, as in others, the highly sophisticated Resource Management Act makes a marked contribution to the advancement of environmental planning and management policy.

The transition from the previous arrangements to the Resource Management Act procedures including the existence (but not the use) of the Environmental Protection and Enhancement Procedures for central government projects not covered by the Act, has lasted longer than anticipated. While many regional and district plans are partially operative, most were not expected to be finalised until 2002 or later (Dixon *et al.*, 1997; PCE, 1998). The Ministry for the Environment employs only a small number of central and regional office staff on EIA. Generally, the number of central and local

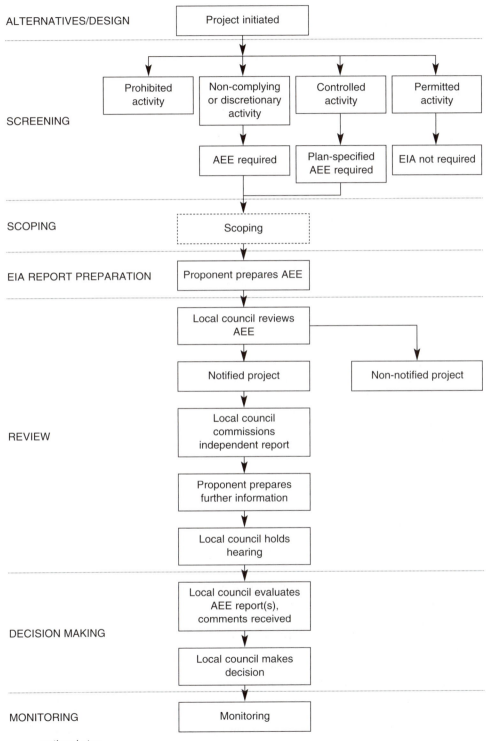

ALTERNATIVES/DESIGN

Project initiated

SCREENING

Prohibited activity

Non-complying or discretionary activity

Controlled activity

Permitted activity

AEE required

Plan-specified AEE required

EIA not required

SCOPING

Scoping

EIA REPORT PREPARATION

Proponent prepares AEE

Local council reviews AEE

Notified project

Non-notified project

REVIEW

Local council commissions independent report

Proponent prepares further information

Local council holds hearing

DECISION MAKING

Local council evaluates AEE report(s), comments received

Local council makes decision

MONITORING

Monitoring

-------- optional step

AEE   assessment of environmental effects

Figure 5.5   Main steps in the New Zealand EIA process

government officers, consultants, academics and environmental campaigners involved in EIA in New Zealand is very small (probably fewer than 50 full-time equivalents).

Several clarifying amendments to the Resource Management Act have been made and a number of reviews have been undertaken (see, for example, Dixon *et al.*, 1997), proposals advanced (for example, McShane, 1998) and rebuttals made (for example, PCE, 1998[7]) (see Chapter 17). Proposals for major amendment were advanced (MfE, 1998) and the consequent amendment bill has been widely discussed in parliament and elsewhere. No further amendments had been enacted in mid-2001 and those recommended by the Local Government and Environment Select Committee were less radical than those advanced earlier.[8]

The South African EIA system

South Africa has a population of 45 million and an area of 1.2 million square kilometres. It is simultaneously both a developed and a developing country. For the vast majority, life is concerned with survival: housing is often no more than a shanty, unemployment is endemic, and ensuring that there is enough to eat means that concern for the environment has very low priority. This has been exacerbated by alienation arising from dispossession of traditional lands (Sowman *et al.*, 1995). For the affluent minority, South Africa offers an enviable standard of living. This minority is mostly well housed, enjoys a benign climate and has ready access to a beautiful country. Its level of environmental concern is understandably higher than the average but there is a tendency to compartmentalise: not to think of a place from which money can be made as being part of the environment. The needs of the poor majority, the interests of a rich minority, and the desire of local and provincial governments to increase their income from taxation, have combined to make the pressure for development almost irresistible. There have, however, been a number of controversial development proposals which have aroused huge indignation and focused attention on the inadequacy of environmental controls. These include the Lake St Lucia sand mining proposal (see, for example, Whyte, 1995), the Saldanha Bay steelworks development and the rebuilding of the Table Mountain cable car station (Fuggle, 1998).

South Africa has a proud history of EIA, despite an historical lack of awareness of the need to consider environmental issues and a consequent lack of political will to implement controls (Sowman *et al.*, 1995).[9] EIA was discussed at professional gatherings as early as 1974 and there was a flowering of EIA publications in the 1970s and early 1980s (see, for example, Shopley and Fuggle, 1984). EIA was seen as a valuable aid to decision making in a 1980 White Paper, but the consequent Environment Conservation Act 1982 made no reference to EIA. The Council for the Environment was set up in 1983 under the provisions of this Act, together with an EIA Committee. This committee was influential in initiating research on, and consultation about, EIA in South Africa. Following an important EIA meeting in 1985, further research and consultation led to the publication of a document on integrated environmental management (IEM) (Council for the Environment, 1989). Sowman *et al.* (1995, p. 51) explained that:

> The term IEM was chosen to indicate an approach that integrates environmental considerations into all stages of the planning and development process and requires post-impact assessment monitoring and management. It was felt that the term EIA was inappropriate as the EIA process was perceived to be too limited in scope, reactive, anti-development, too separate from the planning process, and often the cause of costly delays.

Sowman *et al.* (1995) and Fuggle (1996) indicated that there were several factors which determined the form of South Africa's IEM system, which was developed in considerable isolation from contemporary international thinking on EIA:

- The need to promote economic growth and development to meet the basic needs of the majority of the population.
- The lack of a large cadre of environmental experts.
- The failure of apartheid-imposed technocratic planning and decision making.
- The need for inclusive participatory democracy and empowerment in environmental decisions to counter secretive, non-democratic and highly authoritative traditions, a vocal environmentally concerned middle class and low levels of literacy.

A White Paper on environmental management policy, released in 1998, explained the development of EIA in South Africa lucidly:

> We can characterise the 1980s as 'the decade the environment hit back' with the lives of almost all South Africans touched by major natural disasters such as drought and floods and increasing environmental impacts from industrial development. Following on from this, the 1990s have seen growing awareness of the need for environmental justice and sustainable living throughout society if we are to achieve environmentally sustainable development. (Republic of South Africa, 1998, p. 61)

In the same year as the IEM document was published, the Environment Conservation Act 1989 was promulgated. This contained provisions to give EIA the force of law, though these lay dormant until 1997 (see below). During the 1980s the voluntary undertaking of EIAs, commenced as early as 1971 (Mafune *et al*, 1997) as an input to decision making, increased as a result of rising public clamour. Some provinces passed legislation which enabled administrative authorities to demand EIA where they felt it to be necessary, but these powers were used sparingly and inconsistently.

As a result of consultation arising from practical experience of IEM during the late 1980s and early 1990s, a revised IEM procedure was published in the form of six guideline documents in 1992 (Department of Environmental

Affairs – DEA, 1992). These documents, which are still extensively used in South Africa, dealt with: the IEM procedure; scoping; report requirements; review; environmental characteristics; and terminology. The IEM procedure consisted of three main stages: development and assessment of the proposal; decision; and implementation (Preston *et al.*, 1996). The IEM guidelines formed the basis of several hundred voluntary EIAs in South Africa in which the linkage between EIA and the ongoing environmental management of the implemented project (through environmental management plans, environmental contracts, monitoring and auditing (Preston *et al.*, 1996)) was a key characteristic. In addition, some EIAs were conducted under the provisions of the Minerals Act 1991. The Development Facilitation Act 1995 and the National Water Act 1998 also provided for EIA to be undertaken (Glazewski, 2000, chapter 8).

Despite the generally positive view held of IEM, Ridl (1994, p. 80) felt that in 'the relatively short span of its existence, IEM has not developed a good reputation'. Avis (1994) identified four main problems associated with IEM: its non-enforceable nature; its use of a first world philosophy (EIA) in a third world context; the variations in perceptions between developers and environmentalists; and the reluctance of private landowners to consider alternative sites for development.

In response to continuing pressure to implement the dormant EIA requirements in the Environment Conservation Act 1989, various consultations were undertaken. However, the EIA Regulations, which came into effect between 1 September 1997 and 1 April 1998 (Republic of South Africa, 1997[10]), were radically different from the drafts and thought by many to be hasty and premature. As their name implies, only the EIA procedures for projects were incorporated in the Regulations, not the monitoring, auditing and environmental management provisions, nor the extension of EIA to certain land-use plans and policies which were features of IEM. Some felt the EIA system was little more than a pastiche of the IEM procedure. The decision to legislate was political: the twin aims of achieving tangible progress and overcoming the limitations imposed by severe professional staffing constraints (see below) were uppermost.

The resultant formal requirements of the South African EIA system are shown in Figure 5.6. They require, *inter alia*, that 'independent consultants' be involved in the EIA process. The EIA Regulations were supplemented by EIA guidelines in 1998 (Department of Environmental Affairs and Tourism – DEAT, 1998c). These guidelines, which are closely modelled on the IEM documents, go considerably further than the regulations they are intended to elucidate, particularly in relation to public participation and consultation (see Chapter 16). Granger (1998) highlighted the irony of the EIA regulations being used to require 'EIA' for only a small proportion of the authorisations considered (those for which an environmental impact report (Figure 5.6) is proposed). The belief that the EIA Regulations represented a missed opportunity to legislate the IEM procedure was widespread. As the White Paper on environmental management put it:

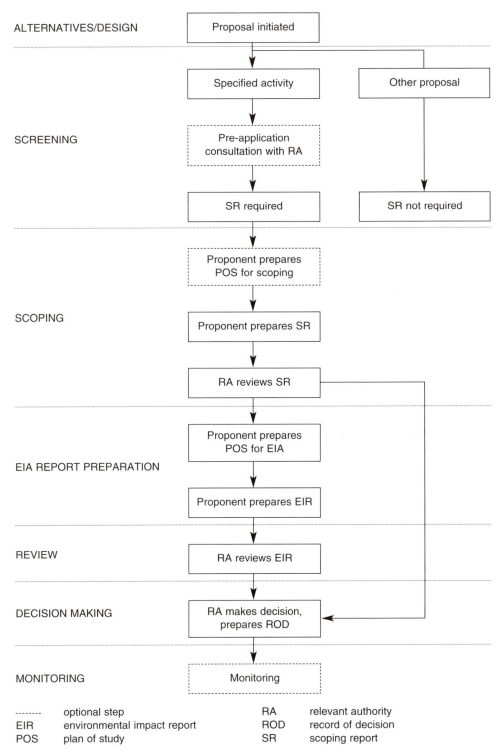

ALTERNATIVES/DESIGN

Proposal initiated

Specified activity

Other proposal

SCREENING

Pre-application consultation with RA

SR required

SR not required

Proponent prepares POS for scoping

SCOPING

Proponent prepares SR

RA reviews SR

Proponent prepares POS for EIA

EIA REPORT PREPARATION

Proponent prepares EIR

REVIEW

RA reviews EIR

DECISION MAKING

RA makes decision, prepares ROD

MONITORING

Monitoring

--------	optional step	RA	relevant authority
EIR	environmental impact report	ROD	record of decision
POS	plan of study	SR	scoping report

Figure 5.6   Main steps in the South African EIA process

The EIA regulations legislated only the scoping and EIA portions of the integrated environmental management (IEM) procedure. This is a major limitation of the current regulations and it has been proposed that the entire IEM procedure should be legislated. (Republic of South Africa, 1998, p. 73)

During the period of transition from minority to majority government in South Africa, the International Development Research Centre made a number of recommendations about environmental management (Whyte, 1995). These led to the setting up by the new government of the Consultative National Environmental Policy Process (CONNEPP) to supplement the activities of the Council for the Environment by bringing together a variety of governmental and non-governmental bodies (Fuggle and Rabie, 1996). CONNEPP succeeded in raising the political profile of the environment (O'Riordan, 1998) by producing a series of widely debated consultative documents culminating in the 1998 White Paper.

This led rapidly to a discussion document on IEM (DEAT, 1998a). Shortly afterwards, a draft environmental management bill (DEAT, 1998b) was issued which proposed IEM procedures for both new and existing activities and for certain land-use plans. These were rushed into legislative form in the National Environmental Management Act 1998 (NEMA) which came into force in January 1999. The remarkable speed with which this pioneering legislation was enacted reflected the political demand for closure of the protracted CONEPP. While NEMA repealed many sections of the Environmental Conservation Act (Glazewski, 2000, chapter 5), the provisions relating to EIA continue in force until NEMA EIA regulations are promulgated. (These, together with some minor amendments to NEMA's EIA provisions (DEAT, 2000a), were expected to come into force during 2002.)

NEMA extends the coverage of EIA to certain projects not included in the 1997 EIA Regulations (for example, mining projects) and to policies, plans and programmes. The EIA provisions of NEMA (see Chapter 5) consist of only two sections. The first deals with general objectives, which faithfully reflect the principles of IEM, for example, to:

Identify, predict and evaluate the actual and potential impact on the environment, socio-economic conditions and cultural heritage, the risks and consequences and alternatives and options for mitigation of activities, with a view to minimising negative impacts, maximising benefits, and promoting compliance with the principles of environmental management. (section 23(2)(b))

The second, lengthy, section of NEMA deals with the implementation of the IEM procedure (Glazewski, 2000, chapter 8). A user guide to NEMA, which mentions EIA, has been issued (DEAT, 1999) and a separate guide on the EIA provisions is being prepared. However, like the US but unlike the Canadian or Australian legislation, NEMA contains little detail: this is to be provided by the forthcoming regulations.

The Constitution of South Africa gives all citizens the fundamental right, now encapsulated in NEMA, to enjoy the environment and provides the right to administrative justice, and hence legal standing, to anyone wishing to intervene in decisions concerning the environment (Peckham, 1997). It also makes the environment a concurrent competence between central and provincial government (Republic of South Africa, 1998). In the case of EIA this means that the provincial government is generally the relevant authority. Indeed, many of the new provincial planning laws have included EIA provisions additional to national requirements (Glazewski, 2000, p. 292). The circumstances in which a province is not the relevant authority include cases where the provincial government is the applicant, when the national government assumes responsibility, and cases involving the compulsory delegation of powers to local authorities which can demonstrate that they have the capacity to discharge them.

The provinces have little formal experience of EIA and have linked it to differing existing duties such as nature conservation. However, no extra funding has been provided to help provincial governments discharge this new responsibility. Like the weak central government Department of Environmental Affairs and Tourism (which has had, at times, only one EIA staff member), all have inadequate staff resources (for example, Western Cape Province had, at one time, only two staff members) and frequent turnover (Duthie, 2001). There is a real danger that South African EIA could become a postiche.

On the other hand, there is a strong EIA consultancy sector in South Africa. This has led to the setting up of a South African chapter of the International Association for Impact Assessment and to the establishment of an international Southern African Institute for Environmental Assessment (similar to the UK Institute for Environmental Management and Assessment). There are substantial EIA training activities and, as mentioned, various guidelines have been issued, including guidance on strategic environmental assessment (DEAT, 2000b).

Notes

1. Amended versions of the Environmental Management Act 1994, of the 1993 Regulations and of the 1994 Decree are available in English. The Ministry of Housing, Spatial Planning and the Environment (VROM) published *The Texts of the Regulations on Environmental Impact Assessment in the Netherlands* in 2000. This document can also be accessed at the VROM web site at www.minvrom.nl

2. Information (in English) about the activities of the EIAC, and about the Dutch EIA system more generally, can be obtained from the Commission's web site at www.eia.nl

3. Information on the Canadian EIA process and many documents and guides can be obtained from the Canadian Environmental Assessment Agency web site at www.ceaa.gc.ca

4. www.ea.gov.au/epbc

5. Personal communication from Nick Gascoigne, Environment Australia. Notifications of referrals (with details) are available on the Environment Australia web site.

6. The more recent reports and much other useful information, including a copy of the Resource Management Act 1991, can be accessed on the Ministry for the Environment web site at www.mfe.govt.nz

7. This report, and other recent documents, can be accessed on the web site of the Office of the Parliamentary Commissioner for the Environment at www.pce.govt.nz This unique body was created in 1986 as part of New Zealand's environmental management revolution.

8. The Committee recommendations, dated May 2001, are posted on the Ministry for the Environment web site at www.mfe.govt.nz

9. This account is an expanded and updated version of Wood (1999b).

10. This and many other documents relating to South African EIA can be accessed on the Department of Environmental Affairs and Tourism web site at www.environment.gov.za

Chapter 6

Legal basis of EIA systems

It is now generally accepted that EIA systems should be based upon clear specific legal provisions. The next section of this chapter explains the reasons for advancing this as an evaluation criterion for EIA systems. It then discusses the requirements for the legal basis of EIA systems in rather more detail and puts forward several more explicit criteria. These requirements are then used to assist in the comparison of the legal bases of the US, UK, Canadian, Dutch, Australian, New Zealand and South African EIA systems.

The legal basis of EIA systems

Following the passage of the National Environmental Policy Act 1969 (NEPA) in the United States, it was not uncommon for governments in other countries to maintain that the main principles of EIA were already provided by their existing legislation. The UK, for example, argued for many years that the principal elements of EIA were satisfactorily furnished by the existing town and country planning system (see Chapter 4). Certainly, some very good EIAs have been undertaken outside the boundaries of a legally enforced EIA system, for example in the UK (Glasson *et al.*, 1999) and in South Africa (Preston *et al.*, 1996). Kennedy (1988) categorised this type of approach as informal-implicit. In it, EIA is adapted to meet the needs of particular situations, an EIA report as such may not be prepared and authorities are not accountable for taking EIA into consideration in decision making.

In the formal-explicit approach, on the other hand, EIA requirements are codified in legislation or regulations, an EIA report must be prepared and authorities are accountable for considering EIA (for example, through judicial review). Kennedy (1988, p. 258) argued that, 'generally speaking, EIA is only integrated in decision-making (that is, it only works) when it is applied in a formal-explicit way'. Sadler (1998, p. 21, emphasis in original) concurred, stating that EIA should have '*a well founded legislative base* with clear purpose, specific requirements and prescribed responsibilities.' This perspective is now generally accepted as evidenced by the growth of EIA legislation in countries throughout the world (Sadler 1996; Donnelly *et al.*, 1998). However, as Glasson *et al.* (1999, p. 39) argued, the existence of mandatory regulations, acts or statutes relating to EIA are 'not necessarily indicative of how thoroughly EIA is carried out.' Brazil and the Philippines were cited as countries that have enacted EIA requirements which have generally been poorly implemented.

A further relevant issue in EIA is the question of how far the detailed operation of the EIA process should be prescribed in laws and regulations, and how much it should be left to the discretion of the relevant authorities. The advantages of a legally specified EIA system may be summarised (Fowler, 1985) as: permanence and evidence of commitment; avoidance of uncertainty; provision of a firm basis for public participation; and enforcement of acceptance of EIA.

On the other hand, the advantages of a largely discretionary EIA system, only the broad details of which are enshrined in law or regulation (Fowler, 1985) are: the desirability of voluntary compliance, the avoidance of judicial involvement; and the retention of discretion. Fowler (1985, p. 205) concluded that 'where a firm political commitment to EIA happens to exist at the time of adoption of a scheme, this is reflected in a legislative base'. He suggested that there has been a gradual shift towards EIA systems 'in which both administrative and judicial supervision is seen as necessary in terms of the efficiency of the overall process'. This appears to hold generally true. Thus the Canadian government, after almost 20 years of discretionary EIA, codified the system in legislation in 1992. Similarly, Australia replaced 25-year-old legislation which provided an outline of EIA requirements with an Act specifying them in detail. The European Commission tightened the EIA Directive, after a decade's experience, to reduce member state discretion, particularly in relation to screening. The New Zealand government, after 17 years' use of discretionary EIA procedures, passed legislation integrating the process into a broader environmental management context in 1991. Similar codification and integration has taken place in South Africa. Two factors driving these legislative changes have been the evolving policy agenda for sustainable development and recognition of the need to rectify acknowledged deficiencies of traditional EA practice (Sadler, 1995).

It is true that some degree of discretion in the operation of the various steps of the EIA process is needed since every eventuality cannot be foreshadowed in laws and regulations. In particular, flexibility is necessary to ensure that the EIA system is focused on the desired outcome of EIA, environmentally sensitive decisions, rather than on ensuring that all the procedural formalities have been completed. Such discretion perhaps takes its most extreme form in the state of Victoria, where the EIA system is now based on guideline procedures quite different from the provisions of the Act underpinning them (Wood, 1993). However, the discretion remaining should not be sufficient to remove reasonable legal certainty, and nor should it enable any participant in the EIA process to gain undue advantage. It is for this reason that 'fast-track' solutions to EIA decision making which automatically permit a development to proceed unless the relevant authorities take appropriate EIA action are generally unsatisfactory. The discretion remaining in an effective EIA system should be broadly acceptable to all parties.

A further issue is whether the EIA system should be independent of existing decision-making procedures or whether it should be integrated into them.

NEPA introduced a completely new procedure which cut across existing decision making. This, not surprisingly, led to confusion and delay in the early years, to considerable duplication of control, and to a desire by other countries to avoid similar problems. This has been manifest in a desire by governments to avoid litigation and by entrenched interests to avoid loss of control over decisions to external agencies. The advantage of separation is the creation of a fresh approach which emphasises the importance of EIA, something which may not be apparent where EIA is integrated into existing procedures if prevailing attitudes among practitioners and decision-makers are not modified. The European Directive on EIA, as amended, is specifically phrased, notwithstanding the US separation model, to allow member states to introduce EIA into existing decision-making procedures. Many of them (including the UK) have followed this route.

There are clearly advantages in both approaches, if they are implemented effectively. Indeed, this distinction between separate and integrated EIA systems may be somewhat arbitrary in practice, since the essential aim of EIA is that decisions on actions are made which take full account of the outcomes of the EIA process. This may, or may not, be achieved in systems utilising either approach.

There is always a danger that, unless the various steps in the EIA process are mandatory, there will be some proponents, consultants, consultees or authorities who will fail, in certain circumstances, to discharge their responsibilities fully. For this reason, each step in the EIA process needs to be specified sufficiently in an Act or in a binding regulation to provide a measure of certainty to the participants in the EIA process. The finer points involved in each stage of the process need not be spelled out in law provided that appropriate additional guidance is made available (for example, in the form of advisory guidelines). It is important, in the interests of certainty, that the specified system is adhered to by all the stakeholders and that accepted procedures are not changed arbitrarily.

While lawyers drafting laws and regulations will always strive to make them unambiguous, others will endeavour to discover loopholes and ambiguities if it is in their clients' interests to do so. Clearly, for an EIA system to function effectively, ambiguities need to be minimised. Where they exist and cause problems in the operation of the EIA system, they should be remedied at the first available opportunity. Herein lies another advantage of specifying some details in the form of regulations or guidance, since they can then be modified without recourse to primary legislation.

EIA is a process which applies to certain types of action, but not to all proposals. The legal requirements relating to EIA should be clearly distinguished from those relating to other types of action so that no confusion exists between different processes. This applies equally to systems in which EIA is separated from other decision-making procedures and those in which it is integrated. The need for differentiation is strongest in integrated EIA systems since the scope for confusion, particularly among proponents, is higher than in separated systems.

In the last analysis, it may be necessary to take enforcement action against one of the participants in the EIA process. This might, for example, be against the responsible authority for not screening the proposal adequately, or for not considering the comments on the EIA report sufficiently in reaching its decision, or against the proponent for not meeting conditions attached to a permission. Such action might be taken by any of the participants in the EIA process, including the public. It is necessary, therefore, that there should be adequate opportunities for the various participants to appeal administratively or to the courts to ensure that the various obligations in the EIA process are properly discharged.

It is important that a clear outline of all the procedures involved in the EIA process be available so that proponents, developers, consultees, the public (and the relevant authorities) can gain an overview of the whole process. This outline should include the time allocated to each stage in the process (a necessary requirement to prevent it from becoming over-lengthy) and any charges involved in it. The various criteria for analysing the legal basis of EIA systems are summarised in Box 6.1. These requirements are used to assist in the review of the legal basis of each of the seven EIA systems which follows.

United States

The US federal EIA system is based upon the broad provisions of legislation – the National Environmental Policy Act 1969 – the brevity of which is matched by its ambiguity. The Environmental Law Institute (1995b) and Caldwell (1998) have argued persuasively that the environmental policy elements of NEPA have been neglected since it grants authority to take action and directs a focus on the future consequences of governmental and non-governmental actions. However, the substantive intent of NEPA, to change the nature of federal decision making, has been gradually eroded by the courts over the years to become a largely procedural requirement relating to environmental impact

---

Box 6.1   Evaluation criteria for the legal basis of EIA systems

Is the EIA system based on clear and specific legal provisions?

- Is each step in the EIA process clearly specified in law or regulation?
- Are the legal provisions sufficiently unambiguous in application?
- Is there a degree of discretion in the provisions which is acceptable to the participants in the EIA process?
- Are the EIA requirements clearly differentiated from other legal provisions?
- Is each step in the EIA process enforceable through the courts or by other means?
- Are time limits for the various steps in the EIA process specified?
- Does a clear outline of procedures and time limits exist for the EIA system as a whole?

statements. Nevertheless, the legal basis of the US EIA system is clearly specified by it. The various EIA requirements of the Act have been clarified over the
years by both the courts and the Council on Environmental Quality (CEQ)
Regulations, themselves based upon legal rulings. The detailed steps in the
process are spelt out in the Regulations, which are widely regarded as providing a model basis for an EIA system, being comprehensive, specific, clear and
surprisingly readable. There is reasonable agreement by proponents, practitioners and environmental groups that the Regulations leave an appropriate
degree of discretion for the EIA process to be applied to the activities of the
very wide range of federal agencies affected by NEPA.

Further guidance has been issued by CEQ to clarify matters not covered
fully in the Regulations.[1] Generally, the various agencies (over 50) have issued
guidelines or regulations to apply the CEQ Regulations specifically to their
own activities and the Environmental Protection Agency (EPA, 1993) has
published a sourcebook for the environmental assessment (EA) process which
includes a set of computer disks. This documentation (primarily intended for
agency staff) forms part of a voluminous literature on NEPA (see, for example,
Environmental Law Institute, 1995a; Fogleman, 1990; Kreske, 1996;
Hildebrand and Cannon, 1993; Clark and Canter, 1997; Mandelker, 2000;
Eccleston, 1999). There is no single official explanation of the EIA system as a
whole beyond the CEQ Regulations themselves.

Each major procedural step in the EIA process can be challenged in the
courts. The fact that such a litigious society as the United States generally generates fewer than 100 court cases per annum on NEPA when its provisions are
applied to at least 50,000 actions must be regarded as a vindication of the
Regulations. Nevertheless, while the legal provisions are generally regarded as
being reasonably unambiguous, the continuing legal actions arguing that EAs
or EISs ought to have been prepared demonstrates the scope they leave for
interpretation.

The Regulations contain provisions relating to the time limits for consultation and participation. They also provide that, subject to certain limitations,
the agency must set time limits if an applicant for the proposed action requests
them (section 1501.8). This provision was the one which the business community found the most attractive (Yost, 1990) but, in practice, it has not been
used a great deal.

The EIA requirements are clearly differentiated from other legal provisions.
Indeed, this separation caused animosity, confusion and delay in the early years
when NEPA was applied retrospectively to projects that were already under
construction. While experience has considerably reduced these problems, some
of the frustration with the complexity of NEPA and other environmental regulations which drove Congressman Hinshaw to enter the holy hypothetical saga
reproduced in Box 6.2 into the record of the House of Representatives still
exists (see, for example, Mandelker, 1993).

Box 6.2   'God and EPA'

In the beginning God created heaven and earth.

He was then faced with a class action lawsuit for failing to file an environmental impact statement with HEPA (Heavenly Environment Protection Agency), an angelically staffed agency dedicated to keeping the Universe pollution free.

God was granted a temporary permit for the heavenly portion of the project, but was issued a cease and desist order on the earthly part, pending further investigation by HEPA.

Upon completion of his construction permit application and environmental impact statement, God appeared before the HEPA Council to answer questions.

When asked why he began these projects in the first place, he simply replied that he liked to be creative.

This was not considered adequate reasoning and he would be required to substantiate this further.

HEPA was unable to see any practical use for earth since 'the earth was void and empty and darkness was upon the face of the deep'.

Then God said: 'Let there be light.'

He should never have brought up this point since one member of the Council was active in the Sierrangel Club and immediately protested, asking, 'how was light to be made? Would there be a strip mining? What about thermal pollution? Air pollution?' God explained the light would come from a huge ball of fire.

Nobody on the Council really understood this, but it was provisionally accepted assuming (1) there would be no smog or smoke resulting from the ball of fire (2) a separate burning permit would be required, and (3) since continuous light would be a waste of energy it should be dark one-half of the time.

So God agreed to divide the light and darkness and he would call the light Day and the darkness Night. (The Council expressed no interest with in-house semantics.)

When asked how the earth would be covered, God said: 'Let there be firmament made amidst the waters, and let it divide the waters from the waters.'

One ecologically radical Council member accused him of double talk, but the Council tabled action since God would be required first to file for a permit from ABLM (Angelic Bureau of Land Management) and further would be required to obtain water permits from appropriate agencies involved.

(continued)

(continued)

The Council asked if there would be only water and firmament and God said: 'Let the earth bring forth the green herb, and such as may seed, and the fruit tree yielding fruit after its kind, which may have seeded itself upon the earth.'

The Council agreed, as long as native seed would be used.

About the future development God also said: 'Let the waters bring forth the creeping creature having life, and the fowl that may fly over the earth.'

Here again, the Council took no formal action since this would require approval of the Game and Fish Commission co-ordinated with the Heavenly Wildlife Federation and Audobongelic Society.

It appeared everything was in order until God stated he wanted to complete the project in six days.

At this time he was advised by the Council that his timing was completely out of the question ... HEPA would require a minimum of 180 days to review the application and environmental impact statement, then there would be public hearings.

It would take 10 to 12 months before a permit could be granted.

God said: 'To Hell with it.'

Source: United States Congressional Record, 1974.

UK

The original regulations incorporating EIA into the town and country planning system (and other statutory procedures) were made under the European Communities Act 1972, the provisions of which do not permit the requirements of European directives to be exceeded when implemented by regulation. However, the Planning and Compensation Act 1991 allows the Secretary of State for Transport, Local Government and the Regions to require EIA for projects needing planning permission additional to those listed in the Directive (see Chapter 4) and this power has been used to extend EIA to a small number of additional projects. The Town and Country Planning (Environmental Impact Assessment) Regulations implement the provisions of the amended European Directive almost to the letter. Like the original Regulations, the amended Regulations, which were laid before parliament prior to coming into effect on 14 March 1999, provide the legal basis for each of the steps shown in Figure 4.1, including scoping should the developer request it. Not only are all the main steps covered by the Regulations but time limits are also specified for each of them. However, no mention is made in the Regulations of monitoring of impacts following project implementation.

The 1999 Regulations removed several ambiguities in the original Regulations regarding the definition of certain types of project: for example, urban development projects now specifically include shopping centres. They also strengthened the EIA system (for example by requiring that supplementary information provided by the developer be subject to the same publicity and consultation arrangements as the original environmental statement). Although EIA in the UK is largely integrated into the town and country planning system, the requirements are clearly distinct from those for normal planning applications, for example, in relation to timescales. Local planning authorities are allowed 16 weeks, rather than the normal eight weeks, to reach a decision on applications involving EIA. The degree of discretion provided by the Regulations (which mirrors that in the existing land-use planning system) appears to be broadly acceptable to most of the main participants (Jones *et al.*, 1998).

The Regulations provide the developer with a right of appeal against an LPA determination that EIA is required. As the EIA system is integrated into town planning procedures, normal appeal provisions against the non-determination of planning applications (and against negative planning decisions) apply. In principle, therefore, there is strong central control over the freedom of LPAs to determine applications involving EIA. While there is no third-party right of administrative appeal in the British planning system, access to the courts is possible where the EIA requirements have not been properly discharged. In practice, there have been only two or three such cases each year, although these have become increasingly influential, leading to changes in procedure.

Circular 2/99 (DETR, 1999b) and government guidance (DoE, 1994a, 1995; DETR, 2000a) together provide clear and detailed guidance on the operation of the procedures. The Regulations contain descriptions of developments and both indicative criteria and *de minimis* (or exclusion) thresholds to be used by LPAs in reaching a judgement about whether EIA is or could be required for Schedule 2 projects. Further explanation, together with advisory criteria and thresholds for Schedule 2 projects, is contained in the advisory Circular.

It is apparent that the Regulations and accompanying guidance contain provisions which clearly and specifically define the basis of the EIA system integrated into British planning procedures.

The Netherlands

The Environmental Management Act 1994, as amended, contains some 43 subsections of detailed requirements relating to: the coverage of EIA; the content of the EIS; the preparation of the EIS; the evaluation of the EIS; the decision-making procedure; transboundary impacts; and post-implementation evaluation. The Act provides for the EIA process to be integrated into existing decision-making procedures. Together with the Inception Memorandum Regulations and the amended EIA Decree,[2] the Act contains provisions relating to each step in the EIA process shown as obligatory in Figure 5.1, although the European Commission has identified shortcomings (see Chapter 5). The one area not provided for in the Act relates to the provision of supplementary information as a result of the review of the EIS by the EIA Commission and the public.

The Act, the Regulations and the Decree have proved not to be unambiguous in application, especially where screening is concerned. Over the years, additional provisions relating to exemptions, screening, treatment of alternatives and transboundary impacts have been added. Some of these were in response to the recommendations of the first report of the Evaluation Committee on the Environmental Protection Act (ECW, 1990) and some to implement the amended European Directive on EIA. Notwithstanding the detailed nature of the legal requirements, numerous requests for clarification have been addressed to the Ministry of Housing, Spatial Planning and the Environment (VROM), to the Ministry of Transport, to the EIA Commission (EIAC), to provincial EIA coordinators and to competent authorities. Some of these requests have concerned the issue of whether economic and social impacts should be included in the EIS. VROM has taken the view that the EIS should be confined mainly to physical environmental impacts, though proponents are free to include economic and social impacts if they choose to do so. Other requests have concerned the Decree provisions relating to activities such as waste disposal. Generally, VROM and EIAC have both acted to clarify the meaning of the legal provisions when necessary. The remaining degree of discretion in the operation of the EIA system is felt to be appropriate by most of the participants in the EIA process.

There is a general right of appeal against the decision to which the EIA process is linked in the Netherlands. This, together with the publication of virtually all the documentary material associated with the EIA process, means that there can be adequate enforcement of the various steps in the EIA process. Some of the stages in the EIA process (for example, screening) are also enforceable through the courts and there have been more than 200 administrative appeals involving EIA in the Netherlands. Cases have related to the granting of exemptions, to screening, to failure to carry out EIA and to the nature of the scoping guidelines for each EIS (the content of these was deemed to be nonbinding and hence not to be open to appeal).

While the EIA requirements are integrated into existing decision-making procedures, they are clearly differentiated from them. EIA can be linked to licences and permits under the Environmental Management Act 1994, to local land-use plans, to the waste policies in provincial environmental management plans and to other sectoral decisions (for example, those concerning airports and motorways) as well as to policy decisions for activities for which EIA is required at the project level.

Several time limits are specified in the Act: nine weeks (including four weeks of public participation) are permitted from receipt of the notification of intent to the issue of the draft guidelines by EIAC and a further four weeks for the competent authority to issue the formal guidelines. The competent authority is allowed six weeks to determine whether the EIS is acceptable, prior to its publication. Once it has been published, EIAC usually has nine weeks to determine whether the EIS meets the legal requirements and the formal guidelines.

The time period for the making of the decision is determined by the regulations relating to the type of decision concerned (licences, land-use plans, etc.) (EIAC, 1998).

Two provinces have approved requirements necessitating the EIA of a single additional project type and another three require EIA for an extensive set of additional activities. The other seven provinces follow the national EIA Decree, as amended in 1999.

There is a large handbook on EIA procedure (updated in 1994), together with an explanatory leaflet (VROM, 1991). In addition, as mentioned in Chapter 5, numerous other documents on more technical aspects of EIA exist. In general, there is probably more official information about EIA in the Netherlands than in any jurisdiction outside North America.

Canada

The Canadian Environmental Assessment Act, proclaimed in 1995, lays down the steps in the EA process in considerable detail in its 82 sections. There are separate parts of the Act dealing with interpretation, purposes, projects to be assessed, the assessment process, transboundary and related effects, access to information, administration, the Canadian Environmental Assessment Agency (CEAA), review and transitional arrangements. Some government officials fear that, in endeavouring to overcome some of the shortcomings arising from the wide discretion provided by the Environmental Assessment and Review Process (EARP) Guidelines, and in attempting to ensure that the courts would be able to enforce the EA process, the authors of the Act (perhaps inevitably) heavily prescribed it. Hazell (1999, p. 153) felt that the Act 'is too complicated and difficult to understand'.

The Act contains provision for numerous regulations to be made. Four sets were promulgated when the Act was proclaimed. These related to the list of Acts and regulations under which an activity could trigger the Act, a list of projects to be excluded from the provisions of the Act, an inclusion list for physical activities to be covered by the Act and a list of projects to be subject to a comprehensive study. Subsequently, as well as the federal authority coordination regulations (see Chapter 5), regulations relating, *inter alia*, to projects outside Canada have been issued.

The provisions of the Act and the regulations have not proved unambiguous. It would have been surprising, given the number of provisions in the Act and the regulations, if no ambiguities had surfaced. However, since the Act was intended to reduce the uncertainty of application inherent in EARP, such ambiguities ought to be limited in number. Indeed, the most significant amendments to the Act during its passage through the House of Commons centred on the reduction of discretion, though much remained (Gibson, 1993). Much of the detailed implementation of the Act is prescribed in regulations, which can be altered relatively easily.

The Act specifies an EA process which is almost entirely distinct from other legal provisions. Where the federal responsible authority proposes the project, or grants money or land for the project, the Act is usually the only specific legislation applicable. Where the responsible authority is involved in a regulatory role, such as issuing a licence, EA is effectively integrated into other legally mandated decision-making processes. Even here, however, the legal EA requirements are quite distinct and clearly differentiated from other provisions.

The various steps in the EA process are open to varying degrees of public participation and to challenge in the courts. Despite the discretion afforded by EARP, Hunt (1990, in Tilleman, 1995, p. 395) was able to state that:

> Given the fact that, in comparison to the United States, Canadian society tends to be relatively nonlitigious, it should come as no great surprise that the courts have played only a modest role in EIA matters.

This was before the landmark challenges of 1989. In total, there were more than 50 court cases under EARP, a tiny proportion of the number in the United States. The authors of the Act were determined to reduce substantially the number of legal challenges by prescribing the EA steps carefully in the legislation. Following a honeymoon period, there had been more than 20 legal cases by the end of 2000, a higher annual rate than under EARP. It thus appears that, while Canadians are becoming more litigious, the Act's draughtsmen were unsuccessful in achieving their objective.

No time limits are specified in the Act. However, section 59(a) states that regulations can be made to prescribe the time periods relating to the EA process and time limits for some (but not all) of the steps in the EA process have been specified in the federal coordination regulations and in ministerial guidelines (see Chapter 18).

A responsible authority's guide to the Act was published in 1994. This consists of a manager's guide, a practitioner's guide and reference guides providing advice on cumulative impacts, the public registry and the assessment of significance (CEAA, 1994). This lengthy guide provides a clear outline of EA procedures and some indication of methods but does not mention time limits. It was intended that the guide would eventually be supplemented by separate sector-specific guides for the EA of particular types of project but, to date, only a draft guide for mining projects (CEAA, 1998b) has been published. A citizen's guide to Canadian EA has been released (CEAA, 1995). In addition, guides on biodiversity and EA (CEAA, 1996), on the preparation of comprehensive study reports (CEAA, 1997b), on panel review procedures (CEAA, 1997c), together with guidance on cumulative impacts (CEAA, 1999a) and various other guidance documents (for example, on alternatives (CEAA, 1998a)) have been published.

Commonwealth of Australia

EIA at the national level in Australia is clearly differentiated from other legal provisions relating, for example, to land-use planning or pollution control. The Environment Protection and Biodiversity Conservation Act 1999 (EPBC Act) clearly specifies the requirements relating to each of the procedural steps in the EIA process. The Environment Protection and Biodiversity Conservation Act Regulations 2000 (No. 181), which can be altered without changing the primary legislation, provide additional details on, for example, the content of EIA reports. The administrative guidelines on significance (Environment Australia, 2000a) are non-statutory and can thus be amended even more easily.

Whereas only one time limit was mentioned in the repealed Environment Protection (Impact of Proposals) Administrative Procedures (Commonwealth of Australia, 1995 – see, for example, Bates, 1995), no fewer than five decisions have time limits attached to them in the EPBC Act. These relate to the Environment Minister's initial decision as to whether approval is required, to the Minister's decision about the type of assessment to be undertaken, to the Minister's preparation of scoping guidelines to the preparation of the assessment report by Environment Australia and to Minister's decision on the action. There are provisions for reporting failure to meet these deadlines.

While the repealed Impact of Proposals Act and the accompanying Administrative Procedures together set down provisions for each stage of the process, they left a number of ambiguities. For example, they contained an open-ended definition of the activities to be covered (Fowler, 1982, p. 19) and they did not specify the triggering mechanism clearly. There was considerable debate about the nature of enforceable obligations on decision-makers to take the Environment Minister's advice following assessment into account, as the Act required (Commonwealth Environment Protection Agency – CEPA, 1994). It would appear that judicial review was the only available means of compelling decision-makers to take appropriate account of the advice (Fowler, 1982, p. 28). However, there were only six or seven judicial reviews and two or three major challenges in the courts to the system. Münchenberg (1994) explained the small number of legal appeals in terms of cost, difficulties in proving standing, uncertainties about the legal status of the Administrative Procedures and the discretion inherent throughout the EIA process. While the general foundations of the EIA system were clearly based in law, considerable legal uncertainty about specific aspects of the EIA process remained (Bates, 1995; Fowler, 1996).

There was thus ambiguity in the application of the legal provisions, particularly in regard to the proposals which should be subject to EIA. The degree of discretion in the EIA system was unacceptable both to project proponents and to environmental groups who both sought greater legal certainty in the application of the EIA procedure (CEPA, 1994). Many of these problems have been overcome by the EPBC Act's provisions but, inevitably, others have been created.

The EPBC Act, rather like the Canadian Environmental Assessment Act 1992, prescribes the EIA process in substantial detail. It is too early to say whether there will be more legal actions under the provision of the EPBC Act than there were under those of the Impact of Proposals Act but this seems to be a likely corollary of the increased detail specified in the primary legislation. The EPBC Act contains provisions for bilateral agreements between the

Commonwealth and the states to accredit state EIA processes. To date, only one (with Tasmania) has been affected and there is considerable scepticism about whether or not the necessary cooperation will be forthcoming (Fowler, 1996; Ogle, 2000; Scanlon and Dyson, 2001). There are also provisions for case-by-case accreditation of state EIA processes and concerns have been expressed about the quality of these (Hughes, 1999; Ogle, 2000; Padgett and Kriwoken, 2001). The principal area of ambiguity relating to the EPBC Act thus appears to relate to Commonwealth-state cooperation, but screening, which was problematic in the past, is likely to continue to cause uncertainty because of the residual discretion the Environment Minister is allowed (see Chapter 9).

As mentioned in Chapter 5, various explanatory memoranda, leaflets, fact-sheets and other documents have been published to provide a clear account of the Commonwealth EIA system (Environment Australia, 1999a,b, 2000a,b,c).

New Zealand

The Resource Management Act 1991 contains the legal provisions relating to the New Zealand EIA system. As explained in Chapter 5, this revolutionary legislation also provides the legal basis for almost all of New Zealand's environmental protection and management measures and, accordingly, it runs to nearly 400 pages in length. The EIA provisions constitute only a small part of the total (around 20 of the 433 sections) and are closely integrated with other resource planning provisions.

The term EIA is not used in the Act, which uses phrases such as 'an assessment of any actual or potential effects ... on the environment' (section 88(4)(b)) to describe EIA. No phrase is used to describe the EIA report in the Act, despite the fact that New Zealand had 15 years' experience of environmental impact reports. Dixon (1993, p. 244) believed the authors of the Act replaced the term 'impact' by 'effect' to signify a fresh approach to EIA as all proposals and plans now come under this scrutiny.

The Act provides a broad indication of the projects to be assessed: land-use and subdivision consents, discharge, water abstraction and coastal permits unless exempted by local authorities in their policy statements and plans. The Act strongly encourages early consultation and provides an indication of the contents of an assessment of environmental effects (AEE) report in the Fourth Schedule (see Chapter 7). Regional authorities, in their policy statements and plans, and territorial (city and district) authorities, in their plans, can supplement the information applicants are required to provide. Time limits are imposed upon these authorities to process applications once they have sufficient information. The results of public participation and consultation on the basis of the AEE report, and the report itself, must be considered in the decision. The Act further provides for monitoring of the impacts of approved proposals where their scale or significance warrants this (Fourth Schedule, para. 1(i)). In short, the Resource Management Act provides a clear framework for EIA in New Zealand but deliberately leaves considerable latitude to local authorities to determine their own specific EIA requirements. Several outlines of the procedures and the tight timetable for each step in the EIA process exist in Ministry for the Environment (MfE) documentation (for example, MfE, 1999a,b, 2001a – see also Chapter 5) and in Morgan and Memon (1993).

South Africa

The regulations requiring compulsory EIA cover both the EIA process and the outcome of that process (Republic of South Africa, 1997). The regulations, which came fully into effect on 1 April 1998, specify the procedures which must be followed to generate the scoping report, the environmental impact report (EIR) and the record of decision report. Most steps in the EIA process are covered: there are provisions relating to the initiation of the EIA process, alternatives, screening, scoping, preparation and submission of the EIR, and decision making.

The EIA Regulations, which provide the relevant authority with consider-able discretion, are proving to be somewhat ambiguous in application. One major problem still to be resolved is the determination of precisely who the rel-evant authority is. The regulations make it clear that the provincial authority is the decision-maker except for certain activities of national significance (such as where there are direct implications for national environmental policy), in which case the Minister of Environmental Affairs and Tourism takes the decision. However, there is a power (which has already been exercised) enabling local authorities to be designated as relevant authorities by the Minister, following a recommendation by the provincial authority. There is also some confusion over the requirement to appoint independent consultants with no financial interest in the implementation of the proposed activity (see Chapter 11). Whether the regulations apply to applications under way but not determined at the time they came into effect is left to the discretion of the provincial authorities. There are, in addition, some ambiguities over whether all the activities specified should be subject to EIA, no matter how minor they may be. In practice, the relevant authorities are using their discretion to exclude many minor projects from the procedures.

It is probable that all the steps of the EIA process specified in the Regulations are enforceable through the courts but, to date, despite increased legal standing, environmental non-governmental organisations have not seized the opportunity to demonstrate this. It is, perhaps, surprising that no challenges on screening and on the preparation of scoping reports have been mounted. This may be because the non-statutory EIA guidelines (Department of Environmental Affairs and Tourism – DEAT, 1998c) provide a clear outline and explanation of the procedures set down in the EIA Regulations. The only time limits spec-ified in the Regulations relate to that for responding to the advertisement of the application and that within which an appeal must be lodged (see Chapter 13).

The EIA requirements are clearly differentiated from other South African legal environmental provisions though there are problems in dovetailing EIA with planning procedures. It was originally intended that the EIA and planning requirements would be integrated but this aim was abandoned because of opposition from the planning profession. There now exist two separate discretionary decision-making procedures. To add to this complexity, many of the new provincial planning laws have included EIA provisions (Glazewski, 2000). While local authorities to which EIA powers are delegated are able to harmonise the two procedures, this is not possible in the majority of cases where the provincial and local governments exercise separate powers. Confusion among governments, developers and the public has been inevitable but the efforts of planners and environmental managers to harmonise their requirements have proved successful in reducing this.

The Development Facilitation Act 1995, which provides for the fast-tracking of certain types of development, also contains permissive integrated environmental management powers, though these have not been exercised to date. As mentioned in Chapter 7, mineral extraction is excluded from the EIA Regulations but quasi-EIA powers are contained in the Minerals Act 1991 (Peckham, 1997; Glazewski, 2000). Every mining authorisation is subject to a process which results in an environmental management programme report that contains comprehensive information on the project. This includes a description of the impact of the proposed mining activity on a standard list of 17 environmental elements which serves as the basis for the management plan that must be approved before consent is granted. Memoranda of understanding are being drawn up between the Department of Environmental Affairs and Tourism and the nine provinces and between the Department and various central government departments to try to ensure uniformity of EIA practice.

It is expected that the EIA regulations made under the provisions of the National Environmental Management Act 1998 (NEMA) will replace the 1997 EIA Regulations in 2002. NEMA lays heavy emphasis on all aspects of integrated environmental management including mitigation or minimisation of negative impacts on the environment, public participation, ongoing environmental management and strategic environmental assessment (Glazewski, 2000). It is therefore to be hoped that the forthcoming regulations (which must be approved by a new Committee for Environmental Coordination) and accompanying guidance dispel the legal uncertainty associated with the 1997 EIA Regulations.

Summary

Table 6.1 summarises the extent to which the seven EIA systems are based on clear and specific legal provisions. All seven systems meet the review criterion. However, the US and, to a lesser extent, the South African, Acts provide a general outline and rely on detailed regulations for their implementation, and the UK EIA system is based almost entirely on regulation, supported by non-statutory guidance. The New Zealand legislation relating to EIA is extraordinarily brief, has not been supported by regulations, and does not even mention

Table 6.1 The legal basis of the EIA systems

Criterion 1: Is the EIA system based on clear and specific legal provisions?

Jurisdiction	Criterion met?	Comment
United States	Yes	Act and regulations clearly define separate EIA system
UK	Yes	Regulations specifically implement amended European Directive on EIA. EIA mainly integrated within town and country planning system, administered by local planning authorities
The Netherlands	Yes	EIA Act and Decrees specifically provide for clearly defined EIA process integrated into other decision-making procedures
Canada	Yes	Act and Regulations clearly define EIA process largely separate from other decision-making procedures, together with powers of Canadian Environmental Assessment Agency
Commonwealth of Australia	Yes	Act and Regulations together outline separate EIA system. Some legal uncertainty, especially on Commonwealth state cooperation and screening
New Zealand	Yes	Act provides clear broad framework for EIA but allows local authorities considerable discretion in operation
South Africa	Yes	EIA Regulations clearly define EIA process separate from, but parallel to, other environmental control procedures

the term EIA. The most detailed legislation is to be found in the Netherlands, Canada and Australia, where sets of regulations relating to particular aspects of the EIA process have been issued to support the very specific EIA Acts. There also exist specialised EIA agencies in the Netherlands, Canada and Australia (a branch) and, to a much lesser extent now, in the United States to provide advice and guidance on EIA both in general and in specific cases.

Five of the seven EIA systems involve procedures which are quite separate from other authorisation systems. In the UK and New Zealand, on the other hand, EIA procedures (while identifiable legally) are firmly integrated into other types of consent procedure. It is not surprising that the legal requirements for EIA in these two countries are expressed much more briefly than in the Acts and regulations specifying the EIA systems in four of the other five jurisdictions (South Africa is the exception). However, given that there is no central body responsible for EIA in these two countries or in South Africa and that there is a much more limited possibility of appeal to the courts than in the United States, it is apparent that there is likely to be rather more discretion in their EIA systems than in the other four.

Notes

1. Council on Environmental Quality (CEQ 1981a,b); Memorandum: Guidance Regarding NEPA Regulations *48 Federal Register* 34263, 28 July 1983. These three documents are reproduced in CEQ annual reports and in Bass *et al.* (2001).
2. The texts of the Act, Regulations and Decree are available from VROM or on its web site at www.minvrom.nl

Chapter 7

Coverage of EIA systems

The coverage of EIA systems relates both to the range of actions subjected to EIA and to the range of impacts regarded as relevant. While it is generally accepted that the impacts of all environmentally significant new and modified projects should be subject to EIA, there is little unanimity about the extension of EIA to higher-tier actions such as programmes, plans and policies or about the definition of the word environment.

The next section of this chapter discusses several criteria which can be employed in the analysis of EIA systems. These criteria are used to assist in the comparison of the legal bases of the US, UK, Dutch, Canadian, Australian, New Zealand and South African EIA systems.

Coverage of actions and impacts

The National Environmental Policy Act 1969 (NEPA) applies to public actions by the federal government. These actions include the granting of permits for private actions. 'Actions' also include the making of plans and the enactment of legislation but one of the main intentions of the Act was clearly to ensure that the projects initiated by federal government were environmentally acceptable. In practice, the overwhelming emphasis of NEPA application has been upon projects, many of them public. From the outset, the word 'environment' was defined to include social and economic impacts, as well as physical environmental impacts (pollution, effects on ecology, etc.). The coverage of later EIA systems has sometimes, but by no means always, followed the precedent set by NEPA.

If the objective of EIA is to ensure that the environmental impacts of significant actions should be assessed prior to implementation, there appears to be little point in distinguishing between public and private actions. There are sufficient examples of both well- and poorly considered projects by public and private proponents in most countries to counter any argument that, say, private but not public proponents should be subject to EIA. That private proponents have not, in general, been subject to EIA at the federal level in the United States is largely a consequence of a system of government which limits federal intervention in state and local matters.

It is clearly important that no significant types of project, whether public or private, should be exempt from EIA unless there is an overwhelming reason for this (for example, national security considerations). It is not uncommon, for example, for a project authorised by specific legislation enacted by a national government to be exempted from EIA requirements, as the European Directive on EIA allows. Clearly, the unjustified use of such legislation could bypass EIA requirements and this has, in practice, taken place on occasion in a number of countries. There may be certain classes of project which, though their environmental impacts are significant, are normally exempt from EIA. For example, the construction of city tower (high-rise) blocks and the conversion of natural areas to intensive agriculture have frequently escaped EIA in many countries.

It is also important that the impacts arising at different stages of the project are assessed. Thus, impacts arising at exploration, construction, operation, modification and decommissioning stages should be considered. Further, impacts under both normal operating and potential accident conditions need to be evaluated.

Following from this is the issue of whether all the types of project which should be subject to the legal provisions of the EIA system are actually assessed or whether, in practice, some actions are in effect exempted. Such avoidance of EIA requirements may result from accident, from the difficulty of undertaking EIA, from the exercise of power by those assessed, or from setting the criteria for applicability too high. (The screening of environmentally sensitive projects is discussed in Chapter 9.)

The argument that the EIA of, say, the construction of a road, takes place too late in the decision-making process to influence crucial choices between different types of transport system and hence their environmental impacts is well established (Therivel et al., 1992; Wood and Djeddour, 1992; Partidario and Clark, 2000). The same argument applies to projects such as housing schemes where the cumulative impacts of several projects can only be adequately covered at the plan-making stage (Wood, 1988a). The need for EIA at strategic level, i.e. at programme, plan and policy tiers, is widely accepted and many jurisdictions are now implementing strategic environmental assessment (SEA) provisions, though many permit at least some SEA to take place. (SEA is dealt with in Chapter 19.)

The definition of 'environment' in EIA has been treated differently in different jurisdictions. The European Directive on EIA, as amended, eschews consideration of social and economic impacts whereas the EIA systems in many other parts of the world, including many developing countries, evaluate impacts other than those upon the physical environment. It is inevitable that in any reasonably democratic decision-making procedure economic and social factors will strongly influence the outcome as a result of the political process. It was the neglect of the physical environment in decision making which was the original stimulus for EIA in the United States, and it was the need to redress this balance that led to the narrow focus of the European Directive on EIA.

The difference in approach to the coverage of impacts between Europe and the United States can probably be best explained by settlement patterns.

Europe is much more densely populated than the United States and Canada and propinquity has led to the need for relatively stringent controls over new development in which economic and social factors are important determinants. EIA in the United States and Canada developed partially to fill a vacuum created by the absence of a strong land-use planning system and consequently embraced a full range of effects. A further factor is the tradition of distrust of government in the United States which has resulted in greater citizen participation and more open government than apply in Europe (Wood, 1989). Since the public finds the distinction between the treatments of the different types of impact of a given proposal to be artificial, there has inevitably been pressure to treat such impacts comprehensively in North America. In the circumstances, consideration of the social and economic impacts of development within the EIA framework was inevitable. A similar situation has arisen in many other jurisdictions, for example in Australia and New Zealand where EIA has evolved to give explicit consideration to the needs of pre-European peoples.

In the last analysis, the issue of whether or not EIA covers impacts other than those on the physical environment is probably not critical, especially as the distinction between them is often a narrow one in practice. It is, however, important that all impacts on the physical environment are encompassed by the EIA system. Thus, impacts on the various environmental media (for example, the air), on living receptors (for example, people, plants) and on the built environment (for example, buildings) should be considered. Further, indirect impacts arising from other types of induced activity (for example, ancillary service development) and the interrelatedness of environmental impacts (for example, emissions of sulphur dioxide affecting the acidity of freshwater) and cumulative impacts (Council on Environmental Quality – CEQ, 1997a; Ross, 1998; European Commission, 1999a) need to be assessed.

The various criteria which can be used in considering the coverage of EIA systems are summarised in Box 7.1. These are used to assist in the review of the coverage of each of the seven EIA systems that follows.

Box 7.1   Evaluation criteria for the coverage of EIA systems

Must the relevant environmental impacts of all significant actions be assessed?

- Does the EIA system apply to all public and private environmentally significant projects?
- Are the provisions applied in practice to all the actions covered in principle?
- Are all significant environmental impacts covered by the EIA system?

United States

The National Environmental Policy Act 1969 (NEPA) applies to federal actions but not to state actions or to most private projects except where they require a federal permit. Section 102(2)(C) of NEPA (Box 2.1) states that all agencies of federal government must:

> include in every recommendation or report on proposals for legislation and other major federal actions significantly affecting the quality of the human environment, a detailed statement ...

The meaning of each phrase has been picked over by countless court deliberations. Bass *et al.* (2001, p. 30) state that actions typically consist of:

- **Policies**. Adoption of official policies, rules [and] regulations ...
- **Plans**. Adoption of formal plans ...
- **Program[me]s**. Adoption of combined actions intended to implement a specific policy or plan ...
- **Projects**. Approval of specific projects ... including federal activities, federally assisted activities and actions approved by federal permit or other regulatory decisions.

Examples of activities which may be subject to NEPA include discharges to wetlands and federal land management activities such as mining, oil and gas development, highway and airport construction, port development and navigation projects, timber harvesting, and so on.

The very large number of environmental assessments (EAs) prepared each year bears eloquent testimony to the fact that NEPA is somewhat broader in application than it might at first appear. While there continue to be legal arguments about whether an EA is required in particular cases, or whether an environmental impact statement (EIS) rather than an EA should be prepared in certain circumstances, it appears to be true that the EIA provisions enshrined in NEPA are generally applied in practice to almost all the actions to which it is addressed. However, not all environmentally significant actions are 'federal' and not all states have EIA laws (Chapter 2) to cover those not caught by NEPA.

If there has been an area of under-application it has been in relation to policies. NEPA was always intended to permit strategic environmental assessment but, until relatively recently, EIA was largely confined to projects. However, while many programmes and plans are now being subjected to NEPA (see Chapter 19), the EIA of policies appears still to be a relatively neglected area.

The types of environmental impact covered by NEPA are broad. The impacts which should be included in an EIS have been summarised by Bass *et al.* (2001, p. 102) as:

- Direct effects.
- Indirect effects.
- Cumulative effects.
- Growth-inducing effects.
- Changes in land-use patterns, population density, or growth rate.

- Conflicts with land-use plans, policies or controls.
- Unavoidable effects.
- Short-term uses of the environment versus long-term productivity.
- Irreversible or irretrievable commitments of resources.
- Energy requirements and conservation potential.
- Natural or depletable resource requirements.
- Effect on urban quality.
- Effect on historical and cultural quality.
- Socio-economic and environmental justice effects.

In practice, as a result of numerous legal challenges, these types of impact are generally covered where they are likely to be significant.

A 1994 executive order on environmental justice (No. 12898) requires that the environmental, human health, economic and social effects of federal actions, on minority and low-income communities be covered in EISs (CEQ, 1997b; Bass, 1998; Environmental Protection Agency – EPA, 1999c). There has also been an increasing emphasis on the treatment of cumulative impacts in EIA (Canter, 1997; CEQ, 1997a; EPA, 1999a) and on the protection of biodiversity (Bear, 1994; EPA, 1999b). Other issues to have been incorporated in some recent EISs include acid precipitation and global climate change. Recently, there has been debate about how the coverage of issues such as cleaner technology and the ramifications of emission and effluent trading programmes should be incorporated into EIA. However, there have been criticisms that some social and cultural impacts are neglected in the EIA process (King, 1998).

To summarise, the federal EIA system is partial in its application to environmentally significant actions but comprehensive in its coverage of federal actions and environmental impacts. It is thus, perhaps, typical of the fragmentation of US environmental policy which it was designed to address.

UK

The UK Planning Regulations and other EIA regulations do not apply to programmes, plans and policies (but see Chapter 19). EIA applies to the various projects listed in the amended European Directive on EIA, subject to the use of screening criteria, no matter under which UK legislation they fall. This list is lengthy and more comprehensive than that in the original Directive. As mentioned in Chapter 6, EIA has been incorporated into the town and country planning system (and other statutory procedures) by means of regulations made under the European Communities Act 1972 which do not allow the requirements of directives to be exceeded. However, the Planning and Compensation Act 1991 allowed the (then) Secretary of State for the Environment to require EIA for other proposed planning projects. This power was exercised in relation to private motorways, motorway service areas, wind generators and coast protection works in 1994 and in relation to sports stadia, leisure centres and golf courses in 1999.

The UK EIA system, then, is not confined to projects approved under the town and country planning procedures. Between 1988 and 1999, numerous additional regulations came into force in the UK to close the loopholes in the implementation of the provisions of the original Directive which gradually became apparent (Bond, 1997). The 1999 Planning Regulations, together with the other project approval systems into which EIA requirements have been integrated, consolidated all the previous regulations and provide for the assessment of most types of project. Exceptions relate to projects for the restructuring of agricultural landholdings (which are included in the amended Directive but which are not adjudged to take place in the UK) and to classified defence projects (which are not included in the Directive). Formal arrangements have been put in place for the EIA of projects approved under the Transport and Works Act 1992 and by specific Act of Parliament.

Nearly all types of public and private project are thus subject to assessment. However, whether a particular project is assessed depends upon the screening criteria and thresholds which apply to the project type. It also depends on the application of those criteria by local planning authorities (LPAs) and by other competent authorities (see Chapter 9).

As in the European Directive, the Regulations refer to aspects of the physical environment: 'population, fauna, flora, soil, water, air, climatic factors, material assets, including the architectural and archaeological heritage, landscape, and the inter-relationship between [these] factors' (Planning Regulations, Schedule 4, Part I, para. 3). Social and economic impacts are not overtly included in these factors. It is, however, open to LPAs to consider these matters in reaching a planning decision if they choose to do so. The definition of effects adopted in the UK should include 'direct effects and any indirect, secondary, cumulative, short, medium, and long-term, permanent and temporary, positive and negative effects' (Planning Regulations, Schedule 4, Part I, para. 4) where this can be 'reasonably required'. In summary, the coverage of environmentally significant types of project requiring planning permission is, in principle, comprehensive, and, under the new Regulations, considerably improved, but some discretion relating to the coverage of certain types of environmental impact remains.

The Netherlands

The EIA provisions of the Environmental Management Act 1994 apply to significant policy plans and spatial plans involving locational decisions, as well as to projects. In practice, the proportion of EIAs carried out on waste management, electricity supply, water supply, minerals extraction and land-use programmes and plans, has been almost 25 per cent of the total. Most of these EIAs involve either site selection for specific projects or alterations to an existing land-use plan as a result of a proposed project (Arts, 1998).

The EIA system applies to all the activities specified in the EIA Decree. The Decree, amended in 1992, 1994 and 1999, contains a list of projects for which EIA is mandatory (Part C). This now includes all the projects listed in Annex I of the amended European Directive (see Chapter 3). Part D of the Annex to the EIA Decree contains a list of activities and decisions for which EIA may be required, and a list of thresholds to be applied. This list corresponds broadly to Annex II of the EIA Directive.

The Dutch EIA system applies to both public and private projects, though most of the EIAs undertaken have been for private developments. The provision for exemptions from EIA which is contained in the Act has been used only occasionally and was partially repealed in 1994. There have been cases where EIA has not been undertaken because of over-liberal interpretation of the guidance and thresholds. However, in general, EIA is carried out if the activity is listed in Part C or if it is a Part D action which is located in a designated sensitive area or which exceeds the relevant threshold size. As mentioned in Chapter 6, the courts have required EIA where these criteria have been met.

The Act does not contain a definition of the 'environment' to be covered in EIA, a weakness criticised by the European Commission (see Chapter 5). The delineation of topics covered tends to be undertaken by the EIA Commission in the scoping process. The coverage of energy, resource, waste disposal and traffic impacts is standard, and indirect and cumulative impacts are usually included. There has been, as mentioned in Chapter 6, some confusion over the relevance of social and economic impacts to EIA, with the Ministry of Housing, Spatial Planning and the Environment anxious to maintain the emphasis on impacts on the physical environment.

Canada

The Canadian Environmental Assessment Act 1982 requires environmental assessment (EA) where the federal responsible authority (RA):

- proposes the project;
- grants financial assistance to the project;
- sells, leases or grants an interest in land to enable a project to be carried out;
- grants a permit or licence for a prescribed project.

'Project' is broadly defined to include construction, operation, modification, decommissioning or other undertaking (and certain physical activities not related to physical works such as dredging) (section 2). Not only are public projects covered by the Act but many private projects require federal money or land or a federal permit and are thus also covered by the provisions of the Act. Each of the provincial governments has its own EA system, as do various aboriginal and other jurisdictions, and many of those private projects not subject to the Act are caught by their requirements (see Chapter 5).

There are, however, some gaps in the coverage of projects. For example, many projects in urban areas escape EA altogether. Crown Corporations and Harbour Commissions are not defined as RAs under the Act (though their activities may nevertheless be subject to EA) and certain types of projects (for example, overseas activities) are excluded (Commissioner of the Environment and Sustainable Development – CESD, 1998[1]). The Minister of the Environment (MoE), in responding to the five-year review of the Act, proposed a bill expanding coverage to include these and other omissions while eliminating some minor projects (Government of Canada, 2001; MoE, 2001). The Act covers only projects, though the Canadian government has committed itself to the environmental assessment of policies, plans and programmes submitted to a minister or going before cabinet by updating the directive on SEA originally issued in 1990 (Canadian Environmental Assessment Agency – CEAA, 1999e).

The coverage of environmental effects in the Act is potentially broad. Not only must direct changes to the biophysical environment be covered, but also 'effects in several socio-economic and cultural areas that flow directly from the environmental effects of the project' (CEAA, 1994, p. 85). These include effects on:

- human health;
- socio-economic conditions;
- physical and cultural heritage;
- traditional aboriginal land and resource uses.

As mentioned in Chapter 5, the Act encourages RAs to 'promote sustainable development' and panels are beginning to demand that proponents provide evidence that their projects will make a positive contribution to sustainability (Gibson, 2000). For example, the Voisey's Bay Mine and Mill Guidelines stated that, in reviewing the EIS, the Panel would consider:

> the extent to which the undertaking may make a positive overall contribution towards the attainment of ecological and community sustainability, both at the local and regional levels. (CEAA, 1997a, section 3.3)

Because the Canadian environmental assessment process relies heavily on the RAs for its implementation, there have been variations in the extent to which the provisions of the Act have been applied fully in practice, despite the power of the Minister of the Environment to demand additional EA (Nikiforuk, 1997; CESD, 1998). There is no action-forcing mechanism in the Act (beyond the public participation and access to information provisions) to ensure compliance by certain departments although redress through the courts is possible (Hazell, 1999). A further difficulty is caused by 'concern about federal intrusion into provincial spheres of responsibility [resulting in a] narrow approach to environmental assessment' (CESD, 1998, p. 6–18). Thus, a federal EA for a bridge permit might consider the effects on fish habitat (a federal responsibility) but not on most other wildlife habitats (a provincial responsibility). In practice, completeness of EA coverage may be achieved by application of provincial EA provisions but the potential scope for confusion is considerable. To overcome these problems, the Minister of the Environment proposed the establishment of a federal environmental assessment coordinator for each EA and the strengthening of federal/provincial cooperation in EA (Government of Canada, 2001; MoE, 2001).

The Act specifies that EA must include coverage of:

the environmental effects of malfunctions or accidents that may occur in connection with the project and any cumulative environmental effects that are likely to result from the project in combination with other projects or activities that have been or will be carried out. (section 16(1)(a))

An evaluation of the significance of these effects must also be made. Canada has long been concerned with cumulative environmental effects (Peterson *et al.*, 1987; Sonntag *et al.*, 1987; CEAA, 1994, 1999a; Ross, 1998), perhaps because so much of its environment is still in a relatively undisturbed condition (CEAA, 1996). Cumulative environmental effects from projects are defined in CEAA's reference guide on the topic as:

the effect on the environment which results from effects of a project when combined with those of other past, existing or imminent projects and activities. These may occur over a certain period of time and distance. (CEAA, 1994, p. 135)

In practice, this exemplary concern with cumulative impacts has been complicated by artificial jurisdictional limitations on the coverage of EA as well as by methodological uncertainties which have left much to be desired. CESD (1998, pp. 6–19) found that 'only Parks Canada is considering cumulative environmental effects on a regular and rigorous basis' in screenings. Baxter *et al.* (1999) found that, of 12 EAs addressing cumulative effects, only 40 per cent met the acceptability criteria they utilised. David Redmond and Associates (1999) reported that cumulative effects were considered in 60 per cent of the screenings they examined. A number of changes to the Act were proposed following the five-year review to endeavour to strengthen cumulative environmental effects assessment, including federal participation in regional studies (Government of Canada, 2001; MoE, 2001).

Commonwealth of Australia

In debating the Impact of Proposals bill in parliament prior to the passing of this (now repealed) Commonwealth EIA legislation, the responsible minister stated that, 'we will not be limiting its scope in terms of the type of proposal that could be the subject of a statement' (Cass, 1974, p. 4082). However, as shown in Chapter 19, the coverage of the Act was, in practice, almost entirely restricted to projects. As in the United States, the coverage of the EIA system in relation to Commonwealth proposals was broad but it excluded many environmentally significant state and private proposals, many of which were subject to state EIA processes (Harvey, 1998). In practice, it proved easy for Commonwealth departments to avoid the somewhat arbitrary EIA procedures. This avoidance of EIA procedures was one of the principal motivations for reform (Hill, 1998).

The intended coverage of the environmental assessment provisions of the Australian Environment Protection and Biodiversity Conservation Act 1999 (EPBC Act) is shown in Box 7.2. By contrast with the previous legislation, the

Box 7.2   Coverage of the Australian EIA system

3(1)   The objects of this Act are:

(a) to provide for the protection of the environment, especially those aspects of the environment that are matters of national environmental significance ...

3(2)   In order to achieve its objects, the Act:

(d) adopts an efficient and timely Commonwealth environmental assessment and approval process that will ensure activities that are likely to have significant impacts on the environment are properly assessed ...

The matters of national environmental significance identified in the Act as triggers for the Commonwealth assessment and approval regime are:

- World heritage properties
- Ramsar wetlands
- Nationally threatened species and ecological communities
- Listed migratory species
- Nuclear actions (including uranium mining)
- Commonwealth marine areas.

523(a)   **action** includes:
(a) a project; and
(b) a development; and
(c) an undertaking; and
(d) an activity or series of activities; and
(e) an alteration to any of the things mentioned in paragraph (a), (b), (c) or (d).

Sources: Environmental Protection and Biodiversity Conservation Act 1999, and Environment Australia, 1999a, p. 5.

EPBC Act contains specific provisions for the conduct of discretionary strategic assessments of actions that may be carried out under a proposed policy, programme or plan and requiring the strategic assessment of Commonwealth-managed fisheries.

The EPBC Act provides for those responsible for an action, a state government, or a Commonwealth agency to refer a proposal to the Environment Minister for a decision on whether the action requires assessment and approval. The Environment Minister may also request a referral and, if this is not forthcoming, deem the referral to be made. By defining matters of national environmental significance (Environment Australia, 2000a) and the circumstances in which Commonwealth agencies are required to seek the approval or advice of the Environment Minister, the EPBC Act clarifies the coverage of actions to be subject to Commonwealth EIA. However, these matters also limit the coverage of the Commonwealth assessment and approval system. The Commonwealth does not always assess all the impacts of an action. Since activities giving rise to impacts of national significance will also cause other impacts, the Environment Minister cannot make a decision until a notice has been received from the relevant state that it has itself assessed these (for example, broader environmental and social) impacts (section 133(5)). There is therefore a danger that certain environmental impacts may be assessed neither by the Commonwealth nor by the states.

While the Impact of Proposals Act was generally interpreted to include all physical environmental impacts, there was considerable debate about the extent to which social impacts were covered. In practice, many EISs dealt with social impacts and the trend was clearly towards fuller treatment. There were many demands that the Australian Commonwealth EIA system should more formally deal with social and cultural impacts, recognising the fact that about 50 per cent of public submissions in relation to EISs dealt with these. The 1992 Intergovernmental Agreement on the Environment attempted to resolve the issue:

> The parties agree that impact assessment in relation to a project, program or policy should include, where appropriate, assessment of environmental, cultural, economic, social and health factors. (Commonwealth Environment Protection Agency – CEPA, 1992, Schedule 3)

However, the treatment of social impacts, and particularly of impacts on indigenous peoples, continued to leave much to be desired (Craig & Ehrlich *et al.*, 1996). Section 528 of the EPBC Act leaves no room for doubt about the meaning of 'environment', which includes:

(a) ecosystems and their constituent parts, including people and communities; and
(b) natural and physical resources; and
(c) the qualities and characteristics of locations, places and areas; and
(d) the social, economic and cultural aspects of a thing mentioned in paragraph (a), (b) or (c).

It is widely agreed in Australia that EIA should be employed as one means of pursuing the nationally agreed goal of ecologically sustainable development (ESD) (Court *et al.*, 1996). The government therefore proposed formalising the coverage of EIA to include cumulative, social and economic impacts (CEPA, 1994; Hill, 1998). Accordingly, under the EPBC Act, the Environment Minister is required to take the principles of ecologically sustainable development into account when deciding whether to approve the taking of an action (section 131(2)). Indeed, one of the objects of the Act is:

> to promote ecologically sustainable development through the conservation and ecologically sustainable use of natural resources. (section 3(1)(b))

It is apparent that, while practice may have left something to be desired in the past, the coverage of environmental impacts under the EPBC Act is potentially comprehensive and the coverage of actions is appropriately confined to those impacting on matters of national environmental significance. It is, however, too early to comment on whether, in practice, all those actions which ought to be assessed are being referred for a decision on whether assessment and approval is required.

New Zealand

The Resource Management Act 1991 covers local government actions comprehensively. It not only applies to projects but also requires the preparation of policy statements and plans which, in turn, determine how the effects of activities are assessed. The Act applies to almost every proposed project, as most development projects require a resource consent in New Zealand. The EIA provisions thus apply to land-use and subdivision consents (which used to be dealt with under the Town and Country Planning Act) and to discharge, water abstraction and coastal permits (which were previously dealt with under a variety of legislative provisions). Some central government projects with major environmental effects could, in principle, be dealt with under the Environmental Protection and Enhancement Procedures where they fall outside the provisions of the Act (see Chapter 5) but no such cases have arisen to date.

There is considerable debate about whether or not the Resource Management Act requires the strategic environmental assessment (SEA) of policy statements and plans. Section 32 of the Act requires policies and plans to be considered in terms of their costs and benefits (including environmental costs and benefits). Section 5 provides an additional power: it stresses that sustainable management means avoiding, remedying or mitigating any adverse effects of activities (including policies and programmes) on the environment (see Box 5.2). However, no SEA is currently being carried out for regional policy statements or for district plans. The need for SEA has never been tested in the Environment Court.

The legal provisions are far from unambiguous. For example, since all resource consent applications, large or small, are controlled under the provisions of the Act, there is ample scope for discretion in the interpretation of terms such as 'major', 'minor' and 'significant', about which no further advice has been issued by the Ministry for the Environment (MfE) and about which the Environment Court has been almost silent to date. Montz and Dixon (1993) suggested that the way in which such terms were interpreted by planners when applications were received would be important in the successful implementation of EIA. This indeed appears to be the case in practice.

The provisions of the Act are enforceable by universal rights of appeal to the Environment Court and, on points of law, beyond the Court to the higher courts. There are seven Environment Court judges, each of whom sits with two part-time commissioners. Court judgements form the basis of planning and environmental law in New Zealand. The decisions of the Environment Court are helping to provide clarification, legal precedents and some consistency in the interpretation of the Act. The independent Parliamentary Commissioner for the Environment (PCE) can, and does, become involved in

a limited number of EIA issues where this is thought likely to result in recommendations of general applicability.

The extent to which the Act applies in practice to consents with minor
environmental effects depends largely on the screening procedures adopted by
local councils. These procedures include the designation of certain types of
development as permitted activities (thus not requiring an assessment of
environmental effects – AEE) and as controlled activities (thus requiring a
limited form of AEE). They also include the identification of persons likely to
be affected by the proposed development who should be informed about it
(Chapter 9).

The definition of the term 'environment' adopted is also very broad as it
includes 'ecosystems and their constituent parts, including people and communities; and all natural and physical resources; and amenity values' together
with relevant 'social, economic, aesthetic and cultural conditions' (section 2).
The comprehensive interpretation of effects is evident from the 'matters to be
considered' listed in the Fourth Schedule to the Act (Box 7.3). EIA in New
Zealand could therefore be construed as encompassing social impacts where
relevant and, in particular, consideration of Maori cultural and community
impacts.

Box 7.3   Content of a New Zealand EIA report and matters to be considered in its preparation

Resource Management Act Fourth Schedule

1. **Matters that should be included in an assessment of effects on the environment** – Subject to the provisions of any policy statement or plan, an assessment of effects on the environment for the purposes of section 88(6)(b) should include:

(a) A description of the proposal;
(b) Where it is likely that an activity will result in any significant adverse effect on the environment, a description of any possible alternative locations or methods for undertaking the activity;
(c) Where an application is made for a discharge permit, a demonstration of how the proposed option is the best practicable option;
(d) An assessment of the actual or potential effect on the environment of the proposed activity;
(e) Where the activity includes the use of hazardous substances and instal-lations, an assessment of any risks to the environment which are likely to arise from such use;
(f) Where the activity includes the discharge of any contaminant, a descrip-tion of:
   (i) The nature of the discharge and the sensitivity of the proposed receiv-ing environment to adverse effects; and

(continued)

(continued)

    (ii)   Any possible alternative methods of discharge, including discharge into any other receiving environment.

(g) A description of the mitigation measures (safeguards and contingency plans where relevant) to be undertaken to help prevent or reduce the actual or potential effect;

(h) An identification of those persons interested in or affected by the proposal, the consultation undertaken, and any response to the views of those consulted;

(i) Where the scale or significance of the activity's effect are such that monitoring is required, a description of how, once the proposal is approved, effects will be monitored and by whom.

2. **Matters that should be considered when preparing an assessment of effects on the environment** – Subject to provisions of any policy statement or plan, any person preparing an assessment of the effects on the environment should consider the following matters:

(a) Any effects on those in the neighbourhood and, where relevant, the wider community including any socio-economic and cultural effects;

(b) Any physical effect on the locality, including any landscape and visual effects;

(c) Any effect on ecosystems, including effects on plants or animals and any physical disturbance of habitats in the vicinity;

(d) Any effect on natural and physical resources having aesthetic, recreational, scientific, historical, or cultural, or other special value for present or future generations;

(e) Any discharge of contaminants into the environment, including any reasonable emission of noise and options for the treatment and disposal of contaminants;

(f) Any risk to the neighbourhood, the wider community, or the environment through natural hazards or the use of hazardous substances or hazardous installations.

Finally, the word 'effect' is also given a broad meaning in section 3 of the Act and includes: positive or adverse, temporary or permanent, past, present or future and cumulative effects regardless of their scale, intensity, duration or frequency. It also includes risk. Cocklin *et al.* (1992) believed that the Act, by encouraging the territorial authorities to integrate EIA within the plan-making process, provides the basis for effective cumulative effects assessment, something which has previously been lacking in New Zealand, as elsewhere. There is evidence that this is occurring in some authorities but that practice still leaves much to be desired (Dixon, 1993; PCE, 1995).

Major amendments to the EIA provisions of the Resource Management Act were proposed in 1998. These included a narrowing of the definition of 'environment' to confine impacts to those on the physical environment, the reining in of the range of issues listed in the Fourth Schedule (possibly by repealing it) and limiting the grounds of appeal to the Environment Court (MfE, 1998). However, these proposed amendments were rejected by the Local Government and Environment Select Committee in May 2001, leaving proposed revisions aimed at reducing the time and costs involved in obtaining resource consents (see Chapter 5).

Despite these proposals, it is apparent that the Mark II New Zealand EIA system, in common with many other systems of long standing, employs considerably broader definitions of actions, of the environment and of effects than, for example, the amended Mark I UK EIA system.

South Africa

The South African EIA system applies to most public and private environmentally significant projects, but not currently to programmes, plans and policies. There was considerable controversy over the designation of projects in the EIA Regulations during consultations prior to their promulgation. The principal omissions are mining-related activities (covered by the Minerals Act 1991), developments within rivers (covered by the National Water Act 1998) and the intensification of various existing land uses. As elsewhere, it is not possible to assess linked developments, and hence cumulative effects, in South Africa. The EIA of changes of certain land uses (from residential to industrial or commercial use and from light to heavy industrial use) has been suspended, in part because of the lack of capacity of the competent authorities (Granger, 1998). The provision for proponents to apply to have insignificant projects exempted from the EIA requirements (see Chapter 9) is being widely employed.

The term 'environmental impact' is not defined in the EIA Regulations (Republic of South Africa, 1997). However, South Africans are acutely conscious of the social impacts of development, as a consequence of the injustices of apartheid. Thus, the Environment Conservation Act 1989 indicated that environmental impact reports should identify affected economic and social interests and estimate the nature and extent of effects on them (section 26(a)). The guidelines on EIA make it clear that 'the effects on human health, socio-economic conditions, physical and cultural resources should be included' (Department of Environmental Affairs and Tourism, 1998c, p. 23). This does not accord with the practice of certain provincial authorities, which have sometimes tended to define the environment narrowly, principally in nature conservation terms (Granger, 1998).

Integrated environmental management (IEM) has always been based upon a broad definition of the environment. Thus, Volume 5 of the IEM guidelines series (Department of Environmental Affairs, 1992) consisted of a checklist of environmental characteristics which included cultural resources, socio-economic characteristics of the affected public, and social and community facilities, as well as cumulative impacts.

However, practice under the IEM procedure often failed to reflect the guidance. Thus, O'Riordan (1998), in analysing IEM experience, stated that 'there is no definable link between social and environmental assessments in EIA in

South Africa.' Nevertheless, despite widely acknowledged weaknesses in social impact assessment, a review of 28 environmental assessment reports prepared between 1971 and 1986 revealed that

> The initial assumption that the reports would focus primarily on biophysical factors and that social factors would be absent from many of the EAs reviewed, was dispelled by the consideration, in each report, of at least some aspect pertaining to social concerns. (Mafune *et al.*, 1997, p. 206)

The National Environmental Management Act 1998 (NEMA) specifically includes social impacts and it applies to policies, programmes and plans, as well

Table 7.1   The coverage of the EIA systems

Criterion 2: Must the relevant environmental impacts of all significant actions be assessed?

Jurisdiction	Criterion met?	Comment
United States	Impacts: Yes Actions: No	Applies only to federal, not state or most private, projects. Comprehensive coverage of impacts of significant federal actions (including some non-project actions)
UK	Yes	Comprehensive coverage of significant projects approved under town and country planning and other processes. Marginal discretion in impact coverage
The Netherlands	Yes	Covers highly significant projects and certain policies, plans and programmes. Indirect and cumulative environmental impacts covered, but not legally specified
Canada	No	Restricted to federal projects and projects requiring federal finance, land or permit. Artificial limitations on otherwise comprehensive coverage of impacts of certain projects possible
Commonwealth of Australia	Partially	Coverage of impacts (which includes social, economic and cultural effects) and actions confined to those affecting matters of national environmental significance
New Zealand	Yes	Act provides for all local authority approved policies, plans and projects to be subject to EIA covering bio-physical environment, social and economic impacts
South Africa	Yes	Comprehensive coverage of impacts of almost all environmentally significant projects

as to projects. The EIA provisions of NEMA (which was implemented when regulations were approved by parliament, expected in 2002) include mining and certain other projects not covered by the Environment Conservation Act 1989. It is to be hoped that the NEMA EIA regulations will ensure that only significant activities are subject to EIA, relieving hard-pressed provincial staff of the need formally to exempt trivial projects.

Summary

The coverage of the seven EIA systems is shown in Table 7.1. The coverage of impacts and projects in the EIA systems in the UK, the Netherlands, New Zealand and South Africa is, at least in principle, comprehensive. However, those in the United States and the Commonwealth of Australia only partially meet the review criterion and that in Canada does not do so.

It is no coincidence that the three jurisdictions failing to meet the review criterion are federal countries. The reason for the failure of their systems to cover all significant projects is largely constitutional. The jurisdiction of the federal government extends only to certain projects – the remainder are subject to state or local control. Certain states, provinces or territories may possess less comprehensive EIA systems than those of, say, California (Bass *et al.*, 1999) and thus some environmentally significant projects may escape scrutiny altogether. This is obviously true in the many American states which have no EIA system (Mandelker, 2000). (However, it could also happen in, say, the Netherlands if the thresholds and criteria are set too high.) Even so, the most significant public and private projects are normally covered in all the systems though it remains to be seen how comprehensive the South African EIA system (which is less devolved) proves to be.

This jurisdictional boundary problem is also the reason that the treatment of certain impacts is limited in Canada. In practice, the coverage of impacts on the physical environment in EIA reports in all the countries generally tends to be reasonably comprehensive.

Note

1. The reports of the Commissioner of the Environment and Sustainable Development are available at www.oag-bvg.gc.ca

Chapter 8

Consideration of alternatives

The consideration of alternatives has been described as 'the heart of the environmental impact statement in the US' (Council on Environmental Quality – CEQ, 1978, Regulation 1502.14). Sadler (1996, p. 55) asserted that this held more generally: 'The identification, analysis and comparison of alternatives to the proposal is the key to creative, proactive, decision relevant assessment.' Nevertheless, the treatment of alternatives has proved to be a contentious area in EIA. For example, although the treatment of alternatives has been strengthened in the amended European EIA Directive (see Chapter 3), it is still not a mandatory requirement that alternatives be considered in the EIA report. This chapter discusses why the treatment of alternatives in EIA is important and advances several evaluation criteria to assist in the review of EIA systems. These criteria are then employed in the analysis and comparison of the EIA systems in the United States, UK, the Netherlands, Canada, Commonwealth of Australia, New Zealand and South Africa.

Consideration of alternatives in EIA systems

The consideration of the alternatives to an action is the first step in the EIA process (see Figure 1.1). Steinemann (2001, p. 3) regarded it one of the most important stages of the process: '[t]he quality of a decision depends on the quality of alternatives from which to choose.' Marriott (1997, p. 51) felt that the comparative analysis of alternatives should be a thorough, systematic and documented process involving all stakeholders and utilising a 'solid platform of evaluation criteria'.

The proponent of an action has a set of aims to be met which can normally be satisfied in a number of alternative ways, each of which has different effects upon the environment. However, the proponent's aims, and therefore his or her preferred alternative, will not always reflect wider societal aims (Glasson *et al.*, 1999, p. 94). The costs of various alternatives will vary for different groups of people and for different environmental components. In EIA, the choice of an alternative which minimises the environmental impact of the action should be an important determinant of any decision to proceed.

Steinemann (2001) made an important distinction between alternative approaches and alternative designs which echoed the Canadian 'alternatives to' and 'alternative means'. For example, electricity provision to a newly developing suburb might involve some mixture of the construction of new generating capacity, the import of electricity from another region and stringent energy conservation measures, all alternative approaches to the provision of electricity, being functionally different from each other. Should it be decided to construct a new generation facility, further alternatives regarding the type of generator (gas-fired, wind-powered, etc.), location and site layout will exist. These are examples of alternative designs, where the objective of providing electricity is achieved in a functionally similar way, by building a generating facility. Steinemann (2001) observed that alternative designs usually dominate sets of alternatives, largely because the opportunity to develop more far-reaching alternatives is often foreclosed by the late stage at which EIA takes place in the project development cycle.

Even alternative designs may be severely limited where a private proponent already owns the project site or where the proposal involves, say, mineral extraction. Many jurisdictions (for example, the Netherlands) insist that the 'no-action alternative' be fully evaluated during the EIA process. The World Bank (1996) has stated that evaluation of the no-action alternative should be routinely included in EIA reports for projects in developing countries that it funds. McCold and Saulsbury (1998) have stressed the merits of, and approaches to, analysing the no-action alternative.

It is at the design stage, before any commitment to any particular action has been made, that it is easiest and cheapest to choose the alternative which best reduces the environmental impacts of an action. Later in the EIA process it may be necessary to consider another alternative if unforeseen impacts are predicted to arise from that chosen. Marriott (1997) suggested that the starting point in the consideration of alternatives should be the establishment of the purpose and need for the project, from which a range of initial alternatives should logically flow. Canter (1996) observed that typical EIAs in the United States focus on between three and five alternatives.

In order for choices between alternatives to be made, the designer needs to have access to environmental expertise and/or to simple evaluative tools (Morris and Therivel, 2001). For example, regular meetings between designers and environmental professionals together with a specific, if brief, evaluation of the impacts of different alternatives (including the option of not proceeding with the action) can assist in making environmentally appropriate choices. Easy-to-use, if unsophisticated, methods such as simple checklists of environmental factors and quick assessments of the magnitude of comparative impacts can helpfully be employed to evaluate alternatives at this very early stage (Jones, 1999). Canter (1996) provided a comprehensive review of methods for comparing and evaluating alternatives, many of which involved the ranking and weighting of alternatives. The World Bank (1996) also indicated that it might be appropriate to employ ranking, rating and scaling techniques in evaluating alternatives. Glasson *et al.* (1999, p. 95) noted that methods for comparing and presenting alternatives ranged from simple qualitative descriptions to the attribution of monetary values to impacts.

Since the proponent's aims will not necessarily reflect wider interests and societal goals, informal consultation with decision-makers, environmental authorities and representatives of the communities affected may be helpful in choosing between alternatives and in alleviating the frequently narrow focus of project alternatives. However, because of the potential sensitivity of many alternatives that may later be discarded, such consultation needs to be handled with great care at such an early stage in the siting process.

Once the decisions regarding broad approach and location have been made, more detailed design of the action can take place (see Figure 1.1). Here, where more resources are committed to the action, it is equally important that the avoidance, mitigation and enhancement of environmental impacts continue to be considered. Indeed, Marriott (1997) has proposed a continuous approach to the consideration of alternatives throughout the EIA process, in which alternatives are frequently screened and refined in order to avoid or minimise potential impacts. The same techniques of specific evaluation, together with the use of appropriate assessment methods, apply as the range of design alternatives narrows and the preferred design emerges.

The importance of timely consultation and public participation throughout the selection and evaluation of alternatives is often stressed (see, for example, World Bank, 1996). The two principal advantages of this approach are that adequate information about the proposed alternatives can be provided and that, as a consequence, consensus on the most acceptable variant of the action is more likely to be reached. Although the benefits of consultation with environmental professionals and public involvement in the development and selection of alternatives are often highlighted, they may be outweighed by the costs of conflict unless participation is carefully managed. Thus, Steinemann (2001, p. 10) noted that 'the public is usually placed in a position to react to alternatives already developed by the agency.'

Sadler (1996) stated that impact significance could be the key to choosing between alternatives but cautioned that significance can be a contentious concept where value judgements and interpretations are influential to the outcome of the decision (see Chapter 9). This step in the EIA process, which involves more detailed environmental evaluation, is much easier both to accomplish and to demonstrate if the environmental impacts of alternative ways of achieving aims and of alternative locations have been assessed in outline earlier.

CEQ surveyed federal agencies in the United States in 1991 to determine how extensively alternatives were actually considered under the National Environmental Policy Act 1969 (NEPA) and to what extent this consideration was influencing decisions. The outcome was interesting:

> The results of the survey indicated that when alternatives are not fully considered in the NEPA process, litigation is more likely, and agencies are less likely to achieve the original goals of the project in an efficient, economical manner. (Bear and Blaug, 1991, p. 18)

It is difficult to assert that a thorough evaluation of the environmental effects of alternatives has been carried out and that the environmental consequences of the detailed design have been fully considered in detailed project design if the alternative chosen for further elaboration is manifestly more damaging to the environment than some of those rejected. It is for this reason that the analysis of alternatives is so important in the EIA process. Since the analysis of alternatives should not only be undertaken properly but also be seen to be undertaken properly, the environmental evaluations undertaken at the alternatives/design stage should be demonstrated in documentary form.

One method of providing an early check that the environmental effects of alternatives really have been fully considered is their inclusion in preliminary documents produced prior to the EIA report (for example, in the notice of intent in the United States). Such documentation should show clear evidence of the mitigation/avoidance of environmental impacts in the initial action designs. For the reasons stated above, this evidence will usually be in the form of an evaluation of the environmental consequences of the alternatives considered. If such evidence is not forthcoming, the proponent can be encouraged to return to this evaluation and, if necessary, to redesign the proposal before too many resources are expended. If documentation does not have to be produced until the scoping stage, or even later in the EIA process, the proponent's commitment to the design will be greater and the chance of cost-effective amelioration may be reduced.

While less satisfactory than the early submission of public documentation, any requirement to discuss the action with the decision-making and/or environmental authorities prior to submission of the EIA report (for example, at the scoping stage or, preferably, at the screening stage of the EIA process) will involve the inspection of design documents. This will provide an opportunity to check that the most environmentally appropriate alternative and design meeting the proponent's aims has been chosen and, if it has not, to require that further iteration of the design process takes place.

As a final vital check that the environmental consequences of alternative approaches, locations and designs to meet the proponent's aims have been considered, the EIA report should contain evidence to this effect. If the documentation proves to be inadequate it may be possible for the proponent to supply supplementary information, but in some instances reconsideration of the proposal may be necessary.

Since the range of possible alternatives to an action may be legion, a choice will need to be made as to the alternatives to be detailed in the EIA report (and other documentation). This choice is usually made on a case-by-case basis, the standard test being that of reasonableness. Thus, in the United States, non-feasible, remote or speculative alternatives need not be analysed (Fogleman, 1990). Canada varies its requirements on the treatment of alternatives according to the type of assessment undertaken. Other jurisdictions use different approaches in attempting to ensure that the evaluation of alternatives is both effective and efficient.

The existence of published advice on the treatment of the environmental impacts of alternatives in the EIA process is beneficial not only to developers but also to consultants, decision-making authorities, environmental authorities, consultees and the public. The various criteria which can be used in

---

Box 8.1   Evaluation criteria for the consideration of alternatives

Must evidence of the consideration, by the proponent, of the environmental impacts of reasonable alternative actions be demonstrated in the EIA process?

- Must clear evidence of the consideration of the environmental impacts of alternatives be apparent in preliminary EIA documentation?
- Must the realistic consideration of the impacts of reasonable alternatives, including the no-action alternative, be evident in the EIA report?
- Does published guidance on the treatment of the impacts of reasonable alternatives exist?
- Does the treatment of alternatives take place effectively and efficiently?

---

evaluating the treatment of alternatives are summarised in Box 8.1. These criteria are employed to assist in the analysis of the treatment of alternatives in each of the seven EIA systems that follows.

United States

The treatment of alternatives lies at the heart of the US EIA system. The National Environmental Policy Act 1969 (NEPA) specifically refers to the coverage of alternatives to the proposed action (section 102(2)(C)(iii)). This is evident throughout the EIA process, commencing with the environmental assessment (EA). The EA, which is supposed not to exceed 15 pages in length, must briefly discuss the feasible alternatives to the proposed action and their environmental impacts (unless there are no 'unresolved conflicts concerning alternative uses of available resources') (Regulations, section 1508.9(b)). In practice, 75 per cent of EAs considerably exceed 15 pages in length but often include adequate discussion of the environmental impacts of both the proposal and of the alternatives to it.

The evaluation of alternatives in the environmental impact statement (EIS) is governed by the 'rule of reason' under which an EIS must consider, analyse and compare a reasonable range of options that could accomplish the agency's objectives. An explanation of why alternatives were eliminated should be included. The Regulations (section 1502.14) state that the range of alternatives to be considered should include:

- alternative ways of meeting the objective;
- the no-action alternative;
- alternatives outside the lead agency's jurisdiction.

The Regulations further require that rigorous evaluation and comparison are required, that the preferred alternative must be identified and that measures to mitigate the environmental impacts of alternatives must be described. Further, the environmentally preferable alternative must be identified in the record of decision for a proposal.

The Regulations, together with the treatment of alternatives in the first seven of 'NEPA's forty most asked questions' (CEQ, 1981a), provide guidance on the consideration of reasonable alternatives. However, considerable scope for uncertainty remains, given the infinite number of alternatives to an action which may be feasible. The question of alternatives has exercised the courts in the United States on a large number of occasions. Criticisms of the treatment of, in particular, the no-action alternative in EISs have been expressed (McCold and Saulsbury, 1998). Following a major study of the treatment of alternatives in EISs, Steinemann (2001, p. 13) reported that:

> Even though the EIS rigorously analyzes the environmental impacts of alternatives and involves the public, it occurs too late. More environmentally sound or publicly acceptable alternatives can be overlooked or eliminated, and not reconsidered, before the rigorous environmental analyses.

CEQ is right to claim that the treatment of alternatives is 'the heart of the environmental impact statement' (Regulations, section 1502.14). There can be no doubt that the US EIA process requires the demonstration, by the federal agency concerned, that the environmental impacts of alternative actions have been considered (though not as fully as they could be) before a decision is reached.

UK

The necessity to integrate environmental factors in the choice between alternatives and in initial project design is a fundamental reason for making the proponent responsible for producing the EIA report in the UK. While the consideration of alternatives is still not a mandatory requirement in the UK, the amended Regulations state that an environmental statement (ES) must include 'an outline of the main alternatives studied by the applicant or appellant and an indication of the main reasons for his choice, taking into account the environmental effects' (Schedule 4, Part II, para. 4).

The Circular makes it clear that, while the Directive and the Regulations do not expressly require the developer to study alternatives, the ES must record the consideration of any alternatives which are studied. It emphasises that:

> consideration of alternatives (including alternative sites, choice of process, and the phasing of construction) is widely regarded as good practice. ... Ideally, EIA should start at the stage of site and process selection, so that the environmental merits of practicable alternatives can be properly considered. Where this is undertaken, the main alternatives considered must be outlined in the ES. (Department of the Environment, Transport and the Regions – DETR, 1999b, para. 83)

However, if the developer states that no alternatives were studied, none needs to be described in the ES.

This strong encouragement to consider alternatives has been a consistent theme in government guidance (Department of the Environment – DoE, 1989, 1994c, 1995; DETR, 2000a). Proponents have been urged to consider strategic alternatives (for example, alternative processes and locations) early enough for them to be feasible options in making choices (DETR, 1997d, p. 49). The reporting of the consideration of alternatives in ESs is important because preliminary EIA documentation does not have to be submitted to the local planning authority or to any environmental authority. Further, because discussion between the proponent and these bodies is not required in the UK EIA system, the ES provides the only formal check on the treatment of alternatives during the EIA process.

The lack of regulatory weight given to the treatment of the environmental impacts of alternatives has been reflected in practice. The Department of the Environment reported that the treatment of alternatives in early ESs was frequently unsatisfactory (DoE, 1991a), a finding endorsed five years later (DoE, 1996). There was substantial improvement in the treatment of alternatives in later ESs, but DoE (1996) indicated that it was still not satisfactory. Blackmore *et al.* (1997, p. 225) confirmed that 'alternatives were either not considered in detail or that coverage was very limited or non-existent' in the ESs they analysed. A review of 100 ESs found that, in 20 cases, no alternative sites, routes or processes were presented, and that in a further 13 cases, alternatives were considered, but without any environmental criteria being applied (DETR, 1997d). Encouragingly, in 34 per cent of the ESs 'it was evident that environmental criteria had had some influence in the selection of the site, route or process' (DETR, 1997d, para. 3.30). It is to be hoped that the introduction of the requirement to report those alternatives studied in the ES will lead to further improvements in practice.

The Netherlands

The Dutch EIA system lays considerable emphasis on the treatment of alternatives, because this largely subsumes the treatment of mitigation. In the Netherlands, alternatives may include measures which would be described as mitigatory elsewhere (for example, technical controls over pollution). Section 7.10 of the Environmental Management Act 1994 specifies the minimum contents of an EIS and lays great emphasis on the coverage of reasonable alternatives. Realistic alternatives to the proposed activity and their environmental consequences must be described both in the EIS and in the non-technical summary. Further, a comparison between the environmental impacts of the proposed development and of the alternatives considered is required. Section 7.10 in effect requires that the no-action alternative be described in the EIS because the future environment in the absence of the proposal or any of the proposed alternatives must be described. In addition, the 'alternative which prevents the adverse effects on the environment or, in so far as this is not possible, reduces them as far as possible using the best means available of protecting the environment' must also be described in the EIS (section 7.10(3)). This 'alternative most favourable to the environment' (or environmentally preferable alternative) is a key element in the Dutch EIA system and requires that impacts be reduced to levels as low as reasonably achievable. Compensation measures must also be considered.

Although the Inception Memorandum Regulations do not specify that the proponent must include mention of alternatives in the notification, some information on alternatives is almost always provided. This provides the EIA Commission (EIAC) with an indication of the alternatives the developer is likely to consider in the EIS. EIAC then suggests, in its recommended guidelines, the additional alternatives which it believes should be studied. These alternatives may include different sites, processes, designs or mitigation methods and are normally adopted by the competent authority in its formal guidelines.

There is a published guide specifically dealing with the treatment of the environmental impacts of alternatives. Some general guidance is also available in the EIA handbook (see Chapter 5) and is provided by van Eck (1994) and van Eck *et al.* (1994). EIAC provides project-by-project guidance on alternatives in its guideline recommendations. A considerable proportion of the guidelines for each project is normally devoted to the treatment of alternatives (see, for example, van Eck *et al.* 1994). EIAC's review criteria (see Chapter 12) lay considerable emphasis on alternatives (de Jong, 1997; Scholten, 1997).

Following receipt of the EIS, the competent authority is obliged to state the grounds on which the decision is based (see Chapter 13). It is therefore compelled to indicate how it has taken into account the effects of the proposed activity, and of its alternatives, on the environment.

Despite differences of view between business participants in the EIA process and environmental groups and improvements in the treatment of alternatives, in practice, alternatives are not always treated as fully as the legislation requires. In particular, certain aspects of the alternative most favourable to the environment are often not pursued by either the proponent or the competent authority. Although many examples of the use of environmentally determined alternatives becoming the main drivers of the project design exist (Verheem *et al.*, 1998), many alternatives appear to involve changes at the margin, rather than radically different approaches. Thus, the Evaluation Committee on the Environmental Protection Act (ECW, 1990, p. 7), in its first report, found that:

> There is a statutory requirement to specify the most environmentally sound alternative to the proposed activity. If the alternative is not worked out in sufficient detail – which often happens – it is impossible to obtain a satisfactory impression of the most viable alternative.

About one-third of EISs need to be supplemented to ensure that alternatives are dealt with in sufficient detail to meet the requirements of, *inter alia*, the European birds and habitats directives (EIAC, 2001a).

Canada

There is considerable emphasis on the treatment of alternatives in the various EA documents referred to in the Canadian Environmental Assessment Act

1992. There is, however, no requirement for the consideration of alternatives and their environmental effects to be included in screening reports. This is left to the discretion of the responsible authority (RA). Section 16(1)(e) of the Act states that any other matter relevant to the screening may be included if the RA requires it, including 'alternatives to the project'.

It is mandatory to consider 'alternative means' in comprehensive study, mediation and panel reports. Section 16(2)(b) of the Act requires that such reports include consideration of 'alternative means of carrying out the projects that are technically and economically feasible and the environmental effects of any such alternative means'. The consideration of 'alternatives to' the project remains discretionary in these reports but inclusion may be required by the Minister of the Environment in consultation with the RA. Box 8.2 summarises the distinction between 'alternative means' and 'alternatives to' in the Canadian system. There is an operational policy statement on the treatment of alternatives which includes two examples and provides some guidance to this rather confusing area (Canadian Environmental Assessment Agency – CEAA, 1998a). In essence, it suggests that RAs should develop criteria to enable choices to be made on environmental, economic and technical grounds.

Gibson (1993) felt that the discretionary provision relating to 'alternatives to' the project was likely to be used rarely and fell short of the comparative

---

Box 8.2   Canadian Environmental Assessment Act provisions for alternatives

Alternatives

The CEAA distinguishes between 'alternative means' and 'alternatives to':

'Alternative means' of carrying out the project are methods of a similar technical character or methods that are functionally the same. 'Alternative means' with respect to a nuclear power plant, for example, includes selecting a different location, building several smaller plants, and expanding an existing nuclear plant. 'Alternative means' that are technically and economically feasible must be considered in a comprehensive study, mediation and panel review, but are discretionary under a screening.

In contrast, 'alternatives to' the project are functionally different ways of achieving the same end. For example, 'alternatives to' the nuclear power plant include importing power, building a hydroelectric dam, conserving energy, and obtaining the energy through renewable sources. Consideration of 'alternatives to' the project is at the discretion of the RA in screening, or of the Minister in consultation with the RA in a comprehensive study, mediation, or panel review.

Source: CEAA, 1994, p. 76.

evaluation of alternatives necessary to force effective integration of environmental concerns into the design stage for new proposals. In practice, while some EA reports have provided broad-ranging coverage of 'alternative means' (if not of 'alternatives to'), most screening reports have failed to mention alternatives at all. Thus, the treatment of alternatives in comprehensive studies has not always been satisfactory. Nikiforuk (1997, p. 15) cited an example where belated public involvement in such an EA resulted in an alternative which protected the coast and saved money. However, Canadian practice in relation to alternatives has been generally satisfactory at the panel review stage. Thus, in relation to the Voisey's Bay Mine and Mill, the Panel Guidelines required the proponent to provide:

> an analysis of alternatives available to the Proponent if the Undertaking does not proceed;
>
> descriptions and assessment of alternative means of carrying out the Undertaking and its key components that are technically and economically feasible. These shall include, but not be limited to, tailings disposal, waste rock disposal, fly-in-fly-out camp or town site, concentrate storage and transportation. Sufficient information should be provided for the reader to understand the reasons, including consideration of environmental and socio-economic impacts, for selecting the preferred alternative and for rejecting others. (CEAA, 1997a, Section 6.1)

Commonwealth of Australia

The treatment of alternatives has always been, in principle, an important element in the Australian EIA system. Thus, in the notice of intention prepared for screening purposes under the repealed Impact of Proposals Act, a list of the alternatives considered had to be prepared. Equally, the content requirements for Commonwealth environmental impact statements included the no-action alternative, 'feasible and prudent' alternatives and the reasons for choosing the preferred alternative. The requirements for the treatment of alternatives under the Environment Protection and Biodiversity Conservation Act 1999 (EPBC Act) are similar.

Box 8.3 shows the information requirements relating to alternatives in EIA reports that are specified in the Environment Protection and Biodiversity Conservation Regulations 2000. A description of any feasible alternatives and their impacts must be included in the preliminary information if no environmental impact statement or public environment report (PER) is to be prepared. Box 8.3 emphasises that rather more information about feasible alternatives and their impacts must be included in draft PERs and EISs. In both cases the no-action alternative must be described and the relative impacts of the alternatives must be contrasted. These requirements are more stringent than those of the earlier legislation. Unfortunately, an alternative proposal which emerges as a result of the EIA process technically becomes a different action under the EPBC Act provisions.

No specific guidance on the treatment of alternatives exists and, in practice, their coverage in EISs and PERs has tended to be weak. In particular, the

Box 8.3 Content requirements for alternatives in Australian EIA reports

Preliminary information

A description of:

(a) any options for how the proposed action may be taken; and
(b) any feasible alternatives to the proposed action, including not taking the action; and
(c) the relative effect of the options and alternatives on the relevant impacts of the action.

Draft public environment report and environmental impact statement

to the extent reasonably practicable, any feasible alternatives to the action, including:

(i) if relevant, the alternative of taking no action;
(ii) a comparative description of the impacts of each alternative on the matters protected by the controlling provisions for the action;
(iii) sufficient detail to make clear why any alternative is preferred to another . . .

Sources: Environment Protection and Biodiversity Conservation Regulations 2000, Schedule 3, Article 6.01; Schedule 4, Article 2.01(g).

no-action alternative has frequently been glossed over in EISs. This is nicely exemplified by the EIA for the Second Sydney Airport. The guidelines for the EIS (see Chapter 10) specified that several alternatives, including the do-nothing option, should be 'examined and discussed'. Because the proposal was so controversial, Environment Australia commissioned an independent review of the EIS. The auditors reported that, notwithstanding the guidelines' requirements, alternatives were not assessed in the draft EIS and that the do-nothing option was not assessed.

Overall, there is little evidence that the environmental impacts of meaningful alternatives have been adequately discussed in Australian Commonwealth EISs in the past, despite the requirement to do so. It remains to be seen whether the stronger requirements of the EPBC Act will lead to an improvement in practice.

New Zealand

New Zealand assessment of environmental effects (AEE) reports are expected to include consideration of alternatives where significant adverse environmental effects are likely. The Fourth Schedule to the Resource Management Act 1991 (see Box 7.3) requires proponents of proposals likely to have a significant impact to provide information on 'any possible alternative locations or methods for undertaking the activity'. In addition, where the discharge of a pollutant is involved, the proponent must describe 'any possible alternative methods of discharge, including discharge into any other receiving environment'. The Act provides probably the first instance of the integration of legislative provisions for EIA and for achieving the best practicable environmental option anywhere in the world (see Box 7.3).

Where a project is deemed by the local council to have major effects and is notified (see Figure 5.5 and Chapter 9) the council may:

> Require an explanation of:
> i) any possible alternative locations or methods for undertaking the activity and the applicant's reasons for making the proposed choice. (section 92(2)(a))

This provision clearly invites councils to ensure that alternatives are fully explored. It is noteworthy, however, that the no-action alternative is not specifically mentioned in relation to project EIA.

The Act states only that information on alternatives and on other matters in the Fourth Schedule should be included 'subject to the provisions of any policy statement or plan'. However, since section 88(6)(b) of the Act states that the assessment 'shall be prepared in accordance with the Fourth Schedule', it is apparent that a proponent ought to provide this information where an AEE report is required unless the territorial or regional council has specified otherwise in its plan or policy statement. It is, or course, open to regional and territorial councils to include more stringent requirements relating to the treatment of alternatives in submitted AEE reports in their policy statements or plans.

There is no published guidance on the treatment of the impacts of alternatives in the New Zealand EIA system. Indeed, it is notable that none of the guidance commissioned (Morgan and Memon, 1993) or published by the Ministry for the Environment (see Chapter 5) stresses alternatives (see, for example, MfE, 1999a) beyond the brief statement that an 'assessment of alternatives should only be required if there is the potential for significant adverse effects' (MfE, 1999b, p. 42).

It is thus apparent that the treatment of alternatives has not been given priority in the implementation of the New Zealand EIA system (Smith, 1996). This is reflected in practice: generally alternatives do not appear to have been treated satisfactorily in EIA. Most applicants justifiably argue that their proposals will not cause significant effects and that the requirement to consider alternatives is therefore inoperative. It is for the local authorities to request information about alternatives where they feel that effects are likely to be significant but this seldom happens (Parliamentary Commissioner for the Environment – PCE, 1995).

South Africa

Alternatives receive considerable attention in the South African EIA system. The Environment Conservation Act 1989 stresses the importance of the treatment of alternatives in environmental impact reports (EIRs) repeatedly (section 26). The regulations require that the scoping report contains 'a description of alternatives identified' (Republic of South Africa, 1997, Regulation 6(l)(d)). The EIA guidelines (Department of Environmental Affairs and Tourism – DEAT, 1998c, p. 23) categorise alternatives as follows:

- Demand alternatives e.g. using energy more efficiently rather than building more generating capacity.
- Activity alternatives e.g. providing public transport rather than increasing the road capacity.
- Location alternatives either for the entire proposal or for components of the proposal e.g. the location of a processing plant for a mine.
- Process alternatives e.g. the re-use of process water in an industrial plant, waste-minimising or energy efficient technology, different mining methods.
- Scheduling alternatives – where a number of measures might play a part in an overall programme but the order in which they are scheduled will contribute to the overall effectiveness of the end result.
- Input alternatives e.g. use of alternative raw materials or energy sources.

They also describe the use of the no-action alternative:

> The option not to act is often used as a base case against which to measure the relative performance of the other alternatives. The relative impacts of the other alternatives are expressed as changes to the base case. (DEAT, 1998c, p. 23)

This emphasis on alternatives in scoping virtually repeats previous integrated environmental management guidance (Department of Environmental Affairs – DEA, 1992, vol. 2). The EIA guidelines (DEAT, 1998c) provide advice on the identification of alternatives and suggest that no more than three alternatives should be investigated.

Where an environmental impact report is to be prepared, the plan of study must include 'a description of the feasible alternatives identified during scoping that may be further investigated' (Republic of South Africa, 1997, Regulation 7(1)(b)). The EIR must contain:

(a) a description of each alternative, including particulars on –
   (i) the extent and significance of each identified environmental impact; and
   (ii) the possibility for mitigation of each identified impact;
(b) a comparative assessment of all the alternatives. (Regulation 8)

The EIA guidelines (DEAT, 1998c) provide no further guidance about the fulfilment of these requirements.

It is widely accepted that the treatment of alternatives in EIA reports in South Africa prepared prior to, and under, the integrated environmental management procedures has left much to be desired (see, for example, Weaver, 1996). In particular, there has been considerable resistance to considering the no-development alternative in EIA reports. Mafune *et al.*'s (1997) survey of 28 early EIA reports found that 9 considered 'alternatives to' a proposal and that 8 of these dealt with the no-action alternative. A further 10 reports contained a discussion of alternative sites and corridors. In total, therefore, 61 per cent of

the sample of reports produced between 1971 and 1986 contained meaningful discussion of one type of alternative or another. There was little evidence
of any trend in treatment of alternatives over time. A study of 34 scoping
reports in North West Province by Potchefstroom University[1] revealed that 50
per cent of the reports contained no reference to the consideration of alternatives and in 10 per cent of cases the only alternative to the project proposal to
be considered was the no-project option. Although practice is believed to have

Table 8.1 The consideration of alternatives in the EIA systems

**Criterion 3: Must evidence of the consideration, by the proponent, of the environmental
impacts of reasonable alternative actions be demonstrated in the EIA process?**

Jurisdiction	Criterion met?	Comment
United States	Yes	Treatment of alternatives required in almost every environmental assessment and lies at 'heart of' environmental impact statement. Practice generally good
UK	Partially	Regulations now require discussion of alternatives (where studied) and guidance has long advised it. Practice improving
The Netherlands	Yes	Alternatives, including the no-action and the environmentally preferable alternatives, must be considered in scoping, the EIA report and the decision. Practice often good
Canada	Yes	Treatment of 'alternative means' in comprehensive study, mediation and panel reports but not in screening reports, discretionary provision for treatment of 'alternatives to' in all EA reports. Practice varies
Commonwealth of Australia	Yes	Alternatives (including the no-action alternative) must be described in preliminary information and fully treated in EISs and public environment reports. Practice often inadequate
New Zealand	Partially	EIA report should contain discussion of alternative locations and methods where significant effects are anticipated. Practice often weak
South Africa	Yes	Treatment of alternatives must be considered in scoping report and in environmental impact report. Practice established but often weak

been improving in more recent years, there is clearly still substantial room for improvement.

The National Environmental Management Act 1998 (NEMA) specifically requires consideration of alternatives (section 24(7)(b)) and of the 'option of not implementing the activity' (section 24(7)(c)). The promulgation of EIA regulations under the provisions of NEMA (expected in 2002) provides an opportunity to strengthen the legal requirements relating to the assessment of realistic alternatives.

Summary

The consideration of alternatives in the seven EIA systems is shown in Table 8.1. It can be seen that the 'alternatives' criterion is met, to a greater or lesser extent, in the United States, the Netherlands, Canada, Australia and South Africa. It is, nevertheless, true that the treatment of alternatives in these jurisdictions often leaves a great deal to be desired. If practice in the United States is at the leading edge, practice in the treatment of alternatives in the Australian federal EIA system and in South Africa has frequently been unsatisfactory, and that in some cases in Canada and the Netherlands has been properly criticised.

The only EIA systems which do not require the treatment of alternatives in EIA reports are those of the UK and New Zealand. The consideration of alternatives is, in effect, still discretionary in the UK because only 'an outline of the main alternatives studied by the applicant' needs to be described. Nevertheless, the official guidance strongly advises that alternatives be described in environmental statements. The New Zealand requirements relating to alternatives have not been interpreted consistently. In practice, some UK and New Zealand EIA reports contain adequate discussion of a reasonable range of alternatives to the proposed action but this is totally absent from others.

Practice in relation to the treatment of the no-action alternative and the environmentally preferable alternative could undoubtedly be improved in all the jurisdictions and in developing countries to enable a better informed comparison to be made with the proposed action and hence to assist in reducing the severity of its environmental impacts.

Note

1. Personal communication by Andries van der Walt of unpublished research by the Environmental Assessment Research Group at Potchefstroom University.

Chapter 9

Screening of actions

The determination of whether or not an EIA report is to be prepared for a particular action normally hinges upon the question of the significance of its environmental impacts. In screening, as in other stages of the EIA process, 'evaluating the significance of environmental effects is perhaps the most critical component of impact analysis' (Sadler (1996, p. 118). Two broad approaches to the establishment of significance during screening may be identified in EIA systems:

1. The compilation of lists of actions and of thresholds and criteria to determine which should be assessed.
2. The establishment of a procedure for the discretionary determination of which actions should be assessed.

In practice, most EIA systems adopt a hybrid approach involving lists, thresholds and the use of discretion, i.e. a balance between objective and subjective approaches to determining the significance of a particular action. In some systems, different types of EIA (with different documentary and participation requirements) are employed for actions with different levels of significance.

This chapter discusses the determination of the potential significance of impacts in the screening of actions and puts forward several evaluation criteria intended to assist in the analysis of EIA systems. The screening provisions of the EIA systems in the United States, UK, the Netherlands, Canada, Commonwealth of Australia, New Zealand and South Africa are then reviewed utilising these criteria.

Screening of significant actions

It is clearly important that some form of screening takes place in any EIA system. Without it, large numbers of actions would be assessed unnecessarily and/or actions with significant adverse impacts would not be assessed. 'A screening mechanism seeks to focus on those projects with potentially significant adverse environmental impacts or whose impacts are not fully known' (Glasson et al., 1999, p. 88). In order to make such choices, the concept of significance is employed to help to determine the level of assessment required during the EIA process. Kjellerup (1999, p. 4) suggested that 'decisions in the screening phase call for clear, easily conducted tests regarding the significance of the proposed project.' Worryingly, considering its centrality in the EIA process, significance is rarely defined or explained (Weston, 2000).

The question of significance has created difficulty in the United States from the outset and has been the most frequent cause of litigation on EIA over the years (see Chapter 2). The US courts have generally ruled that an environmental impact statement was necessary when significant effects were present and where its preparation was reasonable under the circumstances. This type of legal test could, at least in principle, be applied in other jurisdictions. Interestingly, the courts in the United States have rejected specific size or monetary factors as a guide to determining the significance of an action (Bear, 1989). These problems are not confined to the United States. For example, Kjellerup (1999) reported that the Danish EIA system lacked a focused approach to significance interpretation, and that, as a result, the handling of significance varied between different projects, detracting from the purpose of EIA, i.e. the avoidance of harmful environmental impacts.

In practice, both inclusion and exclusion lists are employed in the United States, as elsewhere (Glasson *et al.*, 1999, p. 88). Inclusion lists (for example, Annex I of the European Directive) determine the projects to be subject to EIA, according to factors such as project type and location. Exclusion lists (for example, the quantified *de minimis* thresholds and criteria for different categories of Schedule 2 projects in the UK Regulations) specify the projects not requiring an EIA, due to the insignificant nature of their environmental impacts.

The 1978 Council on Environmental Quality (CEQ) Regulations set forth ten general criteria for the determination of significance on a case-by-case basis (Box 9.1). These intensity criteria are to be applied within the societal and environmental context in which the action would occur. Some of these criteria have been adopted or modified by other jurisdictions. Sadler (1996, p. 121) has recommended a four-step methodology to aid the evaluation of significance of proposed actions. The approach includes consideration of: the nature and extent of potential project impacts; the likely adverse effects on the environment; the magnitude of potential impacts; and the opportunities for impact mitigation.

As mentioned in Chapter 3, the European Directive on EIA lists a limited number of Annex I projects for which assessment is mandatory (save in exceptional circumstances). It then goes on (Annex II) to specify other projects for which criteria and thresholds may be applied to determine whether or not they should be assessed. Because of the unacceptable variations in the member state criteria adopted relating to the nature, size and location of proposals, the amended Directive contains a new Annex III designed to result in much greater harmonisation in screening decisions. Other jurisdictions (for example, New Zealand) have not published lists of criteria or lists of proposals but have left local authorities to establish their own discretionary exclusion procedures.

The United States possesses a *de facto* simplified form of EIA report: the environmental assessment. The use of an additional level of assessment was

Box 9.1   US Council on Environmental Quality significance criteria

1. Is the impact adverse or beneficial?
2. Does the action affect public health or safety?
3. Is the action located in a unique geographic area?
4. Are the effects likely to be highly controversial?
5. Does the proposed action pose highly uncertain or unique or unknown risks?
6. Does the action establish a precedent for future actions with significant effects, or represent a decision in principle about future considerations?
7. Is the action related to other activities with individually insignificant but cumulatively significant impacts?
8. To what degree may the action affect designated or listed and protected sites?
9. To what degree may the action adversely affect endangered or threatened species and habitats?
10. Could the action contravene other environmental legislation?

Source: amended from CEQ Regulations 40 CFR 1508.27(b).

considered but rejected by the European Commission (see Chapter 3) but has been adopted, in a modified form, by other jurisdictions such as Canada, Australia and South Africa.

In practice, there is a variety of methods and procedures for screening actions which mix the approaches described and which permit different levels of scrutiny and of consultation and participation (Jones, 1999). Whichever approach is adopted, it is clearly important that, if screening is to be operated effectively, the proponent should be required to submit information to assist the decision-maker and/or the relevant environmental authorities in determining whether EIA is necessary in any particular case. The environmental assessment nominally performs this function in the United States while the scoping report serves the same function in South Africa. In practice, the EIA process normally only progresses further in a small minority of cases in both countries.

Since proponents require as much certainty as possible in determining whether assessment is likely to be required, clear and detailed information about actions, criteria, thresholds and screening procedures generally should be available. Such guidance is helpful not only to the proponent but also to all the other participants in the EIA process.

To instil confidence in the screening process, an identifiable decision should be made by a publicly accountable body and the reasons for making it should be on the public record. Further involvement by the public and by environmental authorities would require a formal period of public participation and, beyond that, a third-party right of appeal against screening decisions. This would include the right of the proponent to appeal.

Box 9.2   Evaluation criteria for the screening of actions

Must screening of actions for environmental significance take place?

- Is there a legal test of whether the action is likely to affect the environment significantly?
- Is there a clear specification of the type of action to be subject to EIA?
- Do clear criteria/thresholds exist (e.g. size, location)?
- Do different types of EIA exist for different types of action?
- Must documentation be submitted by the proponent to assist in screening?
- Does published guidance about actions, criteria, thresholds and screening procedures exist?
- Is the screening decision made by a publicly accountable body?
- Does consultation and participation take place during screening?
- Is there a right of appeal against screening decisions?
- Does screening function effectively and efficiently?

Whatever screening procedure and level of participation is adopted, it is necessary that it be operated effectively and efficiently. That is, in general, those actions with significant impacts (and only those actions) should be assessed and decisions should be made within a specified period of time without undue expenditure by any of the participants in the EIA process.

The criteria which can be advanced to analyse the treatment of screening are summarised in Box 9.2. These criteria for dealing with the issue of significance of action impacts are now used in reviewing the way this stage is handled in each of the seven EIA systems.

United States

There is no formal legal test in the US EIA system of whether the proposed action is likely to affect the environment significantly. Nor is there a set of criteria or thresholds in existence which permits precision in determining whether an environmental impact statement (EIS) must be prepared. Rather, a three-step process applies under the provisions of the National Environmental Policy Act 1969 (NEPA):

1. Determine whether NEPA applies to the proposed action (includes determination of whether the action is 'categorically excluded').
2. Determine whether the proposed action may 'significantly affect the quality of the human environment' (usually involves preparation of an environmental assessment).
3. Prepare an EIS if significant impacts are anticipated.

In practice, few actions are subject to both an environmental assessment (EA) and an EIS: the EA is bypassed for certain types of proposed action for which agencies have published EIS inclusion criteria in their own NEPA Regulations.

The EA is a supposedly concise public document which is intended to serve three purposes:

1. Provide sufficient evidence and analysis to determine whether an EIS is required.
2. Support an agency's compliance with NEPA when no EIS is required.
3. Facilitate preparation of an EIS when one is required.

It must discuss the need for the proposed action, reasonable alternatives (see Chapter 8), the probable environmental impacts, and the agencies and persons consulted. Agencies must provide notice of the availability of EAs. Scoping of EAs is not a requirement but is undertaken by some agencies. Other federal agencies and the public are supposed to be involved in the preparation of the EA (Regulations, section 1501.4(b)) but the different federal agencies have adopted different rules about consultation and public participation.

The decision to proceed to an EIS or to prepare a finding of no significant impact (FONSI) is taken by the lead agency. The FONSI must succinctly state the reasons for deciding that the action will not have significant effects on the human environment and summarise or attach the EA. Many federal agencies are now preparing mitigated FONSIs, i.e. reducing all the significant impacts of the proposed action to less than significant levels.

There have been many legal challenges to EAs and FONSIs, relating both to their absence and to their quality, indicating that there is considerable dissatisfaction with the screening process. This is not surprising. Some 50,000 EAs are prepared annually by federal agencies (or by contractors or applicants), but agencies do not routinely use EAs to determine whether an EIS is needed (Blaug, 1993). Some EAs are undoubtedly EISs in disguise, perhaps to try to avoid the public scrutiny and possible delay involved in EIS preparation (Weiner, 1997). Despite the length of EAs (see Chapter 8) and their cost (sometimes over $1 million) the majority of federal agencies do not, in practice, involve the public (Blaug, 1993). Since around 100 EAs are prepared for each EIS, this indicates serious shortcomings in US EIA screening.

UK

As described in Chapter 4, the UK Planning Regulations have contained two lists of projects since 1988. These were both extended to cover a wider range of developments by the 1999 Regulations. For Schedule 1 projects (for example, oil refineries and wastewater treatment plants), for which environmental impact assessment is mandatory, it is normally clear from the thresholds whether a particular project requires EIA. For the longer list of Schedule 2 projects (such as quarries, urban developments and golf courses), whether a project will require an environmental assessment depends on the likely significance of its environmental effects. These, in turn, will depend on the environmental sensitivity of the location, the characteristics of the development, and the nature of the potential impact (Department of the Environment, Transport and the Regions – DETR, 1999b).

The types of project potentially subject to EIA, the criteria and thresholds and the procedures to be followed are set out in the Planning Regulations and in Circular 2/99 (DETR, 1999b). Because these criteria are necessarily general, the Regulations set out quantified *de minimis* or exclusive thresholds and criteria for different categories of Schedule 2 projects while the Circular sets out indicative thresholds and criteria (Figure 9.1). EIA is likely to be required for:

- major development of more than local importance;
- development in particularly environmentally sensitive locations;
- development with particularly complex and potentially hazardous environmental effects. (DETR, 1999b, para. 33)

In essence, if the Schedule 2 development is located in a sensitive area, or is a major development of more than local importance, which exceeds an exclusive threshold, it is considered to be a potential EIA development, i.e. one for which EIA may be required. If the development does not meet the indicative criteria and thresholds listed in Annex A of the Circular (DETR, 1999b), EIA is unlikely to be required. All developments in sensitive areas (national parks, national nature reserves, etc. – see DETR, 1999b, Annex B) are potentially subject to EIA and must be screened. Finally, a limited number of developments (usually involving emissions which are potentially hazardous or damaging) may be subject to EIA because of the nature of their impacts (DETR, 1999b, para. 41). Examples of exclusive and indicative criteria and thresholds for certain mineral operations are presented in Box 9.3.

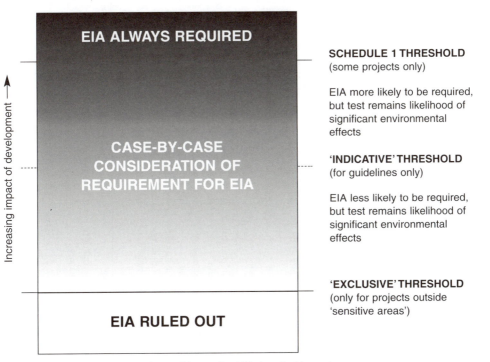

Figure 9.1   An illustrative guide to the UK thresholds system
Source: DETR, 1997a.

Box 9.3   Exclusive and indicative UK screening criteria and thresholds for extractive industry projects

Extractive industry activity	Exclusive thresholds and criteria below which EIA is not required (Regulations, Schedule 2(2))	Indicative thresholds and criteria below which EIA is unlikely to be required (Circular, Annex A)
(a) Quarries, open-cast mining and peat extraction (unless included in Schedule 1) (b) Underground mining	All development except the construction of buildings or other ancillary structures where the new floorspace does not exceed 1,000 square metres	A7. The likelihood of significant effects will tend to depend on the scale and duration of the works, and the likely consequent impact of noise, dust, discharges to water and visual intrusion. All new open-cast mines and underground mines will generally require EIA. For clay, sand and gravel workings, quarries and peat extraction sites, EIA is more likely to be required if they would cover more than 15 hectares or involve the extraction of more than 30,000 tonnes of mineral per year
(c) Extraction of minerals by fluvial dredging	All development	A8. Particular consideration should be given to noise, and any wider impacts on the surrounding hydrology and ecology. EIA is more likely to be required where it is expected that more than 100,000 tonnes of mineral will be extracted per year
(d) Deep drillings, in particular: (i) geothermal drilling (ii) drilling for the storage of nuclear waste material (iii) drilling for water supplies with the exception of drillings for investigating the stability of the soil	(i) In relation to any type of drilling, the area of the works exceeds 1 hectare; or (ii) in relation to geothermal drilling and drilling for the storage of nuclear waste material, the drilling is within 100 metres of any controlled waters	A9. EIA is more likely to be required where the scale of the drilling operations involves developments of a surface site of more than 5 hectares. Regard should be had to the likely wider impacts on surrounding hydrology and ecology. On its own, exploratory deep drilling is unlikely to require EIA. It would not be appropriate to require EIA for exploratory activity simply because it might lead to some form of permanent activity
(e) Surface industrial installations for the extraction of coal, petroleum, natural gas and ores, as well as bituminous shale	The area of the development exceeds 0.5 hectares	A10. The main considerations are likely to be the scale of development, emissions to air, discharges to water, the risk of accident and the arrangements for transporting the fuel. EIA is more likely to be required if the development is on a major scale (site of 10 hectares or more) or where production is expected to be substantial (e.g. more than 100,000 tonnes of petroleum per year)

Source: Planning Regulations Schedule 2; and DETR, 1999b, Annex A.

The UK system is binary: either a project is subject to EIA or it is not. There is no provision for 'simplified' EIA, although, inevitably, the environmental statements (ESs) for certain projects (especially those with potentially complex impacts such as toxic waste incinerators) tend to present a much fuller treatment than those for others (for example, afforestation projects). The vast majority of developments which fall to be approved by local planning authorities (LPAs) within the town and country planning system are minor and not subject to EIA.

LPAs, which are elected bodies, are generally responsible for screening decisions in the first instance. They are required to reach, and record, a formal screening opinion' in relation to all applications for Schedule 1 or Schedule 2 developments not accompanied by an ES. A formal opinion of this kind may be requested by the proponent, in which case the LPA may, 'exceptionally' (DETR, 1999b), refer to the statutory consultees for advice. Where a formal opinion is sought by the proponent, information about the nature, purpose and possible effects of the proposal on the environment must be provided. If the LPA, on receipt of this information, or of a planning application unaccompanied by an ES, determines that an ES is required, there is provision for the developer to appeal to the Secretary of State for the Environment against this screening decision.

The Secretary of State can issue a screening direction that an EIA be undertaken and submitted to the LPA even in the absence of an appeal by the developer. Public pressure or the opinion of statutory consultees could thus succeed (no doubt on rare occasions) at central government level in obtaining an EIA, even if it fails to convince the LPA of the need. There is, however, no formal third-party right of appeal against screening decisions. The Secretary of State also has the reserve power to direct that EIA be undertaken for any development listed in Schedule 2 (Regulation 4(8)).

The original procedures for obtaining opinions and directions were often not used in practice. Of a sample of 24 ESs received by LPAs in the early years of EIA, just over half were initiated by a request from the LPA, with just under one-third being volunteered by the developer without previously notifying the LPA. In a separate sample of 24 LPAs which had not received ESs, half the LPAs appeared not to have considered the question of whether or not to request EIA, the 'decision' often being taken by junior planners having little knowledge of EIA (Department of the Environment – DoE, 1991a). There was widespread agreement among LPAs that further guidance was needed on how to define significant environmental effects for Schedule 2 projects under the original Regulations (DoE, 1991a).

More recent research indicates that planning officers with greater experience of EIA are more likely to request an ES than less experienced planners (a small minority of whom appear to try to avoid EIA altogether) (Weston, 2000). About 55 per cent of developers asked LPAs if an ES would be required, 19 per cent submitted an ES without prior consultation and 23 per cent provided an ES later, when requested to do so. The most important screening criteria in deciding whether to request an ES were the nature of the project and its proximity to a sensitive environmental receptor. The Secretary of State has been issuing about 12 directions a year, with a small majority ruling that no EIA was necessary. Screening could be a complex process, involving qualitative and quantitative approaches, with the outcome largely determined by the 'enthusiasm and commitment of the officers making the decision', relying on professional, political and intuitive judgement (Weston, 2000, p. 194). Developers and their consultants were almost equally divided between those with positive and those with negative approaches to EIA screening. Most planning officers felt that the indicative thresholds approach to screening was appropriate (Weston, 2000).

Overall, it can be seen that, while screening of environmentally significant actions does take place, practice has been variable and not always effective. The 1999 Regulations and Circular provide clearer criteria and thresholds, which have led to a doubling in the number of ESs (Figure 4.2).

The Netherlands

As explained in Chapter 7, Part C of the Annex to the Environmental Impact Assessment Decree contains a list of activities for which EIA is mandatory. This Part contains three columns: in the first the activities for which EIA is required are set down; in the second column threshold values beyond which EIA must be applied are given; and in the third column the project or plan decisions which involve mandatory EIA are specified. The thresholds are usually based on area (for example, an industrial estate of more than 150 hectares) or weight (for example, an industrial waste incinerator for 100 tonnes or more per day). Little discretion is seemingly allowed: implementation of EIA is mandatory if the extent of the proposed activity exceeds the threshold. (Proponents may, if they choose, voluntarily undertake EIA.)

Many of the threshold values were considered by the participants in the EIA process to be arbitrary, unclear and set too high (Commission of the European Communities, 1993). In effect, the Netherlands initially operated a highly sophisticated EIA system for a highly selective set of projects in which too many environmentally significant proposals eluded the EIA net. However, the Decree was amended in 1994 to include all Annex I projects. In addition, further clarification has resulted from the provision of extra information, the making of supplementary policy and, occasionally, from judicial opinion.

The new screening procedure, implemented in 1994 and amended in 1999, supplements the list in Part C. Part D of the Annex to the EIA Decree contains a further list of activities and thresholds (for example, an industrial estate of more than 75 hectares) which meets the requirements of the amended EIA Directive. This Part D list, which broadly corresponds to Annex II of the EIA Directive, is utilised with accompanying criteria relating to the nature of the impacts, the sensitivity of the area, whether impacts are likely to be cumulative, the complexity of the project, etc., to determine whether EIA is necessary. The screening procedure is applied by the competent authority.

There is no provision for different types of EIA but the scoping guidelines issued by the competent authority (see Chapter 10) in effect provide for different, project-specific types of EIA. The inception memorandum (i.e. notification of intent, which should be brief) covers the purpose and goal of the proposal, a general description, the alternatives to be considered, a description of the area and the impacts to be considered. It is used not only for screening purposes, and for the preparation of scoping guidelines, but may sometimes be used as an EIA report in its own right (for example, for dike improvement projects). Information on screening is available both in the EIA Decree and in the EIA handbook (see Chapter 5).

While there is consultation of specified bodies (for example, the regional environmental inspectorate) during screening, there is no public participation. There is a third-party right of appeal to the competent authority and to the courts at the decision stage of the EIA process and such appeals often relate to screening decisions (de Jong, 1997). In practice, there are some differences in the interpretation of Part D criteria by the different provinces. On balance, while too many projects have escaped the EIA process in the past (and some insignificant projects have been assessed), screening is currently working reasonably well in the Netherlands, despite criticisms by the European Commission (see Chapter 5). Nevertheless, the Evaluation Committee on the Environmental Management Act (1996) suggested that there should be greater consultation and participation in screening decisions, but this recommendation was rejected because it was felt that it would prolong the EIA process.

Canada

The Canadian screening situation is complicated by two factors. First, the EA system has two separate, sometimes successive, procedures, each with its own screening steps. Second, one track in the self-directed assessment is called a 'screening', an important purpose of which is supposedly to determine whether additional EA is required.

In the self-directed assessment the responsible authority (RA) determines, utilising project information usually provided in the proponent's application, the exclusion list and the inclusion list (Chapter 6), whether EA is required. If it is required, the RA determines whether this will be in the form of a screening or of a comprehensive study by referring to the comprehensive study list (which consists of types of project, sizes of projects and certain locational considerations). There is no legal test of significance but decisions are potentially subject to judicial review in the federal courts. As mentioned in Chapter 5, over 5,000 screenings are carried out each year. Over 99 per cent of all EAs are screenings, with less than 10 comprehensive studies being initiated each year (Commissioner of the Environment and Sustainable Development – CESD, 1998; Canadian Environmental Assessment Agency – CEAA, 2000). Far more 'comprehensive studies' are initiated each year under provincial EA legislation: (for example, about 40 in Alberta). An additional category of EA document is the class screening report which acts as a model for future projects. This report is intended to present a comprehensive discussion of the generic environmental effects of a class or type of project and to identify known mitigation measures. Only two class screening reports had been completed by the end of 1999, but a further 15 were under way (Minister of the Environment – MoE, 2001), indicating that greater emphasis is being given to this type of screening, as recommended by CESD (1998).

There is no legal requirement for the proponent to provide a document to assist in the decisions to require a screening report or to require a comprehensive study, nor for any form of public participation or administrative appeal. CESD (1998) found that not all federal projects were being assessed since some proponents were not applying for the regional federal permits, that there were uncertainties about the application of the Fisheries Act 1985 by the Department of Fisheries and Oceans and that some unnecessary screenings were taking place. Proposals to strengthen the self-assessment procedures were advanced by the Minister in the bill to revise the Act (Government of Canada, 2001; MoE, 2001).

The decision to require further EA (public review), beyond the self-directed assessment, is taken by the Minister of the Environment, either at the request of the RA or without such a request. The Minister can require public review following the RA's decision upon completion of the screening report (or of the comprehensive study report), while self-directed assessment is taking place or, occasionally, before self-directed assessment has commenced. The criteria for the decision are the likely significant adverse effects of the project, and/or the existence of widespread public concerns. About two panel reviews take place each year, a figure which hardly changed with the coming into force of the Act.

The Minister, after seeking the advice of the RA and CEAA, may decide that mediation will either take place instead of a panel review or be applied during the course of a panel review. The criteria for determining the appropriateness of mediation include:

- What are the potential sources of uncertainty or disagreement? For example, do the disputes involve fundamental opposition to the proposed project, technical issues, the determination of environmental effects and their significance, or the effectiveness of mitigation measures?
- Are these disagreements negotiable? Is there room for compromise and consensus?
- Who are the main parties involved?
- Do the parties agree on the areas of uncertainty or disagreement?
- Are there representatives who can speak on behalf of the interests?
- Are the parties willing to participate in mediation? (CEAA, 1994, p. 117)

As mentioned in Chapter 5, no mediations have been conducted to date because if the parties fail to agree they must proceed to a panel review. Informal experience of mediation has been positive and it has been proposed that the Act be amended to remove this disincentive (Government of Canada, 2001).

Unlike the screening step in the self-directed assessment procedure, there are no clearly specified types of project which should be subject to public review. The Minister has the benefit of the screening or comprehensive study reports and of the RA's decision in making this screening decision. The fact that so few projects are subject to formal public review in Canada makes the decision on referral a highly political decision and the Minister is clearly subject to public and governmental pressure (the decision is sometimes taken by cabinet). Because screening and comprehensive study reports are public documents, the public are able to argue for further EA. Further, the Minister can be, and has been, challenged in the courts, if a public review is not required. There has been much criticism of the uncertainty surrounding decisions requiring panel reviews (see, for example, Nikiforuk, 1997). To reduce this uncertainty among proponents, the Minister has proposed the streamlining of the procedure so that a public review cannot be demanded after a comprehensive study report has been produced. Rather, a decision would have to be taken early in the comprehensive study either to continue or to require a public review (Government of Canada, 2001; MoE, 2001).

Commonwealth of Australia

The repealed Environment Protection (Impact of Proposals) Act 1974 was triggered not by the Environment Minister (as it is in most of the state EIA systems) but by the so-called Action Minister (the Commonwealth Minister responsible for the action or decision). The Action Minister had to determine whether or not a proposal was environmentally significant and only then did Environment Australia (EA) become involved in determining the type of assessment to be undertaken. This frequently took place late in the planning process. As mentioned in Chapter 5, during the quarter-century lifetime of the Act, only 178 of the 4,000-plus referrals required the preparation of an EIS and 37 required the preparation of a public environment report (PER) (introduced in 1987). Only 6 inquiries were required.

The lack of a departmental power of direction and of clear screening criteria and information left the EIA system open to political discretion and this, in turn, led to demands that the Environment Minister should trigger the Act (Australian and New Zealand Environment and Conservation Council, – ANZECC, 1991; Commonwealth Environment Protection Agency – CEPA, 1994; Fowler, 1996). However, while this was firmly resisted by other departments and by industry, business and conservationists alike agreed on the need for greater legal certainty (CEPA, 1994) and on a reduction in the level of discretion which may have been an over-reaction to the legal challenges made to the US National Environmental Policy Act (Fowler, 1982). This need was addressed in the CEPA (1994) proposals by designated lists of proposals that would normally attract the application of EIA, and by giving the Environment Minister the power to trigger the Act. The Minister's consultation paper went further, proposing that 'the Environment Minister will be responsible for deciding whether a proposal or activity may have a significant impact on a matter of national environmental significance' (Hill, 1998, p. 13). This reform was implemented in the Environment Protection and Biodiversity

Conservation Act 1999 (EPBC Act), overcoming one of the greatest weaknesses of the previous Commonwealth EIA system.

Under the provisions of the EPBC Act, the first of two screening stages
involves the referral of a proposed action to the Environment Minister using a
prescribed form (Environment Protection and Biodiversity Conservation
Regulations 2000, Part 4). This referral may be made by the proponent, by a
Commonwealth or state agency, or at the request of the Environment Minister
(who may be alerted by the public) or the Minister may deem the referral to
have been made. There is an online interactive map, a database and question-
naire to help proponents decide whether or not they need to refer a proposal.

The referral must be published on the internet[1] so that the public can comment, and relevant Commonwealth and state agencies must be consulted
about whether or not the proposal is a controlled action, i.e. if it significantly
impacts on matters of national environmental significance. Administrative
guidelines on significance have been issued to assist in the making of this
decision (Environment Australia, 2000a). The Minister must make a binding
decision on whether approval is required within 20 business days
(Environment Australia, 1999a). If so, the Minister must designate a proponent, publish the decision and, if requested, provide reasons for the decision.
Should the relevant state or territory request it, the Minister must reconsider
this screening decision.

For actions in a state (or territory) with a bilateral assessment agreement
with the Commonwealth in place,[2] the state confirms that state assessment will
apply and that an assessment report will be provided to the federal
Environment Minister for consideration in the approval decision. Where no
bilateral assessment agreement is in place, the second screening stage involves
the Environment Minister deciding which of the following assessment
approaches is to apply to the action:

- an accredited state assessment process;
- an assessment on the basis of 'preliminary documentation';
- a public environment report;
- an environmental impact statement;
- a public inquiry (see Figure 5.4).

This decision is made on the basis of the prescribed preliminary information
(Regulations, Part 5) which the proponent must provide.

This preliminary information bears many similarities to the US environmental assessment but it must be provided even if a more detailed form of
assessment is subsequently required. It should contain a description of the proposal, feasible alternatives to the proposed action, a description of the environment affected, an indication of the potential environmental impacts and a
description of mitigation measures to be taken to reduce this (Regulations,
Part 5). While there are no provisions relating to the making public of the preliminary information at this stage in the EIA process, it normally forms part of
the assessment documentation made publicly available by the proponent. The
Minister must consult the appropriate state before taking the decision as to

which assessment approach is required, must take the decision within 20 business days and must publish notice of the decision.

Environment Australia has published guidance on the circumstances in which the different forms of environmental assessment are appropriate (Box 9.4). There is a limited provision for appeal against the decision that approval will be required but not against the level of assessment. These decisions are, however, subject to judicial review.

Box 9.4   Australian criteria for determining types of assessment

Assessment on preliminary documentation is likely to be appropriate when:

- the number and complexity of relative impacts is low and locally confined; or
- the relevant impacts of the controlled action can be predicted with a high degree of confidence; or
- the relevant impacts have been or are being adequately assessed under Commonwealth or State legislation.

Assessment by public environment report is likely to be appropriate when:

- an assessment of the relevant impacts is expected to focus on a relatively small number of key issues; and
- an adequate assessment of these issues will require the collection of new information and/or further analysis of existing information.

Assessment by environmental impact statement is likely to be appropriate when:

- an assessment of the relevant impacts is expected to raise complex issues, or a large number of issues; and
- an adequate assessment of these issues will require the collection of new information and/or further analysis of existing information.

Assessment by public inquiry is likely to be appropriate when:

- the relevant impacts are likely to be relatively high; or
- the relevant impacts, or the management of those impacts, are outside the control of a single proponent; or
- a public inquiry is necessary to desirable to ensure effective and efficient public involvement in the assessment process.

Assessment by accredited process is where a state or territory will manage the assessment or the Commonwealth will do so under other legislation. This assessment approach allows case-by-case accreditation of state, territory or Commonwealth assessment processes in situations where bilateral agreements and declarations do not apply.

Source: adapted from Environment Australia, 2000d.

There is no doubt that, despite concerns about the possible accreditation of inadequate state (Hughes, 1999; Ogle, 2000) and Commonwealth agency EIA processes (Padgett and Kriwoken, 2001), the screening provisions of the EPBC Act are a considerable advance on the previous legislation. Triggering now depends on the nature of the impact, not on the type of action or whether the proponent is a Commonwealth body. There has been pressure, from environmental groups and others, to expand the range of impact triggers (see, for example, Ogle, 2000) and consultation about the inclusion of a 'greenhouse' trigger was under way in mid-2001. A proposal for a National Heritage Places trigger was before parliament in late 2001 and some environmental groups have suggested that all actions involving impacts on World Heritage sites should be subject to an inquiry.

Too many trivial proposals, and insufficient major proposals, were referred under the Impact of Proposals Act. It appears that proponents have erred on the side of caution under the EPBC Act and have referred proposals to gain a definitive decision about whether Commonwealth environmental approval is required or not. While the numbers of referrals seem to be continuing at a similar level, this tendency is expected to diminish as experience is gained, at least for certain types of action. Under previous legislation there was a perceptible trend towards the preparation of notices of intention containing detailed impact mitigation measures by consultants, leading to a reduction in the need for EISs, as in the United States. This is expected to continue under the EPBC Act, with the majority of assessments not completed within accredited state processes being undertaken on the basis of preliminary documentation.

New Zealand

Since, in principle, EIA applies to all resource consent applications, screening is important in eliminating minor or irrelevant projects from further consideration. The Resource Management Act 1991 delegates this task, like most other EIA responsibilities, to the regional and territorial authorities. The Ministry for the Environment (MfE) has no direct role in screening and has not issued specific guidance on screening or recommended screening criteria. Rather, regional and territorial plans provide guidance to applicants about the environmental issues of concern in their areas and councils provide advice about whether an assessment of environmental effects (AEE) is required or not (MfE, 1999a). Figure 5.5 shows that various options exist at the initial screening stage, and indicates the two-phase nature of screening.

The role of mandatory regional policy statements and coastal plans, and discretionary plans on other topics, and of district plans (which must accord with regional policy statements and plans) in setting out the criteria for determining whether resource consents require AEEs in New Zealand is crucial. In particular, rules may be included in district plans which prohibit certain activities and permit others without the need for resource consent. Similarly, controlled activities, and the nature of any assessment required for these, can be specified.

Finally, non-complying and discretionary activities, for which EIA is always required, can be set down in rules. By defining the nature of activities, the limits on the size of controlled activities or permitted activities and/or the nature of their effects, district plans should in effect provide the means of screening projects subject to EIA. However, not only is this not occurring in practice but many minor projects requiring consents still require an AEE if they are not permitted uses.

All applicants are required to complete a prescribed form and attach an assessment of the environmental impacts of their proposal, however minor. All the 53,000 consent applications (MfE, 2001c) submitted each year should therefore be accompanied by an AEE report, however brief. Local authorities may require further information (for example, on alternatives – see Chapter 8) if the proposal is notified. The decision on whether to notify the application is taken by the council on the basis of the AEE report submitted. If it determines that the project is likely to have major effects it is in effect making a screening decision and requiring a fuller AEE of the application. This happens in about 5 per cent of cases or about 3,000 cases per annum (MfE, 2001c).

There is no provision for different types of EIA report. Rather, the Resource Management Act 1991 specifies that any assessment 'shall be in such detail as corresponds with the scale and significance of the actual or potential effects that the activity may have on the environment' (section 88(6)(a)), provided that the requirements of the Fourth Schedule (see Box 7.3) are met. There is thus a continuum of report sizes, from half a page to several hundred pages (Morgan, 1995).

The New Zealand EIA system therefore relies on the judgement of the regional or territorial authority which has discretion over how an activity should be classified. Based on an assessment of the environmental effect, the local authority must decide whether an activity is non-complying, discretionary, controlled or should be permitted without the need for a resource consent. The local authority must, however, require an AEE for every resource consent; it does not have discretion to waive this requirement, although the consent need not be publicly notified.

There is public participation and consultation in the preparation of policy statements and plans (and thus in determining for which projects an AEE report must be provided. People likely to be affected by a proposed activity have to give their written consent before non-notification is permitted (section 94(c)), i.e. they are involved, to some extent, in project screening decisions. There is a right of appeal to the Environment Court on screening as on other issues relating to resource consents. The Parliamentary Commissioner for the Environment (PCE, 1995) and Morgan (2000b) have indicated that, while screening practice is still developing, there are considerable variations in the provision of guidance to applicants by local authorities and in local authority screening decisions.

South Africa

The South African regulations specify a set of activities which must be subject to EIA. Very few thresholds to eliminate minor activities, and no classifications of affected environments to exclude non-sensitive areas, are provided. In principle, therefore, any new activity of the types specified is subject to EIA, as is any upgrading of these activities. In addition, certain land-use changes are deemed likely to have a 'substantial detrimental effect on the environment' (Republic of South Africa, 1997).

This is a different approach from that adopted in the integrated environmental management (IEM) guidelines (Department of Environmental Affairs – DEA, 1992, vol. 1) in which, although few thresholds were specified, a list of sensitive environments was provided. The intention was to use the list of activities and of environments in matrix format to decide if EIA was necessary or not.

The EIA Regulations require the proponent to submit a completed application form to the relevant authority where authorisation is sought (Republic of South Africa, 1997, Regulation 4(1)). There is a requirement for the applicant to advertise the application where directed to do so by the relevant authority (Regulation 4(6)). Relevant authorities almost always require advertisement. The explanatory EIA guidelines indicate that 'it is essential for the applicant or consultant to consult with the relevant authority', *inter alia*, to 'determine whether the proposed activity needs to comply with the regulations' (Department of Environmental Affairs and Tourism – DEAT, 1998c, p. 19), and this is almost universally being done.

In practice, provincial authorities are using considerable discretion in determining whether an activity should be subject to EIA or not. Many trivial activities, especially in urban areas, are being excluded from the EIA process by the use of memoranda of understanding and less formal means. In addition, numerous exemptions are being granted by provincial authorities, on the basis of specific applications, to projects which are clearly scheduled activities but which it is believed will not cause significant impacts (Duthie, 2001). The provincial authorities have agreed that these applications should be accompanied by what is, in effect, a mini-EIA report designed to demonstrate to the provincial authority's satisfaction that impacts are, indeed, insignificant.

In principle, South Africa has a two-stage screening process with stage 1 screening being based upon the completed application form. For perhaps 90 per cent of activities for which a scoping report is prepared no further assessment is deemed to be necessary: authorisation may be granted, granted with conditions, or refused. For the minority of projects, the scoping report provides the basis for stage 2 screening: further EIA (and an environmental impact report) is required. While this decision may be clear from the outset, the significance of certain impacts may become apparent only during the preparation and review of the scoping report.

A study of 75 development applications in North West Province by Potchefstroom University[3] showed that nearly 60 per cent led to scoping reports. The other projects involved either exemption applications or were abandoned before a scoping report was submitted. Decisions were made on three-quarters of the projects for which a scoping report was submitted: the others were either still in progress or were abandoned. Of the 34 projects, a full EIA was required in three cases.

Inevitably, local controversies have arisen over the application of the screening provisions by the provincial authorities. However, the main problem is that too many trivial applications are being processed and only belatedly exempted, wasting limited staff capacity (Duthie, 2001). While there exists the possibility of judicial review of cases where the provincial authorities deem that the EIA Regulations do not apply, or grant an exemption, no consultation or participation takes place and there is no right of appeal against the screening decision. This does not reflect the pre-existing, discretionary IEM procedure, which provided for objections to be lodged by interested and affected parties against the screening decision (DEA, 1992, vol. 1). However, it is widely acknowledged that, in practice, screening proved to be very difficult under the voluntary IEM procedure (DEAT, 1998a).

The Department of Environmental Affairs and Tourism (DEAT, 1998a) has indicated that a set of regional environmental management frameworks is being compiled on the basis of both environmental data and socio-economic planning priorities. These will have statutory standing as an integral part of IEM when the EIA provisions of the National Environment Management Act 1998 (NEMA) (see Chapter 5) come into effect (probably in 2002), providing the information needed to complement the list of activities that the NEMA EIA Regulations will contain. The approval of these regulations presents a very real opportunity to improve both the efficiency and the effectiveness of screening in South Africa.

Summary

Table 9.1 summarises the treatment of screening in the seven EIA systems. It is inconceivable that any EIA system could be operated without some form of screening, so it is not surprising that all the seven systems are adjudged to meet the screening criterion. The EIA systems use a variety of approaches, criteria and thresholds for screening. It is, perhaps, more surprising that, until the coming into effect of the Environment Protection and Biodiversity Conservation Act 1999, the Commonwealth of Australia failed to meet the screening criterion.

It is notable that five of the seven jurisdictions make use of more than one type of EIA document. In the United States the environmental assessment is nominally a screening document but, in practice, it is an EIA report in its own right for thousands of projects each year. The scoping study in South Africa serves a similar purpose. Some of the documents in the two-stage Canadian screening process (and especially screening reports) are not dissimilar. Screening in Australia results in different types of EIA report and in New Zealand in EIA reports of varying length and complexity. In each case, there is a marked tendency to choose less demanding types of assessment.

Table 9.1   The treatment of screening in the EIA systems

Criterion 4: Must screening of actions for environmental significance take place?

Jurisdiction	Criterion met?	Comment
United States	Yes	Use of categorical exclusions, inclusion criteria, and (rarely, in practice) environmental assessments to determine significance of impacts
UK	Yes	Use of lists of projects, indicative criteria and thresholds in screening by LPAs varies
The Netherlands	Yes	Lists of activities, thresholds and criteria in EIA Regulations sometimes allow competent authorities discretion
Canada	Yes	Screening by responsible authority using lists of projects results in 'screening' or comprehensive study. Occasionally, further screening by Minister leads to panel review
Commonwealth of Australia	Yes	Two-stage screening using referral to Environment Minister to determine if assessment required and preliminary information to determine type of assessment (including state accreditation)
New Zealand	Yes	Local authorities must specify types of, and criteria for, actions subject to EIA in their plans. Practice still developing; varies
South Africa	Yes	Two-stage screening process allows for exemptions and discretion in requiring a scoping report which (infrequently) may be followed by environmental impact report

Only the Netherlands and the UK have a single type of EIA report. It is, perhaps, no coincidence that both have strong land-use planning systems under which the environmental impacts of the less significant projects can be assessed. Even here, the Netherlands tends to employ more restrictive thresholds and criteria than the UK, resulting in proportionately fewer EIA reports.

Notes

1. http://www.ea.gov.au/epbc
2. By mid-2001 a bilateral agreement with Tasmania was in existence.
3. Personal communication by Andries van der Walt of unpublished research by the Environmental Assessment Research Group at Potchefstroom University.

Chapter 10

Scoping of impacts

The objective of scoping is to identify the significant issues associated with a proposed action and thus to determine the issues to be addressed in the EIA report. Scoping was not an original requirement of the US National Environmental Policy Act 1969 (NEPA) but was added in 1978 in response to the encyclopaedic nature of many environmental impact statements (EISs). Scoping was intended to ensure that more focused EISs were prepared and, incidentally, has assisted in increasing coordination between proponents in the EIA process and in the agreeing of action-specific timetables.

Scoping has proved to be a successful innovation. Indeed, Weston (2000) has argued that scoping can be regarded as the most important stage in the EIA process. If the EIA is not focused on the important issues, on the one hand delays may result from the need to gather more environmental information whereas, on the other hand, resources may be wasted if minor issues are not eliminated from the assessment resulting in a poor quality EIA report. Mulvihill and Jacobs (1998, p. 351) reported that 'scoping can exert a strong influence in shaping a relevant impact assessment and increasing the probability of a process that satisfies stakeholders.' They felt that scoping was especially important for controversial projects and/or where a diverse range of stakeholders was involved in the EIA process.

Following the lead set by the United States, many other jurisdictions have adopted scoping procedures, though not all have emulated the public participation provisions of the original. This chapter describes approaches to scoping and suggests criteria for use in the evaluation of the treatment of scoping in EIA systems. These criteria are then utilised in the review of the EIA systems in the United States, UK, the Netherlands, Canada, Commonwealth of Australia, New Zealand and South Africa.

Determination of the scope of the EIA

Scoping is intended to focus the EIA on the most important issues, eliminating irrelevant impacts, while ensuring that indirect and secondary effects are not overlooked. It involves identifying the issues and concerns that should form the focus of the study effort and deciding on the appropriate level of study for an EIA (Marriott, 1997, p. 39). The Department of the Environment (DoE, 1995, p. 13) asserted that: 'effectively, scoping is the key to a good quality environmental statement.' Scoping, which is usually initiated after the decision to undertake a full EIA has been taken, is part of a cyclical process. Thus Glasson *et al.* (1999) felt that refining the focus of the assessment on to the most significant impacts should continue throughout the EIA process. Marriott (1997, p. 40) noted that: 'the scoping process should be specifically designed to suit the needs of the particular project or action being proposed.' Mulvihill and Jacobs (1998) also stressed the importance of the scoping process being creative and flexible.

There are several steps involved in the scoping process, one representation of which is shown in Box 10.1. An important starting point is the preliminary identification of impacts. Generally, reference should be made to guidance on procedures and methods which needs to be available if the contribution of the proponent and of the other stakeholders to scoping is to be maximised. Clearly, the use of guidance specific to the type of action proposed is likely to be more helpful than the use of general guidance equally applicable to all types of action or to certain classes of action. (The consideration of a general set of impacts, provided it is reasonably comprehensive is, nevertheless, considerably preferable to the absence of any such requirement.) Previous action-specific guidelines or EIA reports should also be helpful in scoping impacts because they relate to either actions or locations similar to those proposed. Such guidance or guidelines will normally furnish a checklist of impacts to be considered.

The use of checklists and matrices as part of the scoping process is widely recommended (DoE, 1995; Jones, 1999) and other approaches such as networks and flow diagrams can be employed (Carter, 1996). The Council on Environmental Quality (CEQ, 1997a) advanced a framework for scoping in cumulative effects assessment to establish how the environment has already been affected by present and past activities and to identify proposals which might affect the environment in the future. It concluded (p. 11) that: 'expanding environmental impact assessment to incorporate cumulative effects can only be accomplished by the enlightened use of the scoping process.' As Mulvihill

---

Box 10.1   Steps to be considered in scoping

1. Develop a communications plan (decide who to talk to and when).
2. Assemble information that will be the starting point of discussion.
3. Make the information available to those whose views are to be obtained.
4. Find out what issues people are concerned about. (Make a long list.)
5. Look at the issues from a technical or scientific perspective in preparation for further study.
6. Organise information according to issues including grouping, combining and setting priorities (make the long list into a shorter list).
7. Develop a strategy for addressing and resolving each key issue, including information requirements and terms of reference for further studies.

Source: Ministry for the Environment, 1992b, pp. 9, 10.

and Jacobs (1998, p. 351) stated: '[a]s it sets the stage for subsequent steps in the EA process, scoping needs to be a sufficiently broad umbrella that accommodates diverse approaches to identifying, classifying and assessing impacts.'

The set of impacts identified as a result of using scoping techniques will include many which are irrelevant or insignificant and some which have been double-counted but may still exclude some impacts which are potentially important. Sadler (1996, p. 113) confirmed that, in many EIA systems, difficulties are experienced in 'scoping to prioritise issues', i.e. in identifying the impacts critical to the proposal.

This point underlines the importance of consultation and public participation in scoping, first in raising issues for consideration early in the EIA process and, later, in narrowing down the range of issues to be considered in the EIA. In the original American terminology, public participation was an essential element of 'scoping' and is a requirement in many of the older EIA systems (for example, in those of Canada and Australia). Although consultation of environmental authorities and other bodies by the proponent is now discretionary under the provisions of the amended European Directive, public participation is still not a requirement. There is, nonetheless, considerable unanimity of view that consultation with decision-making and environmental authorities, with interest groups such as local voluntary conservation groups, and with the local community should assist in ensuring that all potentially significant impacts are identified. The advantages of public participation within the scoping process were summed up by CEQ (1997b, p. 12):

> By discussing and informing the public of the emerging issues related to the proposed action, agencies may reduce misunderstandings, build cooperative working relationships, educate the public and decision makers, and avoid potential conflicts.

However, consultation and, especially, public participation can prove time-consuming. As (Sadler 1996, p. 113) observed: 'the challenge is to maintain a dialogue with stakeholders without bogging the process down.' Several methods of maintaining this dialogue have been employed, including community-led scoping, public and invitee meetings, questionnaires, surveys, etc. (Jones, 1999). For the scoping process to be efficient and equitable, sufficient information must be made available by the proponent to allow concerned stakeholders to participate meaningfully in the process.

Like screening decisions, scoping decisions frequently hinge on the issue of significance. These decisions often have to be made by individuals with the appropriate levels of knowledge and expertise who are able to say from past experience:

1. what significant effects are likely to arise;
2. how they are likely to impact on the environment;
3. what steps might be taken to deal with them.

In practice, it appears that consultation, previous professional experience and comparison with analogous actions dominate technical or systematic methodological approaches in scoping (see, for example, Weston, 2000).

Different mechanisms are used in different jurisdictions to ensure that the skills of such individuals are brought to bear on the action in question. These vary from the use of appropriate consultants by the proponent, through informal and formal consultations of 'interested parties at all levels of government, and all interested private citizens and organisations' (Bear, 1989, p. 10064), to the use of specialist panels and of representative consultation committees. Weston (2000, p. 200) suggested that because of the subjectivity inherent in defining significance in scoping, decisions as to what to include in an EIA 'are inherently based upon value judgements and are made within a political context.' Ultimately, however, despite stakeholder input into the scoping process, it is the responsibility of the proponent to ensure that the significant issues are assessed (Marriott, 1997).

The most basic means of ensuring that some form of scoping takes place is to require the proponent to consult the decision-maker and/or environmental authorities prior to submission of the EIA report. Consultation of environmental authorities can also reveal useful insights and, further, may ensure that coordination between them is more likely. Such consultation provides an opportunity for the opinions of the relevant authorities about the scope of the EIA to be expressed. It is this form of scoping that is provided for, at the proponent's discretion, in the amended European Directive.

Whether or not such consultation takes place, the proponent may be required to address a general or generic set of impacts in the EIA. That such a requirement has been met can easily be demonstrated in the EIA report. The preparation of action-specific scoping guidelines may not be a formal requirement in some jurisdictions (for example, the UK), whereas in others (for example, South Africa) the proponent will be required to prepare and publicise such guidelines and, in yet others, the decision-making body or environmental authority may be responsible for their preparation (for example, Canada). It is now generally accepted that the preparation of action-specific (rather than generic or general) guidelines by either the proponent or by the decision-making or environmental authorities renders the EIA process markedly more effective.

A danger in such an approach, particularly where public consultation takes place, is that irrelevant impacts will be incorporated in the guidelines. Some mechanism (such as round-table discussions) to eliminate such impacts, and hence to permit the EIA to focus on the relevant issues, is necessary. A record of the scoping process, including any discussions should be available for public inspection. This record should, preferably, demonstrate that the various relevant authorities, and the public, have indeed expressed their views about the scope of the EIA and that these views have been considered by the proponent.

Whatever scoping procedure (and level of participation is adopted), it is obviously necessary that it operate effectively and efficiently. Sadler (1996, p. 115) has defined an effective scoping process as one that 'results in a closure on priority issues and establishes the basis for a focused decision relevant assessment.' Box 10.2 summarises a set of evaluation criteria which can be employed to review the treatment of scoping in EIA systems. These criteria are now utilised in analysing the scoping process in each of the seven EIA systems.

Box 10.2   Evaluation criteria for the scoping of impacts

Must scoping of the environmental impacts of actions take place and specific guidelines be produced?

- Must the proponent consult the environmental authority early in the EIA process?
- Must the proponent prepare information as a basis for scoping?
- Is scoping mandatory in each case?
- Must a general or generic set of impacts be addressed in the EIA?
- Must action-specific scoping guidelines be prepared?
- Are irrelevant impacts screened out?
- Does published guidance on scoping procedures and methods exist?
- Is consultation and participation required in scoping?
- Is there a right of appeal against scoping decisions?
- Does scoping function efficiently and effectively?

United States

The National Environmental Policy Act 1969 is as silent about scoping as it is about almost all the EIA procedural stages. Scoping was introduced in the 1978 CEQ Regulations as a result of experience with the previous guidelines, which did not require scoping. To determine the scope of environmental impact statements (EISs), agencies must consider types of action, alternatives and impacts. The first formal step in EIS preparation is the publication of a notice of intent which must contain a description of the agency's proposed scoping process, including any scoping meetings (which are recommended by, but are not a requirement of, the Regulations). The open scoping process is intended to obtain the views of other agencies and the public regarding the topics to be included in the EIS. The Regulations (section 1501.7(a),(b)) state that the objectives of scoping include:

- determining which significant issues should be analysed in depth in the EIS;
- identifying and eliminating issues which are insignificant or which have been dealt with elsewhere;
- allocating responsibilities among agencies;
- identifying relevant environmental review procedures, documents and consultation requirements;
- setting page and time limits.

There is no prescribed list of impacts which must be included in EISs beyond the specification that direct, indirect, connected, similar and cumulative actions must be considered, together with alternatives and mitigation measures (Bass *et al.*, 2001). A record of the scoping process must be kept since the Regulations (section 1502.9(a)) state that draft EISs 'shall be prepared in accordance with the scope decided upon in the scoping process'. CEQ (1981b) has issued guidance on scoping which not only advocates the use of public meetings and other methods of ensuring participation but suggests that a scoping report be prepared. This non-mandatory document should be a record of the decisions made during the scoping process and contain a summary of the issues to be evaluated in the EIS and of the views of those participating in the scoping process.

There is no formal right of appeal against scoping decisions but it is customary for additional impacts to be addressed in the draft EIS if it later becomes apparent that they are likely to be significant. On the whole, the formalised scoping process works well, providing an agreed list of contents to be covered in the draft EIS. There is, however, a tendency to include impacts of questionable significance, rather than to exclude them, partly because of the fear of legal challenge. Further, while some public scoping meetings are well attended, others fail to attract a single participant. Notwithstanding its shortcomings, scoping is generally regarded as a valuable addition to the EIS preparation process in the United States (Weiner, 1997) and is also sometimes used in the preparation of environmental assessments (EAs).

UK

There is no requirement in the UK for the proponent to consult the local planning authority (LPA) prior to submission of the environmental statement (ES), or to undertake any form of scoping. However, the amended Regulations allow a developer to request a formal pre-application scoping opinion from the LPA: 'A person who is minded to make an EIA application may ask the relevant planning authority to state in writing their opinion as to the information to be provided' (Regulation 10(1)). In this case, the developer must provide a location plan and 'a brief description of the nature and purpose of the proposal and its possible environmental effects, giving a broad indication of their likely scale' (Department of the Environment, Transport and the Regions – DETR, 1999b, para. 55). The LPA is under a statutory requirement to consult the various EIA statutory consultees and to provide an opinion within five weeks; where it fails to do so, the developer may apply to the Secretary of State for a scoping direction instead.

These provisions codify previous government advice, since the Department of the Environment (DoE) has consistently advised developers to consult LPAs about the coverage of environmental statements (DoE, 1995; DETR, 2000a). Consultation of statutory consultees and, in some instances, of the public during scoping has also been recommended:

> While developers are under no obligation to publicise their proposals before submitting a planning application, consultation with local amenity groups and with the general public can be useful in identifying key environmental issues, and may put the developer in a better position to modify the project in ways which would mitigate adverse effects and recognise local environmental concerns. (DETR, 2000a, para. 39: originally DoE, 1989)

The statutory minimum content of an ES consists of a description of the development, a description of mitigation measures, the data necessary to identify and assess the main effects, an outline of the main alternatives considered and a non-technical summary. Further information, describing the environment and the likely significant effects, is to be included to the extent that is reasonably required for the effects to be assessed (Planning Regulations, Schedule 4). The Circular (DETR, 1999b, para. 91) recommends these topics, together with other information in Annex A, and elsewhere in the Circular, as a starting point. The UK guide to EIA procedures contains a six-page checklist of issues that might need to be covered by an ES, including the risk of accidents (DETR, 2000a, Appendix 5). Like the information provided in the DoE guide to the preparation of ESs (DoE, 1995), this checklist has no statutory standing. No official guidance on generic sets of impacts for particular types of projects has been issued.

In a study of a sample of 40 cases where ESs were submitted, about two-thirds of developers or their consultants undertook early voluntary consultations with the LPA, and another 25 per cent discussed scoping prior to submission of the ES. LPAs were generally able to influence the scope of the ES during these discussions, nearly three-quarters of them suggesting topics for inclusion. Discussions with the statutory consultees, prior to submission of the ES, took place in about 60 per cent of cases. The public (mainly in the form of major public interest and local action groups) was involved in scoping in about 40 per cent of cases. The various consultees influenced the scope of the ES by adding important topics to the developer's initial list in about one-third of cases (Jones et al., 1998). These data indicate an improvement over the years since an earlier study found that suggestions by voluntary groups to either the LPA or the developer/consultant had very little influence on the scope of the ES (DoE, 1991a).

In a more recent survey of LPAs, Weston (2000) echoed and expanded upon these findings. Two-thirds of the LPAs sampled reported that developers consulted them in all the EIA cases they dealt with and another quarter in 75 per cent of the LPAs' EIA cases. The scoping methods used by a sample of 33 consultancies centred on consultation with the LPA and other consultees, on previous experience and on professional judgement. Twenty-five per cent used published guidance and only 12 per cent used matrices, computer models and other methodologies to identify likely significant impacts. The LPAs sampled felt that the amended Regulations would have little effect on current practice though, interestingly, about 20 per cent believed that they would result in an increase in the number of topics and 10 per cent in a reduction (Weston, 2000).

It is clear that informal scoping arrangements between the developer/consultant and the LPA and, to a lesser extent, with consultees are working reasonably well and that the amended Regulations should codify existing practice. However, it is also clear that public involvement in scoping in the UK is less satisfactory.

The Netherlands

Scoping has been a requirement of the Dutch EIA system since its introduction in 1987. The Environmental Management Act 1994, section 7.15, requires project- (or plan-) specific guidelines to be prepared for each EIA. The notification of intent prepared by the proponent alerts the competent authority that an EIA is to be undertaken and that guidelines are to be prepared. In turn, the competent authority must publish the notification of intent and apprise the EIA Commission (EIAC), which must produce its recommendations on guidelines within nine weeks of publication of the notification. To make the preparation of scoping guidelines easier, the content of the inception memorandum (a notification of intent, usually 10–30 pages in length) was specified in greater detail following the report of the Evaluation Committee on the Environmental Protection Act (ECW, 1990). As mentioned in Chapter 6, the guidelines must be issued by the competent authority within 13 weeks (unless the authority is also the proponent).

EIAC sets up a small group of independent persons from its panel known to have relevant expertise and provides administrative support (EIAC, 1998). This group takes into account the results of consultations with the proponent and the various relevant agencies, and the representations of the public (particularly regarding specific local conditions), about the content of the EIS. The statutory consultees normally comment mainly on the characteristics of the proposed site but public responses, which vary greatly, often focus more on the merits of the proposal than on the scope of the EIA.

EIAC then makes recommendations on the environmental effects to be described, the objectives of the proposal, the relevant planning policies, environmental designations and environmental standards for the area, the environmental aspects that should be addressed, the specific local conditions which must be described and the alternatives that should be given attention (EIAC, 1998). EIAC occasionally makes recommendations about the methods to be employed (see van Eck *et al.* (1994) for examples of guidelines).

Generally, previous guidelines for similar projects provide a model for the action-specific guidelines. EIAC has prepared standard text for certain sections of all guidelines (EIAC, 1998). However, with the exception of guidelines for the EIA of motorways issued in 1994, no general or generic guidelines have been prepared.

EIAC's advice, which is made public, provides the competent authority with a draft of the scoping guidelines. Indeed, guidelines written by EIAC are sometimes adopted by the competent authority more-or-less word for word, though competent authorities often make major amendments (EIAC, 1998). During the 1990s EIAC produced about 80 sets of guidelines each year, a total of 976 having been published by the end of 2000 (EIAC, 2001a).

Advice on scoping is contained in the EIA handbook (see Chapter 5) and there is specific guidance on scoping and guidelines. The scoping guidelines have tended to be somewhat general (despite lengths of 20–30 pages or more) and often do not eliminate potentially irrelevant topics, leading to criticism of the best efforts of the influential EIAC (EIAC, 1998). However, guidelines rarely neglect relevant impacts and are becoming more focused. The main problems appear to be the elimination of irrelevant impacts and encouraging all the competent authorities (rather than the current 50 per cent) to take full responsibility for, and to commit themselves to, the guidelines issued in their name.

Canada

Scoping is mandatory in Canada. The Canadian Environmental Assessment Act 1992 requires the responsible authority (RA) to determine the scope of the screening or comprehensive study, while the Minister of the Environment (MoE) determines the scope of panel reviews and mediations when fixing the terms of reference (section 16 (3)). No scoping procedure is specified. The Canadian Environmental Assessment Agency's guide to the Act (CEAA, 1994, section 1.4) suggests that there are four aspects to scoping:

1. Determining the scope of the EA (i.e. the relevant components of the project and of the environment).
2. Determining the factors to be considered (i.e. the relevant environmental impacts).
3. Determining the parties interested in the project and their concerns.
4. Determining the appropriate level of effort and analysis, given the project, the likely nature of its environmental effects and the public's concerns.

The proponent must consult the RA (if the two are not one and the same) to enable scoping of both the project and the impacts to take place.

The EA must address the environmental effects specified in the Act (Box 10.3) and informal, action-specific, scoping guidelines may be prepared, but there is no legal requirement for them to be published. The intention is that only potentially significant effects will be considered in the EA and that the appropriate level of analysis should be determined. As in the preparation of the EA reports, the amount of effort devoted to scoping depends on the scale of the EA and is therefore much greater for a comprehensive study than for most screenings. General guidance on scoping procedures (CEAA, 1994) has been supplemented by guidance on scoping in comprehensive studies (CEAA, 1997b) and in panel reviews (CEAA, 1997c). However, apart from the draft guidance prepared on mining projects (CEAA, 1998b), no other sectoral CEAA scoping guidance has been published. There is no requirement for consultation and participation and no right of appeal (save through the courts) against scoping decisions. However, as at the screening stage, it is in the interests of the RA to consult with the public likely to be affected and seek to overcome their concerns. Thus, some scoping reports are made available to the public.

In practice, David Redmond and Associates (1999) found that RAs used matrices, checklists and other formal methods in two-thirds of the screenings they analysed. There has been criticism of CEAA's guidance by RAs and

---

Box 10.3   Factors to be considered in Canadian EA reports

All EA reports related to screenings (including class screenings) comprehensive studies, mediations or panel reviews must address the following factors:

- The environmental effects of the project, including:
    - the environmental effects of malfunctions or accidents that may occur in connection with the project;
    - any cumulative environmental effects that are likely to result from the project in combination with other projects or activities that have been or will be carried out.
- The significance of the environmental effects.
- Public comments, if applicable or received in accordance with the regulations.
- Technically and economically feasible measures that would mitigate any significant adverse environmental effects of the project.
- Any other matter relevant to the assessment that the responsible authority (RA) may require, such as the need for and alternatives to the project.

The RA must also identify those environmental effects (including any directly related to human health, heritage, socio-economic conditions and other factors) relevant to the assessment requiring further investigation.

In addition to the above factors, a comprehensive study, mediation or panel report must also consider:

- The purpose of the project.
- Technically and economically feasible alternative means of carrying out the project as well as the environmental effects of these alternative means.
- Effects on the capacity of those renewable resources likely to be significantly affected by the project to meet present and future needs.
- The need for, and the requirements of, any follow-up programme.

Source: amended from CEAA, 1994, p. 75.

---

concern has been expressed that scoping has sometimes resulted in too narrow a range of impacts. This related particularly to EAs triggered by the issuance of a licence or permit (i.e. to regulatory activity) and was due to 'the general nature of the Agency guidelines; lack of departmental guidelines; and concern about federal intrusion in areas of provincial jurisdiction' (Commissioner of the Environment and Sustainable Development – CESD, 1998, pp. 6–16). Furthermore, 'different federal authorities may have different opinions on how a project or an assessment should be scoped' (pp. 6–23). The Commissioner, following his review of EA procedures, recommended that federal authorities should develop their own scoping guidelines for typical projects, in conjunction with the Agency. This should be one product of the increased cooperation proposed by the Minister of the Environment (MoE) in the bill amending the Act (Government of Canada, 2001; MoE, 2001).

The scoping stage in panel review relates to the preparation of EISs and is based upon a screening report or a comprehensive study report when one is available. (Scoping does not formally apply in the mediation procedure in which the issues are explicit from the outset.) Once the panel has been appointed and given its terms of reference by the Minister (following consultation with the RA), it hears the views of those involved regarding the scope of the review, prepares draft project-specific guidelines, and revises them in the light of comments. These guidelines are frequently extremely demanding of the proponent. Those for Voisey's Bay Mine and Mill, for example, ran to over 40 pages (CEAA, 1997a) and included requirements for:

> The assessment of impacts and the proposed mitigation measures shall include, but not be limited to, effects of the Undertaking on the following matters. . . .
> (f) Social and cultural patterns, including:
>   (i) cultural and spiritual life of the communities, including language loss or retention;
>   (ii) patterns of social organisation at the household and community level, including the organisation of work, mutual aid, and sharing; and,
>   (iii) social relations between residents and non-residents, between men and women, among generations, and between aboriginal and non-aboriginal persons. (CEAA, 1997a, section 9.2)

There is now considerable experience of scoping for EISs in Canada and, in general, the panel scoping procedure has worked effectively (see, for example, the description of the scoping of the Great Whale Project EIA by Mulvihill and Jacobs (1998)). Public involvement in scoping is one of the reasons that confidence in panel review in Canada has generally been high. It is clear, however, that scoping practice in relation to screenings and comprehensive study reports is often considerably less satisfactory.

Commonwealth of Australia

Scoping has long been a requirement of the Australian EIA process, but without formal procedures for consultation and public participation. It is, perhaps, significant that the Australian scoping requirements were not written into the original administrative procedures, but are now used by all the states as well as by the Commonwealth as a means of improving the quality of EIA reports (Australian and New Zealand Environment and Conservation Council – ANZECC, 1991). The Commonwealth Environment Protection Agency (CEPA, 1994) proposed the introduction of a formal public scoping process. It advanced a strong case for the transition from agency to public scoping:

> Proponents benefit from public scoping by having a properly targeted environmental impact assessment process which is more assured of canvassing all important issues up front. Early public involvement in the assessment process can also help reduce controversy. The Environment Protection Agency benefits by identifying and accessing local knowledge, and is therefore in a better position to accurately assess the environmental impacts of projects. (CEPA, 1994, p. 32)

This reform was, however, not mentioned in the Minister's consultation paper (Hill, 1998) and the scoping provisions of the Environment Protection and Biodiversity Conservation Act 1999 (EPBC Act) are similar to those previously employed.

The broad contents of environmental impact statements and of public environment reports (PERs) are specified (Environment Protection and Biodiversity Conservation Regulations 2000, Schedule 4). However, the Australian procedures also require the preparation, by Environment Australia, of guidelines on the content of a draft EIS or PER which include the relevant impacts of the action. These guidelines must be prepared within 20 business days of deciding that an EIS or PER is needed (or of the end of any period allowed to comment on draft guidelines). They should address the issues listed in Schedule 4 to the Regulations and ensure that the assessment:

(a) assesses all the relevant impacts of the action; and
(b) provides enough information about the action and its relevant impacts to allow the Minister to make an informal decision whether to approve the action. (Regulations, Schedule 1, para. 4.03)

In practice, these project-specific guidelines have usually been discussed with the proponent before being issued by Environment Australia.

While there is no formal requirement to do so, the EPBC Act permits other bodies and the public to be consulted on the content of the guidelines. This is expected to occur routinely as agency consultation has become normal and public participation increasingly took place under the previous legislation (Harvey, 1998). However, it left much to be desired (Craig & Ehrlich *et al.*, 1996). The EPBC Act requires the notification of the finalised guidelines on the internet and the previous practice of reproducing the guidelines in the draft EIS is expected to continue. In practice, draft guidelines are also published on the Environment Australia web site.[1]

Guidelines have frequently been detailed (for example, they ran to 26 pages in the Second Sydney Airport EIA). Those for the proposed national low level radioactive waste repository EIS were 36 pages in length (Environment Australia, 2001). An extract from the guidelines dealing with environmental impacts and risks in the operational phase is reproduced in Box 10.4. Project-specific guidelines have been criticised for being insufficiently focused and for failure to consider fully the no-go option (although this was not the case in the Second Sydney Airport guidelines – see Chapter 8). In practice, guidelines have tended to be encyclopaedic and scoping has not eliminated irrelevant or unimportant impacts from consideration. There is no guidance on methods of, or procedures for, scoping beyond reference to guidelines for previous projects. Notwithstanding these difficulties, the utility of project-specific scoping guidelines (and, to a lesser extent, of participation in their preparation) is universally accepted.

Box 10.4   EIS guidelines for Australian radioactive waste repository

Risks of contamination to the environment and the consequences need to be considered in detail, including:

- Atmospheric emissions, including intentional and unintentional releases of radionuclides and a discussion of any releases in terms of anticipated volumes, dispersion, approved discharge limits and impacts.
- Aqueous emissions, including intentional and unintentional releases directly to the environment and/or to the water management system, anticipated volumes, discharge limits, and anticipated impacts on surface/ground water and subsequent uses of this resource.
- Cumulative risks to humans and the environment during the operational and surveillance phases of the project including
  - Analysis of exposure pathways to the environment and humans.
  - Accumulation through environmental pathways and the food chain.
  - Risks associated with characterisation, encapsulation, conditioning, quality assurance and certification of waste packages.
  - Implications of proposal for existing health status of any nearby communities, or surrounding land uses, review and assessment of health risks, background information on likely effects and levels of ionising radiation.
  - Assessment of overall environmental risks from all elements of the proposal.

Source: Environment Australia, 2001, pp. 19–20.

New Zealand

The Resource Management Act 1991 not only specifies that the details of any assessment should correspond to the scale of the effects of the action but that it must be prepared in accordance with the Fourth Schedule (see Chapter 9). The Fourth Schedule was a late addition to the Act, inserted following pressure from environmental groups and the Parliamentary Commissioner for the Environment (PCE), who were concerned that too much discretion in establishing assessment procedures was being left to local authorities. It was intended to bring together the requirements of the Act relating to the contents of an assessment of environmental effects in the form of a checklist. Councils are supposed to provide advice to applicants about the matters that should be covered in assessment of environmental effects (AEE) documents (Ministry for the Environment – MfE, 1999a). Anxiety was expressed by the Commissioner and others (Morgan, 1995) that some local authorities might use the Fourth Schedule as an inflexible listing of required information, rather than as a guide to obtaining information about relevant significant impacts.

While the Fourth Schedule is intended to provide a guide as to which information might be furnished by a developer in each case, it is notable that the requirements include the treatment of alternatives, the risk of accidents and monitoring, together with a strong suggestion that early consultation should take place. However, scoping is not mandatory. This omission is surprising in view of the role of scoping in the Environmental Protection and Enhancement Procedures (MfE, 1987), the stressing of scoping in recent guidance (MfE, 1999a) and the published guidance devoted to scoping (MfE, 1992b) (see Box 10.1). For notified projects (see Chapter 9) the local council may 'require an explanation of (ii) the consultation undertaken by the applicant' (section 92(2)(a)). In effect, an applicant is being instructed that it would be unwise to neglect consultation about the proposal and the scoping of its effects.

Regional and district plans should give guidance as to scoping and consultation requirements (MfE, 1999a). It would be almost impossible for the consent authority to take account of Maori interests, as required by the Act, without early consultation with Maori representatives (i.e. scoping). If affected parties are to give their written agreement to the proposal they must be consulted by the applicant prior to the submission of the application. However, applicants are usually consulted after the assessment of environmental effects has been completed so their views are not incorporated during scoping.

Morgan (1995) has reported mixed experience with local authority involvement in scoping. Some councils have given a clear invitation to applicants to discuss proposed actions with them, whereas others have left applicants to make the initial approach or relied upon the content requirements in the Fourth Schedule. While there is some very good practice, some applicants have not informed affected parties about the environmental effects of their proposals (PCE, 1995). As in other jurisdictions, applicants and local authorities have generally found early discussions to be helpful in modifying project design to reduce impacts.

In summary, the preparation of scoping guidelines is not obligatory in AEE report preparation despite the experience of scoping gained over the years for larger projects in New Zealand. However, scoping (and consultation and participation) is virtually compulsory for notified projects.

South Africa

Scoping has historically been a strong feature of EIA in South Africa. Such heavy emphasis is placed on this stage that the EIA Regulations permit the relevant authority to request a plan of study for scoping. This may be little more than a record of initial discussions between the applicant and the relevant authority but, if required, it must include, *inter alia*, 'a description of the proposed method of identifying the environmental issues and alternatives' (Republic of South Africa, 1997, Regulation 5(2)(e)). It must also include a timetable setting out when the various tasks involved in scoping are to be completed and when the relevant authority is to be consulted. The explanatory EIA guidelines (Department of Environmental Affairs and Tourism – DEAT, 1998c) provide additional detail about this preliminary scoping document, which is produced in perhaps half the scoping studies undertaken.

The relevant authority may request further information or, if the information contained in the plan of study is judged to be 'accurate, unbiased and credible' (DEAT, 1998c, p. 22), request the applicant and/or consultant to submit a scoping report. The scoping report (which, as mentioned in Chapter 9, is the final document produced by the developer in perhaps 90 per cent of unexempted cases) must include:

(a) a brief project description;
(b) a brief description of how the environment may be affected:
(c) a description of environmental issues identified;
(d) a description of all alternatives identified; and
(e) an appendix containing a description of the public participation process followed, including a list of interested parties and their comments. (Republic of South Africa, 1997, Regulation 6)

The EIA Regulations tend to focus on environmental issues, rather than directly on impacts. The EIA guidelines provide an explanation:

'**Environmental issues**' may either be:
- definable impacts (e.g. air pollution),
- the cause of an impact (e.g. burning mine-dumps), or
- a generally expressed concern (e.g. social disruption of communities) which need[s] to be translated into specific impacts to be investigated. At this stage a broad range of issues are being identified by all the interested parties. Some of these issues may be significant, insignificant or inaccurate. It is however important to focus on the relevant and important issues and to eliminate the insignificant or inaccurate issues from the investigation. (DEAT, 1998c, p. 23)

The provisions for the advertisement of applications prior to scoping (see Chapter 9) and for a plan of study, and the emphasis placed on scoping in the EIA guidelines, including the strongly recommended public review of scoping reports (DEAT, 1998c), indicate that scoping is still considered to be crucial in South Africa. Although there is no formal requirement to do so, relevant authorities require the production of draft scoping reports for consultation and participation. Scoping reports are expected to contain details of all those consulted and their comments. South African scoping produces a set of activity-specific guidelines for each project that is to proceed to an EIA. Indeed, applicants are encouraged to submit their plan of study for the environmental impact report (EIR) as part of the scoping report.

The integrated environmental management (IEM) guidelines on scoping concluded that 'scoping may well be considered to be the critical stage in the IEM procedure' (Department of Environmental Affairs – DEA, 1992, vol. 2, p. 20). Under both the IEM and the EIA Regulations procedures, many consultants placed about 30 per cent of their resources into the scoping stage, but frequently did not involve specialists in the EIA until the environmental issues had been clearly identified (see, for example, Weaver *et al.*, 1999). Thus, the scoping stage has often involved elements that belong to the EIA report preparation phase elsewhere. Many scoping reports have not only identified impacts but incorporated the evaluation of impacts and included specialist studies. Research in North West Province by Potchefstroom University[2] indicated that specialist studies were used in 35 per cent of scoping reports. It appears that consultants often provide more information than required by the 1997 Regulations because both applicants and authorities are anxious to avoid the resource implications of proceeding to a full EIA process if this can possibly be avoided. As with environmental assessments in the United States, it is difficult to monitor and control the quality of the information in these quasi-EIRs because their content is not specified in the Regulations.

Scoping has not always resulted in the elimination of irrelevant impacts. In a study of 28 EIAs undertaken between 1971 and 1986, it was found that scoping was documented in nearly 80 per cent of the cases but that only in four cases was this a comprehensive, inclusive, activity (Mafune *et al.*, 1997).

The National Environmental Management Act 1998 (NEMA) is silent on the scoping stage of the EIA process beyond stating that EIA procedures must, as a minimum, ensure:

> public information and participation, independent review and conflict resolution in all phases of the investigation and assessment of impacts. (section 24(7)(d))

It is inconceivable that the NEMA EIA regulations, expected in 2002, will not make provision to build on the extensive South African experience of scoping.

Summary

The treatment of scoping in the seven jurisdictions varies, with five of the EIA systems meeting the evaluation criterion (Table 10.1). Scoping is a formal requirement for full EIA reports in the United States and Canada. It is a general requirement for EIA reports in the Netherlands, Australia and South Africa. In each of these jurisdictions, scoping involves the preparation of action-specific guidelines and must incorporate some environmental agency and public participation.

While not a formal requirement in New Zealand, scoping is very strongly encouraged for notified projects under the Resource Management Act 1991 and local authorities can set up their own scoping procedures. The UK makes reference only to scoping at the proponent's discretion in its legal provisions, though scoping is strongly advised. In practice, scoping (in the form of discussion between the proponent and the LPA) frequently takes place but the resulting guidelines are seldom made public. Consultation of environmental authorities and the public in the UK during scoping is less common.

Table 10.1 The treatment of scoping in the EIA systems

Criterion 5: Must scoping of the environmental impacts of actions take place and specific guidelines be produced?

Jurisdiction	Criterion met?	Comment
United States	Yes	Public scoping is used to produce specific guidelines for EISs. Scoping is sometimes used in EAs
UK	Partially	Regulations require participation by planning authorities and statutory consultees if developer opts for scoping. Guidance strongly advises. Frequently occurs
The Netherlands	Yes	Public scoping process, involving EIA Commission, produces action-specific guidelines for EISs
Canada	Yes	Scoping must be undertaken for self-directed assessments: no requirement for publication. Action-specific EIS guidelines issued following consultation in panel reviews
Commonwealth of Australia	Yes	Project-specific scoping guidelines specified by Environment Australia, usually following consultation and participation
New Zealand	Partially	Scoping not obligatory in Act, but virtually compulsory for notified projects. Practice varies
South Africa	Yes	Scoping report must contain specified information: infrequently leads to further assessment

It is now widely accepted that scoping helps to ensure that the relevant environmental impacts are covered in EIA reports (if not that scoping helps to eliminate irrelevant impacts). It has also been found in the more mature EIA systems that scoping ensures that the various parties can participate early in the EIA process. It can therefore only be a matter of time before New Zealand (which already has considerable experience of scoping for many projects) and the UK (which also has considerable scoping experience) adopt formal requirements for mandatory scoping, including the preparation of action-specific guidelines.

Notes

1. www.ea.gov.au/epbc
2. Personal communication by Andries van der Walt of unpublished research by the Environmental Assessment Research Group at Potchefstroom University.

Chapter 11

EIA report preparation

If the treatment of the environmental impacts of alternatives is at the heart of the environmental impact statement (Chapter 8), the EIS is itself at the heart of the EIA process. There can be no meaningful EIA without the preparation of a report or reports documenting the findings relating to the predicted impacts of the proposal upon the environment. Canter (1996) suggested that the preparation of EIA reports is perhaps the most important activity in EIA, as their findings are utilised by decision-makers, government agencies and the public alike. In effect, the EIA report is the face of the EIA process. Despite an enormous literature on EIA methods, few jurisdictions specify how the findings presented in EIA reports should be derived. They do, however, normally specify the minimum content of the EIA report and frequently indicate procedures which must be followed in the preparation of the report (for example, the making available of information by the relevant authorities).

The next section of this chapter is concerned with ensuring that content requirements for EIA reports are achieved in EIA systems. The chapter then discusses EIA report preparation requirements in EIA systems and puts forward a set of evaluation criteria. These criteria are used to assist in the review of EIA report preparation procedures and practice in the United States, UK, the Netherlands, Canada, Commonwealth of Australia, New Zealand and South Africa.

Content of EIA reports

Virtually every EIA system possesses a requirement that an EIA report must describe the proposed action and the environment affected, forecast the significant impacts likely to result from the implementation of the action, and present a non-technical summary. There is also generally a provision that EIA reports contain other material, such as discussions of the alternatives considered (see Chapter 8) and of mitigation measures (Chapter 15). The preparation of this information requires the use of a wide variety of methods and techniques.[1]

The EIA process is cyclical (see Figure 1.1) and the nature of the action is continually refined as its design progresses. Design work is costly and, because approval is not certain when EIA is undertaken, there is a temptation for the proponent to prepare EIA reports on the basis of designs which are insufficiently detailed to allow forecasts to be prepared with accuracy. The decision-making and environmental authorities, however, should be seeking a realistic estimate of impacts which may necessitate more detailed design (and more expense) than the proponent originally contemplated. Whatever degree of detail is finally determined to be appropriate, the EIA report represents no more than a record of the impacts forecast to arise from the proposal as developed at a particular point in time. Because the impacts arising from the proposal are likely to change throughout its development, this record should be made as late as possible, i.e. it should represent the nature of the proposal immediately prior to the submission of the EIA report rather than at the initial design phase.

Glasson *et al.* (1999) stressed that it is important for an EIA report to be comprehensive. The Department of the Environment (DoE, 1995, p. 43) suggested a logical sequence of stages for presenting the findings relating to each impact (for example, on air quality) evaluated in an EIA:

- potential impacts;
- the existing baseline conditions;
- predicted impacts, giving a measure of their nature, extent and magnitude;
- the scope for mitigating adverse effects; and
- a statement evaluating the significance of unavoidable impacts.

Various checklists designed to assist in project description have been advanced (for example, that in Box 11.1). Such information can be provided in a variety of forms: written text, tables, process diagrams, flowcharts, maps, sketches, photo-montage, etc. It can be obtained by utilising design data, published emission data (for example, for air pollutants), published accident data, advice from expert authorities (for example, air pollution controllers), consultancy advice, EIA reports for similar proposals and visits to the sites of similar projects.

Only by carefully and systematically describing the initial or baseline environmental conditions is it possible to present an accurate and convincing picture of the likely effects that the development will have on its environment. It is important to devote sufficient effort to this part of the EIA process, as the accuracy and plausibility of much of the remainder of the EIA report depends upon it. Wherever possible, existing data should be utilised to indicate the principal physical features (for example, geology), existing and proposed land use; the main air, water and land quality characteristics, existing vegetation and wildlife, and existing land-use and other policies, plans and standards for the area (Morris and Therivel, 2001). Such data will often be held by various environmental authorities and need to be readily available to proponents.

Additional information may need to be gathered by observation and measurement, but only after the purpose of this information has been carefully considered. In addition to having a clear objective in EIA report preparation, any specific pre-project baseline studies should also provide the basis for post-project monitoring (Beanlands, 1988; Jones, 1999). Data on the existing environment should, of course, be collected early enough to use it as an input

Box 11.1   A checklist for project description

Nature and purpose of the development

- Function of the proposal, with economic and operational context.
- Alternatives considered (if appropriate).

Characteristics of the proposed site

- Location; Size; Summary of topography, landscape & natural or manmade features.

Characteristics of the proposed development

- Size; Site layout; Shape; Character; Landscape proposals (including grading).
- Car parking; Entrances and exits; Access to public transport.
- Provision for pedestrians and cyclists; Provision for utilities.
- Any other relevant information (including emissions to air, water and land).

Phasing of the development

- *Construction phase*
- Nature and phasing of construction; Frequency, duration and location of intrusive operations.
- Timing, location and extent of mitigation measures; Use and transport of raw materials.
- Number of workers or visitors.
- *Operational phase*
- Processes, raw materials; Emissions (air, water, noise, vibration, lighting, etc.).
- Number of employees or other users; Traffic generation.
- *Likely expansion or secondary development*
  - To be covered so far as the effects of such development can be anticipated at the time the ES is prepared.
- *Decommissioning/Closure stages*

Source: DoE, 1995, p. 40.

into the design process. Only information directly relevant to the forecasting of impacts should be included in the EIA report and, even then, much of it may be most appropriately presented in the form of appendices. Information on the likely magnitude of the impacts of the proposed action on the environment should be presented in the EIA report in as precise, objective and value-free manner as possible. Clearly, it is necessary to distinguish between the nature, extent and magnitude of an impact (for example, forecast dust levels will vary with distance from the source and disappear when emissions cease).

The forecasts of impact magnitude also need to take full account of likely changes in baseline conditions in the absence of the action and of the effect of mitigation measures (Beanlands and Duinker, 1983; Morris and Therivel, 2001).

The timescale and probability of occurrence of predictions should also be stated (Canter, 1996). Forecasting (or prediction) techniques[2] include formal mathematical models (for example, plume diffusion models), physical models (for example, wind tunnel simulators), laboratory models (for example, exposure of plants to pollutants), computer simulation, analogy with similar projects, photomontage, etc. In general, the simplest and least data-demanding forecasting techniques should be employed (Canter, 1996; Ortolano, 1997; Morris and Therivel, 2001). In order to permit external ver-ification (see Chapter 12) and auditing (Chapter 14) the limitations of the data and methods employed, any assumptions made during the forecasting process, together with the confidence which can be placed in the forecasts generated, should be stated. Uncertainty in these forecasts arises not only from the prob-lematic nature of many impacts but also from inaccurate information, from subsequent changes to the design of the project, from simplifications inherent in models and from errors in their use (Wood, G., 1999, 2000).

Such uncertainty may be very important in decision making (Chapter 13), especially where a choice has to be made between closely matched alternatives. Given this element of uncertainty, Glasson *et al.* (1999, p. 138) observed that the forecasts in EIA reports often appeared more absolute than was justified. Post-implementation monitoring schemes are intended to identify and, if necessary, to manage any variations between projected and actual impacts (Canter, 1996; Morris and Therivel, 2001; see also Chapter 14). In addition to acknowledging the uncertainty associated with impact forecasts, it is also important to explain why certain impacts, especially if they were identified during the scoping process, have not been subsequently included in the EIA report. Such explanations can only contribute to the transparency associated with the EIA report.

Whereas forecasting the magnitude of impacts is a matter of determining the quantitative effects, the significance of an impact is a matter requiring value judgement. Determining the magnitude of an impact may provide some indi-cation of its possible significance, but there is not necessarily a direct relation-ship between the two factors. For example, a road traffic noise level of 69 dB(A) (L_{10} 18 hour) in a forest may have a significance quite different from the same noise level in an urban area. Because value judgements will inevitably be made in presenting forecasts in the EIA report, it is essential that statements about significance are clearly distinguished from those about magnitude (Hyman and Stiftel, 1988; Glasson *et al.*, 1999).

It is important that emotive phraseology is avoided and that a consistent vocabulary is employed to describe the significance of impacts in EIA reports. Equally, the basis on which value judgements are made should be clearly explained since, while there may be agreement about the magnitude of impacts, different participants in the EIA process are unlikely to agree about their significance.

Numerous methods of dealing with the significance of impacts have been identified (Thompson, 1990). Some formal methods have incorporated scoring and weighting in which an attempt has been made to quantify significance, and these have rightly been criticised for internalising value judgements (Bisset, 1988). As explained in Chapter 9, there exist few agreed criteria for defining significance. One approach to significance determination is shown is shown in Figure 11.1.

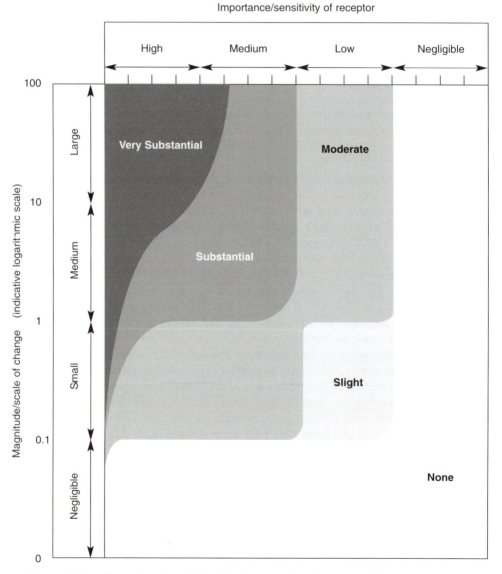

Figure 11.1   Determination of the significance of environmental impacts
Source: Terence O'Rourke, 1999, p. 35.

Where thresholds or standards (for example, ambient noise standards) are exceeded, significance is clearly established. The same is true for certain impacts where, for example, species loss or health damage may be universally regarded as significant. However, the uncertainty associated with certain impact predictions sometimes compounds the difficulty of evaluating their significance (Sadler, 1996). Here, as elsewhere, the use of consultation and participation methods is often appropriate, for example the harnessing of expert opinions (scientific or professional judgement). The organisation of a panel of professionals, perhaps operating on an iterative basis (the Delphi approach), can be helpful in establishing agreement about the significance of impacts. Analogy with similar actions is also frequently employed. The use of public opinion, perhaps by setting up a representative panel, is also effective in establishing significance.

The Department of the Environment (1995) advised that the most significant impacts of the proposal should be given most attention in the text of the EIA report. It advocated assessing the geographical level of importance of the issue considered (international, national, regional or county-wide, district-wide and local), then the significance of the impact (major, minor and not significant) and finally the nature of the impact (adverse/beneficial, cumulative, short/long term, permanent/temporary, reversible/irreversible and direct/indirect). It was suggested that this information, together with an estimate of uncertainty, should then be summarised in a table. However, Glasson *et al.* (1999, p. 141) reported that, in practice, 'much, if not most, current evaluation of significance in EIA is simple and often pragmatic, drawing on experience and expert opinion rather than on complex and sophisticated analysis.'

Preparation of EIA reports

While different EIA systems have different EIA report content requirements, it is clearly important that such provisions be specified precisely. It is also important that procedures be put in place so that proponents can gain access to information about the environment (and the action) held by decision-makers and environmental authorities. A further requirement, given the need for checks and balances in any system driven by the proponent (especially where the proponent is responsible for EIA report preparation), is that checks to reduce the likelihood of inadequate or biased EIA reports exist.

It is ironical, given the volume of literature on EIA methods (see above), that few jurisdictions specify the methods or techniques to be employed in EIA report preparation. In the UK the Department of the Environment (1995) has recommended a systematic methodological approach to the preparation of EIA reports, commencing with careful planning of the preparation of the environmental statement (ES), which in turn can speed up the decision-making process. It suggested that preparation of an ES should include three concurrent stages: determining the content of the ES; establishing a programme and timetable; and assembling the project team. It advised that the EIA report should be accessible to specialists, decision-makers and the public alike. Stressing that the length of an ES does not determine its quality, DoE counselled that it should usually be between 50 and 150 pages long, with appendices if necessary.

It is obviously important that the content of the EIA report be communicated as clearly as possible. Glasson *et al.* (1999) maintained that an EIA report should be: directed at its target audience; well written; avoid technical jargon; state any assumptions on which impact predictions were based; quantify impacts wherever possible; give an indication of the probability of the impact occurring; be honest and unbiased; and be kept as brief as possible. It was stressed that the presentation of an EIA report could have a great influence on how it was received: 'Good presentation can convey concern for the environment, a rigorous approach to the impact analysis and a positive attitude to the public' (p. 177).

Kreske (1996, p. 231) felt that there was no single correct way to prepare an EIS. He suggested that flexibility in environmental impact statement (EIS) preparation was necessary given the variety of proposed actions, environmental issues for a given area, public concerns and political agendas. He also recommended an interdisciplinary and integrative approach to enhance communication between the EIS team members.

The non-technical summary is frequently used to disseminate the findings of the EIA report to the general public at low cost. Since it is often the only document to be examined and since it is frequently read separately from the EIA report, it is important that it should be clear, concise, objective and well written. A good summary accurately reflects the text of the EIA report, presents the main conclusions about alternatives, mitigation, etc., and explains how they were reached. A summary table setting out the main forecast impacts and their significance is often a valuable means of detailing many of the main findings of the EIA report (DoE, 1995).

The non-technical summary is better prepared by assembling summaries of the various sections of the EIA report as they are written and then editing them into a consistent whole than hurriedly as the final step in EIA report preparation. While public relations and document design expertise can be helpful in the production of the summary (as they can be in the presentation of the main EIA report), it is important that any tendency to distort the document by failing to reflect the assessment results accurately should be resisted.

Various methods of checking the content of the EIA report before it is published exist. Perhaps the simplest involves review by the responsible authority before approval to release the EIA report is given. This type of check exists in, for example, the West Australian EIA system (Wood and Bailey, 1994). Another is the swift review of form and content used by the US Environmental Protection Agency prior to formally acknowledging receipt of an EIS. A further approach is to involve a consultative group in the preparation of the various parts of the EIA report and to reach agreement on its content section by section, as in Victoria (Wood, 1993). Some jurisdictions, however, rely on the diffusion of best practice and sanctions later in the EIA process (for example, preparation and submission of further information) as checks on the quality of

EIA reports. This is the formal situation in the UK. It is clear that diffusion of best practice will be much speedier where adequate checks on EIA report quality exist.

Many EIA reports are prepared, in whole or in part, by consultants. Needless to say, the standard of competence among consultancies varies and, especially where EIA has been introduced only recently, some may lack the appropriate range of professional skills. While, over time, reputations for competence in EIA report preparation will be established, there is a constant danger of consultancies being selected by price or suffering from bias (for example, where lucrative design contracts may follow approval of the proposal) unless some form of accreditation of EIA consultants or code of practice is introduced. Some jurisdictions (for example, Flanders in Belgium) have formally introduced such requirements. In South Africa, the consultancy preparing the EIA report is barred from further involvement in the implementation of the project. Elsewhere, several professional associations have either set up voluntary accreditation schemes or are actively considering doing so. In others, voluntary codes of practice are in operation (for example, Australia) or under consideration. There is a non-mandatory registration scheme in existence in the UK. Glasson *et al.* (1999) noted that it is sometimes a requirement for consultancies tendering to undertake an EIA in the UK to include evidence that the environmental statement produced will be reviewed by acknowledged independent experts.

A more radical approach to quality assurance in EIA report preparation is to require that it be prepared not by the proponent but by the decision-maker or the lead agency. This is the original National Environmental Policy Act 1969 model for permit applications to federal agencies, but it takes its most developed form in California. Here, the relevant agency often asks for tenders for environmental impact report preparation from an approved list of consultants. While the proponent pays, the client is the relevant agency (Bass *et al.*, 1999). There are clear advantages to this approach, but any separation of the project design and EIA teams runs counter to the aim of EIA: the consideration and minimisation of impacts as early as possible in the evolution of the proposal.

As at the other stages in the EIA process, the existence of clear and readily accessible guidelines on EIA report preparation, content and form is advantageous. Such guidance (together with scoping guidelines) is helpful in facilitating not only the writing of an effective EIA report but also in increasing the efficiency with which the lengthiest stage of the EIA process is discharged. Such guidance is helpful to proponents and consultants in preparing the EIA report and to decision-making authorities, environmental authorities, interest groups and the public in their subsequent review.

Again, as at other stages in the EIA process, the involvement of consultees and the public in EIA report preparation (not necessarily by means of consultative committees) will lead to improved quality or at least to improved acceptability. This criterion and others which can be advanced to assist in the analysis of the treatment of EIA report preparation in EIA systems are summarised in Box 11.2. The various criteria are used in the analysis of EIA report preparation in each of the seven EIA systems which now follows.

Box 11.2   Evaluation criteria for the preparation of EIA reports

Must EIA reports meet prescribed content requirements and do checks to prevent the release of inadequate EIA reports exist?

- Must EIA reports describe actions and environments affected, forecast impacts, indicate significance and contain a non-technical summary?
- Must information held by the relevant authorities about the environment or type of action be made available to the proponent?
- Does published guidance on EIA report preparation exist?
- Must specified EIA methods or techniques be employed?
- Does accreditation of EIA consultants exist?
- Do checks on the content, form, objectivity and accuracy of the information presented occur before publication of the EIA report?
- Is consultation and participation required in EIA report preparation?
- Does EIA report preparation function efficiently and effectively?

United States

The required contents of an EIS, which are specified in the Council on Environmental Quality's (CEQ's) National Environmental Policy Act 1969 Regulations, can be summarised as follows:

- cover sheet
- summary (not normally exceeding 15 pages in length)
- table of contents
- statement of purpose and need
- alternatives, including the proposed action
- affected environment
- environmental consequences, including mitigation measures
- list of preparers
- list of agencies and organisations consulted
- list of all federal permits
- appendices
- index.

The environmental consequences section of an EIS is intended to form the scientific and analytical basis for the comparison of alternatives. The discussion of environmental consequences must include the environmental effects of the alternatives, including cumulative (CEQ, 1997a), growth-inducing, historical, cultural, socio-economic and environmental justice (CEQ, 1997b) impacts (Bass *et al.*, 2001, p. 102). Conflicts between the proposed action and any relevant land-use plans, policies and controls for the area must be included as environmental effects.

An EIS may contain appendices which present relevant background material of an analytical nature (Regulations, section 1502.18) but the total length of the text of an EIS should not exceed 150 pages or, in the case of an unusually complex proposal, 300 pages. In practice, the average length of the text of EISs is about 200 pages (with ancillary material running to another 200 pages for draft EISs and 400 pages for final EISs).[3] There is no accreditation of EIA consultants and, although many federal agencies still possess sufficient competent in-house staff, consultants are increasingly employed to prepare EISs. It is not uncommon for a 'draft' draft EIS to be prepared, prior to publication of the document, for internal review purposes (Kreske, 1996). This preliminary EIS may occasionally be circulated to a limited number of consultees (including the Environmental Protection Agency). The draft EIS is subject to formal consultee and public scrutiny before the final EIS is prepared (see Chapter 16).

The final EIS should include all the substantive comments on the draft EIS (or summaries of them). It must also include responses to the comments received during the review of the draft EIS (Regulations, section 1503.4(b)). The final EIS may contain modified proposed actions or alternatives, it may develop and evaluate new alternatives, it may supplement, improve or modify analyses, include any necessary corrections or contain an explanation of why no further response is necessary.

The advice on EIS preparation contained in the Regulations and in other guidance (CEQ, 1981a) requires the documents to be analytical (and not encyclopaedic), to focus on significant impacts, to emphasise alternatives, to avoid *post hoc* rationalisation, to be interdisciplinary (i.e. to ensure the integrated use of the natural sciences, social sciences and the design arts (Regulations, section 1502.6)), to be concise, to be written in plain language and to use appropriate graphics.[4] There is no requirement to use specified EIA methods or techniques. Equally, most of the agency regulations contain minimal substantive (as opposed to procedural) guidance (Malik and Bartlett, 1993).

UK

There is no prescribed form of an environmental statement, except that it must contain a description of the development, a description of mitigation measures, the data necessary to identify and assess the main effects, an outline of the main alternatives studied (see Box 11.3) and a non-technical summary (Planning Regulations, Schedule 4, Part II). Further information, describing the environment and the likely significant effects, is to be included to the extent that is 'reasonably required' for the effects to be assessed and which the applicant can reasonably be required to compile (Regulation 2(1)).

While the developer is responsible for the content of the ES finally submitted and for the assessment methods employed, the Regulations enable the developer to collect relevant existing information from the statutory consultees (for example, English Nature) who are under a duty to provide it. Where a local planning authority is informed in writing that an ES is in preparation, it must notify the statutory consultees so that they can be ready to provide the developer with information if requested to do so.

Box 11.3   Mandatory content of a UK environmental statement

1. A description of the development comprising information on the site, design and size of the development.
2. A description of the measures envisaged in order to avoid, reduce and, if possible, remedy significant adverse effects.
3. The data required to identify and assess the main effects which the development is likely to have on the environment.
4. An outline of the main alternatives studied by the applicant or appellant and an indication of the main reasons for his choice, taking into account the environmental effects.
5. A non-technical summary of the information provided under paragraphs 1 to 4 of this Part.

Source: Planning Regulations, Schedule 4, Part II.

As mentioned in Chapter 10, developers and their consultants approach the statutory consultees in the vast majority of cases. On the basis of 40 case studies, Jones *et al.* (1998) reported that consultations, which took place in over 90 per cent of cases, often consisted of requests for information concerning, for example, statutory designations. The proportion of EIAs in which information is sought appears to have increased since an earlier study indicated that rather more than two-thirds of developers and consultants sought relevant information from statutory consultees and other public bodies (Department of the Environment, 1991a).

There exists a good practice guide to the preparation of ESs (DoE, 1995). This provides general guidance about the outline structure and content of any ES (Box 11.4). It contains useful annexes on impacts on human beings, noise and vibration, traffic and transport, land use and other topic areas and recommends the inclusion of a section on relevant policies and plans (DoE, 1995). Neither the Department of the Environment, Transport and the Regions Circular on EIA (DETR, 1999b) nor the DoE guidance on ES preparation (DoE, 1995) specify EIA methods or techniques, though the 1995 guide to ES preparation provides an indication of appropriate prediction methods for the various topic areas covered. Nevertheless, there remains among developers and consultants a sense of inadequate guidance on ES structure, methods and techniques (Jones *et al.*, 1998).

In a study of a sample of 40 submitted ESs, 85 per cent were prepared with assistance from outside consultants (Jones *et al.*, 1998). This represents an increase over earlier years since DoE (1991a) reported that consultants were retained in just over half the EIA cases studied. There is no formal accreditation of EIA consultants in the UK, though the Institute of Environmental Management and Assessment operates a voluntary scheme. There is no formal check on the content, form and accuracy of the ES prior to its release because, as mentioned in Chapter 10, there is no requirement for the proponent to

Box 11.4   Outline structure for a UK environmental statement

1. A non-technical summary, which may also be available as a separate document.
2. Method statement.
3. Statement of key issues.
4. Description of the project.
5. Description of environmental conditions within and surrounding the site.
6. Assessment of environmental effects by topic area.*
   (Each chapter should identify baseline conditions, potential impacts, the scope for amelioration and/or mitigation of impacts, and provide a description of unavoidable impacts.)
7. Appendices of technical data.

* The specified topic areas are:
- Human beings
- Flora
- Fauna
- Soil
- Water
- Air
- Climate
- The landscape
- The interaction between any of the foregoing
- Material assets
- The cultural heritage.

Source: DoE, 1995, pp. 33–4.

consult the relevant authorities prior to submission of the ES, although this is strongly advised. There is a discernible trend for certain clients to demand that their consultants utilise an independent panel to verify the quality of ESs before finalisation.

In summary, ESs have to meet limited content requirements but there are no formal checks to prevent the release of inadequate EIA reports. In practice, checks on ES content and adequacy are made at the review stage in the UK.

The Netherlands

Section 7.10 of the Environmental Management Act 1994 specifies the minimum contents of an environmental impact statement. As well as the detailed discussion of alternatives mentioned in Chapter 8, these include descriptions of relevant environmental plans and standards, of the methodology utilised, and of the shortcomings caused by lack of information, together with a non-technical summary. There must also be, of course, descriptions of the proposed action and of the environment affected and an analysis of the consequences of the action. There is no specific mention of mitigation measures (these are subsumed within alternatives) or of the significance of environmental impacts. The European Commission has identified weaknesses in the legal requirements relating to the content of EISs (see Chapter 5).

Information held by the competent authority must be made available to the proponent who prepares the EIS in accordance with that authority's guidelines for the action. As mentioned in Chapter 10, the guidelines may occasionally specify the type of method to be used (for example, a specific type of noise model), especially where there are overlapping legal requirements. However, most guidelines leave the choice of method to the proponent. The general handbook on EIA (see Chapter 5) contains information on methods, as do the specific guidance documents for each type of impact. There is no shortage of written advice on EIA methods in the Netherlands; however, much of this guidance is not widely used and some of it is now dated. The quality of consultants, who are not accredited, is generally high. Nevertheless, EIS preparation is a lengthy process in the Netherlands: about 50 per cent of EISs take more than a year.

As shown in Figure 5.1, the competent authority evaluates the acceptability of the EIS before it is made available for public review. This check ensures that the requirements of the Act and the recommendations in the guidelines are met. It used to be customary for the EIA Commission (EIAC) to be involved unofficially in this process (ECW, 1990). However, the Commission now becomes involved in this pre-review process only if the EIS deals with a new type of activity or if a special requirement was highlighted during scoping (EIAC, 1998). Six weeks are permitted for the competent authority to decide if the proponent should provide further information. Where such a request is made, it must be made public.

While a few proponents still see EIS preparation merely as a paper exercise, their number is probably diminishing. There is a noticeable improvement in the efficiency of preparation and in quality when proponents or their consultants are preparing a second or subsequent EIS rather than their first EIS (ECW, 1996). However, despite many examples of good practice, the EIA report preparation process in the Netherlands often still leaves much to be desired.

Canada

The content requirements for environmental assessment (EA) reports are set out in Box 10.3. Curiously, in view of the comprehensive nature of these requirements, no non-technical summary is specified (though this is suggested in guidance on the preparation of comprehensive study reports (Canadian Environmental Assessment Agency – CEAA, 1997b). There is a provision in the Canadian Environmental Assessment Act 1992 (section 12(3)) for expert federal departments to make information available upon request by the responsible authority (RA), the mediator or the review panel. EA reports are normally prepared by proponents but are the responsibility of the RA. It is recommended that CEAA check draft comprehensive study reports to help identify and correct major omissions or deficiencies before seeking public input (CEAA, 1997b). There is no mandatory consultation and participation in the preparation of EIA reports in Canada (though in practice this is encouraged and often achieved), nor are consultants accredited.

The responsible authority's guide to the Act provides considerable advice on the preparation of EA reports. It contains sections on screening, scoping, assessing environmental effects, mitigating environmental effects, determining the significance of adverse environmental effects, preparing and reviewing the EA report, decision making and follow-up. These sections are supplemented by reference guides on:

- addressing cumulative environmental effects;
- the public registry;
- determining whether a project is likely to cause significant adverse impacts; and
- assessing environmental effects on physical and cultural heritage resources. (CEAA, 1994)

There is a separate reference guide on the federal coordination regulations. In addition, there are supplementary good practice guides to the preparation of comprehensive study reports (CEAA, 1997b) and to the procedures for assessment by a review panel (1997c). In addition, there is guidance on biodiversity and EIA (CEAA, 1996); cumulative effects assessment (1999a) and various operational policy statements (for example, CEAA, 1998a).

These guides are principally procedural but provide references to the literature and sometimes include suggested methods. In the main body of the RA's guide it is suggested, for example, that project/environment interactions can be identified by such means as overlay maps, matrix tables and expert groups (CEAA, 1994, p. 83).

The determination of the significance of environmental effects is especially important in the Canadian EA system since this directly affects whether the RA can take a decision on the project that would allow the project to proceed (for the self-directed assessment) or whether further review is needed through mediation or a panel review. This determination 'must be based on sound scientific evidence and analysis (including traditional ecological knowledge)' (CEAA, 1994, p. 89), but 'public input can play an important role in the determination' (p. 90). Criteria for determining the significance of effects include magnitude, geographical extent, duration and frequency, irreversibility and ecological context but not the crucial issue of controversy. An example of the factors used in determining whether environmental effects are adverse is shown in Box 11.5.

The length of screening reports varies between 2 and 200 pages, but the vast majority are very brief. Comprehensive study reports tend to be shorter in length than EISs,[5] normally running to around 200 pages. David Redmond and Associates (1999) found that the competence and training of EA staff in various organisations were not impediments to the quality of screenings but

Box 11.5 Canadian criteria for defining adverse effects

Factors used in determining whether or not environmental effects are adverse

Environmental changes:

- Negative effects on the health of biota including plants, animals and fish.
- Threat to rare or endangered species.
- Reductions in species diversity or disruption of food webs.
- Loss of, or damage to, habitats, including habitat fragmentation.
- Discharge or release of persistent and/or toxic chemicals, microbiological agents, nutrients (e.g. nitrogen, phosphorous), radiation or thermal energy (e.g. cooling wastewater).
- Population declines, particularly in top predator, large, or long-lived species.
- The removal of resource materials (e.g. peat, coal) from the environment.
- Transformation of natural landscapes.
- Obstruction of migration, or passage of wildlife.
- Negative effects on the quality and/or quantity of the biophysical environment (e.g. surface water, groundwater, soil, land and air).

Effects on people resulting from environmental changes:

- Negative effects on human health, well-being or quality of life.
- Increase in unemployment or shrinkage in the economy.
- Reduction of the quality or quantity of recreational opportunities or amenities.
- Detrimental change in the current use of lands and resources for traditional purposes by aboriginal persons.
- Negative effects on historical, archaeological, palaeontological or architectural resources.
- Decreased aesthetic appeal or changes in visual amenities (e.g. views).
- Loss of, or damage to, commercial species or resources.
- Foreclosure of future resource use or production.
- Loss of, or damage to valued, rare, or endangered species or their habitats.

Source: CEAA, 1994, p. 92.

that lack of time and other resource limitations might well be constraints. Worryingly, the Commissioner of the Environment and Sustainable Development (CESD), in his study of nearly 200 screenings found that 'in some cases, the official carrying out the assessment or receiving a third-party assessment did not even visit the site' (CESD, 1998, pp. 6–17). The Minister of the Environment (2001) proposed amendments to the Act to increase CEAA's role in improving the quality of EA (Government of Canada, 2001).

Commonwealth of Australia

If assessment is undertaken on the basis of preliminary documentation, no further EIA report preparation is necessary although the proponent must take account of any public comments received. The Commonwealth assessment report can then be prepared. However, environmental impact statements and public environment reports (PERs) must meet identical specific prescribed content requirements (Environment Protection and Biodiversity Conservation Regulations 2000, Schedule 4). These requirements include:

1. The background of the action.
2. A description of the action.
3. A description of the relevant impacts of the action.
4. Proposed safeguards and mitigation measures associated with the action.
5. Other approvals and conditions relating to the action.
6. Environmental record of person proposing to take the action.
7. Information sources employed in the assessment of the action.

Interestingly, there is no provision for a non-technical summary, despite this being specified under the previous legislation. Using these requirements as a basis, the detailed content of each EIS or PER is specified in the project-specific guidelines published by Environment Australia, usually following consultation and participation (see Chapter 10). These include a requirement that the EIS or PER should comprise three elements: an executive summary, the main text, and appendices. They suggest a series of points for inclusion in the summary.

The previous Administrative Procedures (Commonwealth of Australia, 1995, para. 4.5) required that: '[t]he proponent shall consult with the Department throughout the preparation of the environmental impact statement'. This consultation was a crucial element in ensuring that the draft EIS or PER addressed the issues listed in the project-specific guidelines in an adequate manner. In practice, Environment Australia normally received draft chapters from the proponent. Again, permissive arrangements for the consultation of other agencies and bodies during the preparation of the EIS or PER existed under the previous legislation (Commonwealth of Australia, 1995, para. 4.6). This generally worked well, as liaison between the proponent and Environment Australia and other agencies was close. Occasionally, public groups were consulted by the proponent's consultants during report preparation. This consultative (rather than participative) process meant that EIS preparation did not generally taken an excessive amount of time.

These arrangements do not formally apply under the Environment Protection and Biodiversity Conservation Act 1999 (EPBC Act) but they continue informally. However, there are provisions for the proponent to report on consultations undertaken, or expected to be undertaken, during the preparation of the draft EIS or PER (see Chapter 16). In addition, Environment Australia retains the right to approve the publication of the draft EIS (section 103(1)(b)) (or the draft PER) which allows it to withhold permission to release the document until it is satisfied with its quality. This mechanism, which involves vetting the preliminary EIS or PER against the guidelines, has provided an effective latent weapon in the past, and Environment Australia has been reasonably content with the quality of published EISs and PERs. However, the quality of EISs has been criticised by environmental groups and others (see, for example, James, 1995) who suggested that there was a need for improved information systems and more rigorous predictive modelling.

No formal guidance on EIA report preparation, or specification of particular assessment methods to be utilised, has been issued. There is no accreditation of the consultants who are employed to prepare most EIA reports (though this was sought by some environmental groups during the passage of the EPBC Act). It remains to be seen how the new arrangements will impinge on EIS and PER preparation.

New Zealand

As mentioned in Chapter 9, it is apparent that the document (or documents) reporting on the assessment of environmental effects (which is, confusingly, not named in the Resource Management Act 1991) may vary in length from half a page to hundreds of pages, depending on the circumstances. The contents of this assessment are outlined in the Fourth Schedule to the Act. It is perhaps surprising that the Act is silent on the need for a non-technical summary (a requirement in almost every other EIA system). Provision of such a summary is, however, suggested by the Ministry for the Environment in its advice on scoping (MfE, 1992b, p. 26).

The Ministry (MfE, 1999a) has suggested that the assessment information to be provided should be indicated clearly in regional or district plans and that further help could be obtained by discussing the proposal with councils to gain their initial reaction. Because, in many situations, the level of expected impact will be low and the mitigation measures easily defined, the guidance suggests that the applicant should usually be able to furnish the necessary information. However, where the proposal may have 'significant effects on the environment', applicants may need to involve specialist advisers to help them with either a few aspects of their proposals or with the whole assessment of environmental effects (AEE) (MfE, 1999a). (There is no accreditation of EIA consultants in New Zealand.) These cases, which should generally be flagged in the regional or district plan, will usually trigger the requirement that they be notified and that information on alternatives and consultation be provided (Chapters 8 and 10). It is clearly in a developer's interests to provide such information at the outset, to avoid the delays involved in furnishing it upon request once the competent authority has received the application.

Guidance is available to applicants on the preparation of AEE reports in the form of a practical guide (Morgan and Memon, 1993), a scoping guide (MfE, 1992b) and various other documents (see Chapter 5). However, although methodologies such as checklists, matrices and evaluation systems are described in the practical guide, none is specifically recommended. The

Ministry's own guidance (MfE, 1999a) is equally silent about methods. Several seminars and workshops for practitioners were organised in the early period of implementation in which the Ministry provided advice in the form of papers. Following criticisms of practice (see, for example, Parliamentary Commissioner for the Environment – PCE, 1995), the Ministry was given a budget to produce further guidance (for example, MFE 1999a,b) and to become actively involved in the training of practitioners and local politicians.

In addition to a hypothetical gold mine assessment presented as an example of how to undertake an AEE (Morgan and Memon, 1993), the Ministry issued (MfE, 1992a) a fictional case study to provide brief model advice to developers in their approach to the assessment of the environmental effects of a tourism development. A model AEE report for a very minor development (of the type that would not warrant the retention of consultants) is presented in the Ministry guidance to applicants (MfE, 1999a). Such guidance is very important, given that there are no formal requirements for environmental information to be made available to the applicant, or for any formal check to be made on the quality of the EIA report before it is made public (Smith, 1996; Morgan, 2000b).

However, given the discretion that local councils have in specifying EIA requirements, applicants face considerable uncertainty unless the relevant policy statements and plans (or other local guidance) provide the necessary indication as to the appropriate content of AEE reports. Dixon (1993), Dixon and Fookes (1995), Morgan (1995), the PCE (1995) and McShane (1998) have noted that, in practice, the lack of EIA expertise among both local authority staff and applicants has caused considerable difficulties and that many of the Act's requirements, for example in relation to cumulative impact assessment (see Chapter 7) and alternatives (Chapter 8), are often not being met.

Morgan (2000b) reported that applicants needed project-specific advice in undertaking their assessments, and that this was frequently provided orally by councils at pre-application meetings. Nevertheless, many consultants have slavishly followed the topics listed in the first paragraph of the Fourth Schedule as a format for their assessment of environmental effects (see Chapter 10) (Morgan, 2000b). The quality of many AEE reports appears to be poor (Morgan, 1995, 2000b; PCE, 1995). In particular, for small projects:

> Typically, assessments have described the proposed activity, perhaps noting some of the main potential effects to be expected, then jump to comments on the significance of effects, based on personal judgement and the approval of neighbours. (Morgan, 2000b, p. 101)

South Africa

The South African EIA Regulations say little about the preparation of the EIA report (or environmental impact report – EIR) but they require that a plan of study for an EIA with clearly specified contents be submitted (Republic of South Africa, 1997, Regulation 7(1)). This plan of study must be accepted by the relevant authority prior to submission of the EIR. It is clear that the relevant authorities are using their approval of the plan of study to require the preparation and circulation of a draft EIR (despite the absence of any formal requirement that checks be made on the contents of EIRs before they are released), adequate consultation and public participation, and response to comments.

As mentioned in Chapter 8, the EIA Regulations state that descriptions of each alternative and a comparative assessment of the alternatives must be presented in the EIR, together with appendices containing descriptions of:

(i)   the environment concerned;
(ii) the activity to be undertaken;
(iii) the public participation process followed including a list of interested parties and their comments;
(iv) any media coverage given to the proposed activity; and
(v)   any other information included in the accepted plan of study. (Regulation 8)

The EIA Regulations require the majority of the information usually demanded in any EIA process to be supplied in either the main text or the appendices of the EIR.

The EIA guidelines (Department of the Environmental Affairs and Tourism – DEAT, 1998c) provide some elucidation. They indicate (DEAT, 1998c, pp. 27–8) that the assessment of impacts should involve a synthesis of the:

• nature of the impact
• extent
• duration
• intensity
• probability.

The guidelines also suggest that the significance of impacts should be determined, that mitigation measures should be set down, that the key issues should be reassessed and that a comparative assessment of the feasible alternatives should be presented. This is a more conventional recipe for EIA report content, which goes well beyond the requirements of the Regulations. There is no mention of a non-technical summary in either the Regulations or the guidelines, despite this being suggested in the Environment Conservation Act 1989, section 26(a)(vii). As in most other EIA systems, no particular EIA methods or techniques are specified.

The South African regulations contain the unusual requirement that consultants, who must be appointed, have 'no financial or other interest in the undertaking of the proposed activity' (Regulation 3(1)(c)). While many practitioners in South Africa have welcomed this provision (see, for example, Granger, 1998), others feel that multidisciplinary firms are penalised, that the value of developers employing in-house specialists may be lost and that the integration of the design process and the mitigation of impacts is hampered. There is concern that some consultants may not act independently but produce inexpensive 'sweetheart reviews' of the work of others with little critical analysis, or that they may regard success (and hence repeat business), rather than objectivity, as their goal. The accreditation of EIA consultants is being actively considered to try, *inter alia*, to overcome this problem (Duthie, 2001).

The integrated environmental management (IEM) guidelines were rather more prescriptive than the EIA Regulations or the EIA guidelines (Department of Environmental Affairs, 1992, vols 3, 4). Since South Africa has a large and competent EIA consultancy sector some EIA report preparation practice under the voluntary IEM procedure was exemplary. It has been, for example, common practice to produce draft EIA reports and to use separate consultants, sometimes selected by the 'interested and affected parties', to review specialist studies prior to the preparation of the EIR. Weaver *et al.* (2000) suggested that perhaps 10 per cent of the total cost of specialist studies should be budgeted for such independent reviews. They cited the Maputo Iron and Steel EIA as a case where the interested and affected parties had a direct input into the appointment of specialists, following scoping. The summary of the draft EIR for this controversial project, which was accompanied by three specialist reports, was 60 pages in length (Gibb Africa, 1998).

On the other hand, despite the strength of the EIA consultancy sector, there is a shortage of experienced EIA specialists and, in particular, of practitioners trained in the social sciences in South Africa. Consultants have often been appointed too late, have had insufficient expertise, or have undertaken work at less than cost to ensure that lucrative engineering design fees are secured later. This has led, unfortunately, to numerous examples of poor EIA practice (Ridl, 1994; Gasson and Todeschini, 1997; Duthie, 2001). In particular, South African EIA reports have contained few quantified predictions and they have sometimes failed adequately to take account of the comments made during their preparation. Weaver (1996, p. 115) has indicated that key problems included 'not indicating confidence levels of data used in predicting impacts' and 'not specifying [the] significance of residual impacts'.

The National Environmental Management Act 1998 (NEMA) does not specify the content of the EIR beyond demanding that there be 'investigation, assessment and communication of the potential impact of activities' (section 24(7)). The detailed requirements will be set out in the NEMA EIA regulations (expected in 2002). That these requirements will be demanding may be gauged from the specification, in the Act (section 24(7)(e)), that 'reporting on gaps in knowledge, the adequacy of predictive methods and underlying assumptions, and uncertainties encountered in compiling the required information' must be undertaken. The intention is clearly that EIA under NEMA should build on best previous IEM and EIA practice.

Summary

The performance of the seven systems against the EIA report content criterion is shown in Table 11.1. While, in practice, their performance varies substantially within as well as between jurisdictions, the United States, the Netherlands, Canada and Australia meet the criterion. UK and South African

Table 11.1   The treatment of EIA report preparation in the EIA systems

Criterion 6: Must EIA reports meet prescribed content requirements and do checks to prevent the release of inadequate EIA reports exist?

Jurisdiction	Criterion met?	Comment
United States	Yes	Draft EISs are subject to formal checks on required contents prior to publication by Environmental Protection Agency and others
UK	Content: Yes Checks: No	Regulations prescribe content but no formal requirement for proponent to consult or for checks on environmental statement before release
The Netherlands	Yes	EIS checked against guidelines and EIA Act by competent authority before release for public consultation
Canada	Partially	Content prescribed in Act. Few checks on screening reports but checks made by federal authorities to limit inadequacy of comprehensive study reports and EISs
Commonwealth of Australia	Yes	EIA reports checked against project-specific guidelines and vetted by Environment Australia before release
New Zealand	No	Act provides strong guidance as to content but no formal checks on adequacy of EIA reports before release exist
South Africa	Partially	Acceptance of plan of study ensures that content requirements are met but no formal checks on adequacy exist, despite informal use of draft environmental impact reports

EIA reports must meet the content requirements specified in the respective regulations but no checks are made to prevent the release of inadequate EIA reports. Similarly, in Canada, there is no formal check on the content of the numerous screening reports, despite the existence of provisions relating to EISs. The EIA system in New Zealand contains no formal provision as to the content of EIA reports, nor are checks made on their content prior to release.

It is probably no coincidence that neither the UK nor the New Zealand EIA system requires compulsory scoping. Scoping guidelines provide a useful set of criteria against which to judge the content of an EIA report and they are frequently utilised in the other other five jurisdictions for this purpose. However, the checks probably have more to do with the cooperation which scoping engenders between the proponent and environmental and decision-making authorities. Such checks sometimes take place informally in both the UK and New Zealand but the major impediment to the release of inadequate EIA reports in both these countries is the lack of experience of many local authority personnel.

More formal checks on EIA quality in the UK and New Zealand may come either from the adoption of scoping or from the acquisition of experience. In Canada, the number of screening reports, many of which are very brief, militates against pre-release checks, but these could easily be made a requirement. The same is true in South Africa.

Notes

1. See, for example: Shopley and Fuggle (1984); Bisset (1988); Ortolano (1997); Canter, (1996); Glasson *et al.* (1999) and Petts (1999). A distinction is often drawn between 'methods' which, as Bisset (1988, p. 47) averred, 'have been described alternatively as methodologies, technologies, approaches, manuals, guidelines and even procedures', and 'techniques'. Methods are frequently comprehensive or aggregative, and attempt to address several tasks within the EIA process. Techniques, on the other hand, usually focus on a single task, such as the forecasting of noise levels or the organising of a public discussion about impacts. In practice, however, the terms are often used interchangeably. Most EIA methods and techniques have evolved from work in other fields: land-use planning, cost–benefit analysis; multiple-objective decision making; checklists, matrix and network analysis; and modelling and simulation (Hyman and Stiftel, 1988, chapter 1). Their use normally requires the involvement of a range of professionals working as an interdisciplinary team. In general, methods or techniques need to be selected to deal with the tasks involved at each step in the EIA process. It is the choice of an appropriate set of techniques which is the hallmark of good EIA practice, rather than the use of a comprehensive 'method'. Each technique chosen should be appropriate to the EIA task considered, replicable by other users, consistent in use on different alternatives and cost-effective in its use of data, financial, time and personnel reserves.
2. Forecasting and prediction both involve an estimate of the impacts likely to arise from the action. Strictly speaking, a prediction is a forecast with a very high probability (usually certainty). Most estimates in EIA involve an element of uncertainty and should therefore be termed forecasts. In practice, the terms are used interchangeably.
3. EPA Office of Federal Activities web site at: es.epa.gov/oeca/ofa/length.html
4. These requirements also apply, to a lesser extent, to the preparation of environmental assessments.
5. It is of interest to note that the term EIS is never used in the Act. However, both the RA's guide (CEAA, 1994) and the guide to panel reviews (CEAA, 1997c) specify detailed requirements relating to EISs.

Chapter 12

EIA report review

If there is only one point in an EIA process where formal consultation and participation take place, it is during the review of the EIA report. Indeed, in some jurisdictions, public review is virtually synonymous with public participation. This is not to say that all jurisdictions provide for public participation once the EIA report has been prepared (that in Hong Kong, for example, does not always do so: Wood and Coppell, 1999). However, nearly all do. The public review of EIA reports provides an invaluable check on their quality, especially where such checks have not been applied earlier in the EIA process. This chapter advances a set of evaluation criteria for the treatment of the review of EIA reports in EIA systems. These criteria are then employed in the analysis of EIA report review procedures in the United States, UK, the Netherlands, Canada, Commonwealth of Australia, New Zealand and South Africa.

Review of EIA reports

Sadler (1996, p. 122) suggested that the purpose of the review of an EIA report 'is to verify the document is an adequate assessment and is sufficient for the purpose of decision making'. Such reviews, which involved setting the boundaries for the review process, carefully selecting the reviewers, using input from public involvement and identifying review criteria, were seen as a form of quality control. Sadler (1996, p. 124) noted that a review process could involve the following stages:

1. Identify deficiencies in the EIA report.
2. Identify the shortcomings that are crucial to restricting informed decision making.
3. Recommend to the responsible authority how and when shortcomings in the EIA report should be remedied to assist decision making.

The approach to the formal review of EIA reports is handled differently in different EIA systems. In the US model, for example, the draft environmental impact statement (EIS) is used as the basis for consultation and participation and is duly succeeded by a final EIS. The power to require a supplementary EIS also exists. The US Environmental Protection Agency reviews all EISs and publishes its opinions about both the adequacy of the EIS and the environmental impact of the proposed action using a set of general criteria (see below). This

'EIA report – review – further EIA report' pattern has been emulated in other EIA systems (for example, in the Commonwealth of Australia) where the comments by consultees and the public are published as part of the review process.

In yet other systems, no formal provision exists for the proponent to respond to public comments in this way (for example, in the UK). Clearly, while the treatment of EIA review varies in different jurisdictions, the fundamental requirement of this stage in the EIA process is that those bodies with responsibilities and expertise, and the public, should be able to comment upon the EIA report and the action it describes. This stage exists in almost every EIA process. A mechanism is also needed for the consideration of such comments, by the proponent and by the decision-making/environmental authorities, before any decision on the action is made.

One of the most difficult areas in the review of EIA reports, as in the preparation of EIA reports, is ensuring objectivity since the organisation charged with responsibility for formal review (if any) may have a vested interest in the decision about the proposal. There are various methods of ensuring objectivity, including the use of review criteria, the accreditation of EIA report review consultants, the setting up of an independent review body, the publication of the results of the review and the involvement of consultees and the public. It is obviously best to utilise the services of skilled professionals in the review process, whether within the decision-making/environmental authorities, within retained consultancies, or within consultee organisations, including public interest groups. However, staffing resources may not permit this and, since members of the public cannot rely on such expertise to aid their participation, professionalism alone cannot ensure objectivity.

The existence of review criteria can provide a useful focus for the review of EIA reports. Action-specific scoping guidelines, where they are prepared, provide a valuable checklist for review. Another checklist is normally provided by the set of statutory requirements for EIA reports contained in legislation or regulations (or sometimes in formal guidance). Very few jurisdictions, however, have published formal criteria to assist in the review of EIA reports. Guidelines about reviewing have, however, been issued in New Zealand and in the UK which contain some review criteria (see below).

Several sets of criteria intended for use in the review of EIA reports have been published independently (see, for example, Beanlands and Duinker, 1983; Elkin and Smith, 1988). Lee and Colley's 1992 review procedure (in Lee *et al.*, 1999) requires at least two reviewers to utilise criteria relating to the extent to which specific tasks are completed and reported and builds up through a series of stages to an agreed overall evaluation of the EIA report as broadly adequate or inadequate. Such criteria can be used to meet various objectives. Where the treatment of a particular EIA task, or group of tasks, is clearly adjudged inadequate, further information can be requested. The Department of the Environment (DoE, 1996) produced related criteria and the European Commission (2001b) has published review criteria applicable to any member state, and more generally (see Chapter 3). Box 12.1 presents a set of review criteria developed to meet New Zealand EIA requirements.

Box 12.1   Review sheet for evaluating New Zealand assessment of environmental effects (AEE) reports

1.0	*Description of the proposed activity and its setting*	☐

1.1   How clearly, preferably in non-technical language, is the proposal described?

1.2   How clearly, preferably in non-technical language, is the environmental setting described?

To what extent . . .

1.3   . . . are the likely direct links between the proposal and the environment clearly identified in the description? (e.g. discharges to the environment; use of local resources, labour, etc.)

1.4   . . . is a distinction made (if appropriate) between the construction, operation, and/or decommissioning phases of the proposed activity?

1.5   . . . does the AEE refer to any environmental assessment provisions of the district and/or regional plans? (this would include noting that there are no particular provisions, if that was the case)

2.0	**The approach to, and coverage of, the assessment of environmental effects . . .**	☐

2.1   Should alternative locations, or methods, for the proposed activity be   Y/N
considered in the AEE (i.e. significant effects are likely)? [If yes, go to
2.1.1 If no, skip to 2.2]

2.1.1   To what extent have alternatives been considered in the AEE?

To what extent . . .

2.2   . . . is there evidence of the early and meaningful involvement of affected people, groups and communities in the assessment of environmental effects of the proposal?

2.3   Is there evidence that a rational approach to scoping, and especially to impact identification, has been used in the EIA?

2.4   . . . is the coverage of the AEE appropriate for the type and scale of proposal (i.e. there are no obvious, unexplained gaps in coverage)? (see Notes)

2.5   . . . is there an appropriate balance between the biophysical (e.g. effects on water or air quality) and the social and cultural impacts of the proposal (e.g. effects on the neighbourhood or the wider community, and/or health and safety issues)?

(continued)

(continued)

2.6   ... has baseline data been collected? [If not, skip to 2.7]

2.6.1   ... has baseline data collection been directed by the scoping process? (as opposed to a wide, unfocussed collection strategy)

2.7   ... is there evidence of careful selection of indicator variables, both for impact prediction and for monitoring, should the latter be necessary?

2.8   If risk assessment is appropriate for the proposal, to what extent has it been provided in the AEE?

2.9   If hazard assessment is appropriate for the proposal, to what extent has it been provided in the AEE?

3.0   Prediction, mitigation, and monitoring of effects.

To what extent ...

3.1   ... are clear and sound predictions made about possible impacts? [If no predictions are made, skip to 3.10]

3.2   ... (assuming predictions are made), is the basis of the prediction clearly stated (including methods, supporting data etc., as appropriate)?

3.3   ... do the predictions generally provide sufficient information about the nature, severity, likelihood and spatial extent of the impacts such that the implications of the impact can be understood?

3.4   ... is there an appropriate balance between adverse and beneficial impacts?

3.5   ... do the predictions take account of indirect impacts?

3.6   ... do the predictions take account of cumulative impacts?

3.7   ... do the predictions take account of long term impacts?

3.8   ... does the AEE consider possible mitigation measures for the likely impacts?

3.9   ... does the AEE seek to link, and integrate, impacts on different parts of the environment, to provide an overall picture of the impact of the proposal?

3.10   Is monitoring appropriate for the proposal? [If yes, go to next question If no, skip to 4.1]

3.10.1   To what extent has monitoring been dealt with, to a level appropriate for the proposal?

(continued)

(continued)

4.0	**Significance evaluation**	☐

4.1 To what extent is there evidence in the AEE of a systematic approach to evaluating the significance of the identified impacts/effects of the proposed activity?

4.2 How well, overall, are the attitudes of the affected individuals, groups and communities towards the identified impacts recognised in the AEE?

4.3 To what extent does the AEE avoid undue reliance on the valued judgements of the impact assessors?

4.4 Have technical methods been used in the AEE to evaluate the social significance of the identified impacts? [If yes, go to 4.4.1 if no, skip to 5.1]   Y/N

4.4.1 If technical methods have been used to evaluate the social significance of predicted impacts, to what extent have the methods been clearly explained?

4.4.2 If technical methods have been used to evaluate the social significance of predicted impacts, to what extent have the people affected by the proposal been involved in the evaluation process?

5.0	**Communication of impact information**	☐

To what extent ...

5.1 ... is the EIA clearly and simply organised, providing a coherent and useful study?

5.2 ... is the impact information summarised in a form that non-technical people can understand?

5.3 ... is the overall impact of the proposed activity on the environment (including reference to both beneficial and adverse impacts) clearly set out, in an understandable form?

5.4 ... is the discussion of the predicted impacts free from obvious bias (e.g. emphasis on benefits, downplaying negative aspects)?

5.5 ... is the assessment free of superfluous material (that can hide important information)?

5.6 ... have photographs and/or other graphics been used to aid the understanding of information in the assessment?

Overall AEE grade: ☐

(continued)

(continued)
Overall comments
1. Is further information about the proposal and/or its environmental effects required?
   Look back over the grades and, using your impression of the assessment, consider: have the essential issues been addressed, and in a way, and to the extent, that the information is **relevant** *(to the public, as well as council staff, and other interested parties) and can be* **relied on***? What level of uncertainty still exists about important aspects of the activity, and/or potential effects on the environment?*
   Decision: Y/N
   If yes, indicate the information needed:

2. Will the consent application require public notification on the basis of the AEE?
   Is it clear who might be affected by the proposed activity, and have their values and concerns been recognised in the assessment (or through other channels)? How significant (to local people, to interest groups, to iwi, to other potentially affected parties) are the likely effects?
   Decision: Y/N
   If yes, indicate why:

Source: Morgan, 2000a.

It is by no means uncommon for consultants to be retained to undertake, or to assist in, the review of EIA reports. The competence of such reviewers is an important issue, especially as they tend to be engaged to review the EIA reports for the more complex and significant proposals. Generally, review consultants are drawn from the same consultancies that prepare EIA reports, and the desirability of some form of accreditation (Chapter 11) applies equally. Glasson *et al.* (1999) suggested that, because consultancies are in competition with each other, they might not be entirely objective in their approach to EIA report review.

The appointment of an independent panel selected from acknowledged experts in the field to review EIA reports has two advantages. First, it should provide a means of reducing any bias in the relevant authority's decision on the action. Second, it should ensure that the quality of EIA reports improves over time, since its opinions, whether adverse or positive, should be both public and influential. This is certainly the case with the review panels in Canada and the Netherlands, for example.

In order that a review of the EIA report may be seen to have taken place, the outcome should be made public. This is most clearly seen to be the case where a formal review is published (as in the Netherlands or the Commonwealth of Australia, for example). In addition, the comments arising from reviews of the EIA report by the consultees and by the public (correspondence, notes of telephone conversations, minutes of meetings, etc.) should be placed in the public domain (either by publication – with or without editing – or by allowing access to the decision-making authority's files).

One of the more significant checks on the preparation of inadequate EIA reports by proponents is the right to demand, and the duty to provide, further information (including any necessary corrections) following submission. Ideally, such information should be requested after the decision-making or environmental authority has reviewed the EIA report and after there has been a response from consultees and the public, to ensure that a full range of expertise and opinion is brought to bear on the adequacy of the information in the EIA report. An account of how the comments received at the review stage were (or were not) incorporated in the final EIA report should be provided.

The approaches to ensuring the provision of further information vary from a right to request further information but no power to enforce its provision or to 'stop the clock' while it is furnished (as in the UK) to the formal preparation of a final EIA report in response to comments on the draft report (for example, in the United States and in Australia). Buckley (1998a) noted that stringent review can increase not only the quality of the EIA report but also the quality of the EIA process as a whole. Clearly, if the review process is effective, it should lead to improved EIA report quality over time. There is some evidence that this is so (for example, in many European countries – see Chapter 3) but the quality of EIA reports still leaves much to be desired (see below).

As at the other stages of the EIA process, the existence of published advice on the procedures employed in either formal or informal EIA review and on the methods which may be used in reviewing EIA reports is invaluable not only to those engaged in EIA review (the decision-making and environmental authorities, consultees and the public) but also to those involved in EIA report preparation. Such guidance might include a checklist or a set of review criteria (see above).

As mentioned above, if there is one step in the EIA process where public participation takes place, it is following publication of the EIA report. It is preferable that this participation occurs prior to requesting further information from the proponent. If public participation has already taken place during scoping, the public will be better able to comment constructively on the quality of the EIA report. Suitable provision for participation, including adequate availability of copies of the EIA report over a realistic time period, etc., is necessary (see Chapter 16). Similarly, where further information is submitted by the proponent, this too should be open to public inspection and comment.

If a formal review of the quality of the EIA report is carried out and made public, especially if this is in the form of a recommendation to the decision-maker, some form of appeal against the findings of the review should exist, even if this is of an informal nature. In many jurisdictions, where no formal review is published prior to the making of the decision, such a demand can be combined with an appeal against the decision itself.

Although adequate time must be allowed for EIA report review to take place, it is nevertheless important that the report review stage of the EIA process is carried out effectively and efficiently. In other words, the EIA report should be thoroughly reviewed by an appropriate range of participants and further information should be provided without undue demands upon resources or time. Balances are needed to ensure that requests for further information are coordinated (and preferably all made at once), reasonable and not deliberately used as a delaying tactic. These can be introduced by the use of a 'single request' provision, by the use of time limits and by administrative or legal appeal. The various criteria discussed above are summarised in Box 12.2. These evaluation criteria are now used to assist in the analysis of EIA report review procedures in each of the seven EIA systems.

United States

There are comprehensive arrangements under the National Environmental Policy Act for the review of draft environmental impact statements. The (NEPA) lead agency must circulate the draft EIS for review to prescribed agencies. Not only must agencies with any jurisdiction over, or special expertise with regard to, the proposal respond but other federal, state, tribal and local agencies and the public must also be invited to comment. Generally, 45 days are allowed for comment on a draft EIS.

It is customary for consultees and the public to check the draft EIS against the outcome of scoping (which is often summarised in the EIS) (Kreske, 1996). In addition, the Environmental Protection Agency (EPA) employs its own formal review summaries (see below). EPA also conducts detailed reviews of final EISs, especially where significant issues are raised at the draft EIS stage.

Box 12.2   Evaluation criteria for the review of EIA reports

Must EIA reports be publicly reviewed and the proponent respond to the points raised?

- Must a review of the EIA report take place?
- Do checks on the objectivity of the EIA report review exist?
- Do review criteria to determine EIA report adequacy exist?
- Does an independent review body with appropriate expertise exist?
- Must the findings of the EIA report review be published?
- Can the proponent be asked to respond to comments and to provide more information following review?
- Must a draft and final EIA report be prepared?
- Does published guidance on EIA review procedures and methods exist?
- Is consultation and participation required in EIA report review?
- Is consultation and participation required where further information is submitted?
- Is there some form of appeal against review decisions?
- Does EIA report review function effectively and efficiently?

For projects that it rates as environmentally unsatisfactory, EPA may refer the issue to the Council on Environmental Quality (CEQ – see Chapter 13). In addition, EPA will informally review environmental assessments that it receives if so requested by a lead agency. It also informally reviews 'draft' draft and final EISs if requested.

As mentioned in Chapter 11, the comments of EPA and other agencies reviewing the draft EISs must be made available to the public in the final EIS together with responses to them. There is a 30-day waiting period between the publication by EPA of a notice in the Federal Register that the final EIS has been filed and the issuing of the record of decision (see Figure 2.2). The final EIS must be circulated to all federal agencies with jurisdiction or expertise, to the project applicant (where there is one), to persons requesting to be notified and to persons who submitted comments on the draft EIS.

There is no formal general guidance on EIS review but EPA (1984) has issued policy guidance concerning the quality of the draft EIS and the acceptability of the proposed action (summarised in Box 12.3). It has also issued review guidance on the treatment of cumulative impacts, ecological impacts and environmental justice in EISs and environmental assessments (EPA, 1999a,b,c).

EPA records a significant improvement in EIS quality between its published review of the draft and its unpublished review of the final document. Thus, during the period 1987–99, EPA recorded ratings of 3, of environmental objections, or of environmentally unsatisfactory to about 17 per cent of draft EISs (the action rating is based upon the 'worst' alternative). During the same period, the percentage of final EISs so graded fell to 5 per cent (based upon the 'chosen' alternative). The percentage of EISs for which environmental concerns was recorded also fell substantially between the draft and final documents.[1] Notwithstanding these improvements in the environmental acceptability of the actions described in EISs, it appears that the quality of the EISs themselves has not improved over the years. Tzoumis and Finegold (2000, p. 576) found that there 'appears to be very little learning from previous years', and recommended that there should be more coordination and sharing of knowledge about NEPA within agencies.

Generally, peer review of EISs is thorough although there have been some criticisms of lack of consistency in review findings from the different EPA regional offices and of the time taken for EIS review. Overall, however, the process seems to work reasonably efficiently and there is no doubt that it is effective in ensuring that the concerns of the various relevant agencies and, perhaps to a lesser extent, the public are met.

UK

There is a requirement for the environmental statement (ES) to be made available for consultative and public review in the UK, but not for the preparation of any formal review report by the local planning authority (LPA) as a separ-

Box 12.3   US Environmental Protection Agency review criteria

Criteria for draft EIS adequacy

Category 1: Adequate
• The EIS's treatment of impacts and alternatives is adequate.

Category 2: Insufficient information
• The draft EIS does not contain sufficient information to fully assess all reasonable alternatives.

Category 3: Inadequate
• The deficiencies in impact analysis or alternatives are of such magnitude that a revised draft EIS should be recirculated. The action is a potential candidate for CEQ referral.

Criteria for environmental effects of proposed federal actions

LO: Lack of objection
• EPA review has not identified any potential impacts requiring changes to the proposal.

EC: Environmental concerns
• EPA review has identified environmental impacts that should be avoided to fully protect the environment.

EO: Environmental objections
• EPA review has identified significant impacts that must be avoided to fully protect the environment.

EU: Environmentally unsatisfactory
• EPA review has identified adverse impacts of sufficient magnitude that the action should not proceed as proposed. The proposal will be referred to CEQ unless the unsatisfactory impacts are mitigated.

Source: adapted from EPA, 1984.

ate stage in the EIA process. The LPA review of the ES is normally in two stages. The first stage is an early evaluation of the ES to see whether more information should be requested:

> Local planning authorities should satisfy themselves in every case that submitted statements contain the information specified in Part II of Schedule 4 to the Regulations and the relevant information set out in Part I. (Department of the Environment, Transport and the Regions – DETR 1999b, para. 109)

The second stage is a fuller review once the results of consultation and participation have been received. This latter stage is normally completed immediately prior to decision making. The responses from consultees and public participants, together with the LPA's own review, are usually open to public inspection, thus ensuring that some element of objectivity is seen to apply.

Government guidance on the review of ESs has been published, providing a framework for reviewing the content of an ES and advice on evaluating the treatment of individual environmental effects (DoE, 1994a). This document is intended to help LPAs to determine whether the ES is adequate and to evaluate the information outlined in it to reach a decision. Review criteria based on Appendix 5 of the procedural guide (DETR, 2000a: originally DoE, 1989), and tables to aid evaluation, are provided.

Notwithstanding the availability of this guidance, LPAs review ESs without the benefit of formal review criteria beyond Schedule 4 to the Planning Regulations or of any specialised review body. Where they undertook their own reviews, the LPAs studied by Jones et al. (1998) relied on the Regulations in a quarter of cases, on consultations in two-fifths of cases and on combinations of consultation and the use of guidance documentation in a quarter of cases. Kreuser and Hammersley (1999, p. 369) confirmed that, because planners felt unable to evaluate ESs critically, they 'place great reliance on the consultees to review, verify and summarise at least parts of ESs'.

The LPA may commission consultants to review the ES but there is no provision to charge the costs to the developer. There is, as mentioned in Chapter 11, some voluntary accreditation of EIA consultants. The Institute of Environmental Management and Assessment is not only involved in this process but also provides an ES review service to LPAs utilising a modified version of the review criteria advanced by Lee et al. (1999). A study of 40 ESs showed that nearly one-quarter had been reviewed by external consultancies (Jones et al., 1998). Glasson et al. (1999, p. 233) suggested that about 10–20 per cent of ESs are reviewed externally.

As a result of its own review and the responses from statutory consultees and the public, LPAs may request further information (for example, on how certain objections are to be overcome). However, DETR has indicated that this power should be used sparingly:

> Authorities should only use their powers … when they consider that further information is necessary to complete the ES and thus enable them to give proper consideration to the likely environmental effects of the proposed development. … Authorities should not … simply … obtain clarification or non-substantial information. (DETR, 1999b, para. 111)

A planning application accompanied by an ES must be advertised by the LPA and copies of the ES must be made available to the public for inspection and to a set of statutory consultees. Additional copies must be made available for purchase at a price reflecting the cost of printing and distribution (Regulation 18). The Planning Regulations impose a time limit upon LPAs to process the review phase of the EIA process, and there is no mechanism for 'stopping the clock' while further information (including any corrections) is provided. Such information must be made available to the consultees and the public for further comment (see, for example, Read, 1997).

Should proponents choose to do so, they can decline to provide the information requested and subsequently appeal to the Secretary of State for non-determination of the planning application. However, this is frequently a lengthy process. There is no provision for the proponent to prepare a formal response to comments, for example in the form of a final EIA report. In practice, LPAs appear to request additional information in about two-thirds of EIA cases (Glasson *et al.*, 1999, p. 233). For example, further information or evidence was requested in relation to over two-thirds of the 40 ESs referred to above. This further information was generally provided by the developer (Jones *et al.*, 1998).

In the early years of formal EIA the quality of the majority of ESs was unsatisfactory. DoE (1991a) found that about 60 per cent of a sample of 24 ESs were unsatisfactory, but the LPAs concerned took a more sanguine view as a result of their own evaluations, with only about 25 per cent being adjudged to be unsatisfactory.

The situation has improved subsequently. Jones *et al.* (1998) found that over two-thirds of a sample of ESs submitted during 1990–92 were satisfactory but that, anomalously, this declined to rather more than a half in 1992–94. Lee and Brown (1992) found similar improvements over time, although ES quality varied with project type, the length of the ES, the size of the project, the experience of the developer, and whether a consultant was used. These results were confirmed by DoE (1996) and Glasson *et al.* (1997), who reported that 60 per cent of post-1991 ESs were satisfactory, compared with 36 per cent of pre-1991 ESs (see also Glasson *et al.*, 1999). It is notable that the quality of ESs for major roads improved markedly following the publication of project-specific guidance by the Department of Transport in 1993 (Glasson *et al.*, 1999). DoE (1996) reported that statutory consultees felt that the quality of ESs was improving, though still wanting. However, LPAs remain substantially less critical of ES quality than academic commentators (Jones *et al.*, 1998).

Although these findings suggest progress, there remains considerable scope for improvements in ES quality in the UK. It is apparent, however, that the review process is public and that, in practice, proponents generally respond to the points raised.

The Netherlands

Once the competent authority has accepted the EIS, the public must be notified and the EIS is made public together with the draft decision on the proposal. The public review period of at least four weeks for the EIS coincides with the public review period for the application to which the EIS relates. There is provision (Environmental Management Act 1994, section 7.24) for a public hearing if this is requested, as it usually is. The various statutory consultees are also asked to comment on the EIS. The results of the consultation process are then passed to the EIA Commission (EIAC) which receives the EIS at the time of its publication by the competent authority.

EIAC checks the EIS against the legislation and the regulations and against the scoping guidelines to see whether it is essentially complete during the five weeks following the period of public participation. It makes a judgement about whether the EIS is adequate for decision-making purposes (by asking, for example, if the predictions are likely to be accurate, if all the reasonable alternatives have been analysed and if the EIS is sufficiently objective). The published scoping guidelines, the requirements of the Act and a brief set of operational review criteria are used as the basis for its check both of completeness and of adequacy (Sielcken *et al.*, 1996; de Jong, 1997; Scholten, 1997). EIAC must first identify and list omissions and mistakes in the EIS and then evaluate, using four operational criteria (Box 12.4), which of these is so crucial that it may affect the decision. It must take the opinions of the statutory consultees and the public into account in framing its recommendations. It endeavours to utilise the latest detailed information about, for example, land contamination in its comments and to organise these according to their significance to the competent authority's decision. If EIAC finds that substantial information is lacking, it must recommend to the competent authority how this deficiency can be overcome in as short a period of time as possible (Scholten and Bonte, 1994; Scholten and van Eck, 1994; EIAC, 1998).

Box 12.4   Dutch EIAC operational review criteria

1. **The description of the proposed activity (the preferred alternative) and its environmental impacts is adequate.**

Example: Integral EIS Amsterdam Airport area

In the EIS the impacts of night flights (sleep disturbance) were calculated according to the assumptions made in the standard applied, i.e. with closed bedroom windows and a night-time period of just seven hours (between 23.00 pm and 06.00 am). The Commission was of the opinion that an assessment based on a partially opened bedroom window and a night-time period longer than seven hours would give a more realistic picture of the actual impacts, and recommended that this be provided in a supplementary report. This was carried out, and resulted in the night-time flight regime being extended from 06.00 am to 07.00 am.

2. **The description of the alternative most favourable to the environment (AMFE) and its environmental impacts is adequate.**

Example: Combined heat and power station UNA in Diemen

The possibilities for discharging cooling water from this power station are severely constrained. The temperature of the Amsterdam–Rhine Canal is already adversely affected by current high discharges of cooling water. The alternative body of surface water, the lake IJmeer, which could be used for cooling is ecologically valuable. The EIS contained no alternative making use

(continued)

(continued)
of both bodies of water; nor was there an environmental alternative to discharging chlorine, used to remove algae, into the lake. The Commission recommended that a supplement be prepared which describes these options as full alternatives. This supplement formed the basis for the discharge licence which prescribes both of these measures.

3. **The description of the other alternatives and their environmental impacts is adequate.**
Example: A58 Motorway, Schoondijke–Sluis
The guidelines explicitly required that a new alternative route be developed. This was not taken up in the EIS, because it was claimed to be unrealistic. The Commission judged the argumentation to be unsound. The supplement prepared by the Department for Public Works and Water Management revealed that the alternative was realistic, but did not perform significantly better than the other alternatives in terms of environmental impact.

4. **The comparison of alternatives is adequate.**
Example: Choice of a sanitation sludge disposal system for the northern provinces
Multiple criteria analysis (MCA) was used in the EIS to carry out a comparison between different treatment techniques, but the criteria used to evaluate the environmental impacts were not stated, making verification of the comparisons impossible. The criteria were later provided and explained in a supplementary report. According to these criteria, the differences revealed by the MCA were so small relative to the margins of uncertainty that no firm conclusions could be drawn from the resulting ranking of alternatives.

Source: adapted from Scholten, 1997, p. 25.

EIAC's review findings are published. These findings (which are supposed to be, but are not always, confined to the contents of the EIS rather than to the advisability or otherwise of the proposal), are usually accepted by the competent authority. If the findings are negative the competent authority will almost always require supplementary information (see Figure 5.1). This further information is treated in the same way as the original EIS and is reviewed by EIAC. It may, depending on its importance, be subject to review by the public (most competent authorities arrange for this to occur). There is now informal Ministry (VROM) guidance on the treatment of supplementary information.

EIAC undertook 49 reviews of EISs during 2000 and 661 by the end of 2000 (EIAC, 2001a). Notwithstanding the preparation of scoping guidelines for each proposed activity, and improvements in EIS quality, the Commission has found substantial deficiencies in the content of EISs necessitating the supply of supplementary information in about one-third of cases each year (Scholten and Bonte, 1994, p. 5; Mostert, 1995). In addition, supplementary information, mostly of a relatively minor nature, is requested in other cases.

The principal areas of weakness in practice relate to treatment of alternatives, miscalculation of extent of impacts, overlooking of impacts, use of outdated models or inadequate baseline information (Scholten and van Eck, 1994; VROM, 1994a) and to the treatment of issues raised by the European Habitats and Birds Directives.

In general, the EIA review process is working reasonably well. The quality of EISs is improving as a result of the experience of developers, the competence of consultancies, the high level of expertise within some competent authorities (for example, the Ministry of Transport and the Province of South Holland), the increasing sophistication of some non-governmental organisations and the use of specialist panel secretaries by EIAC. While there is no right of appeal against EIAC's review findings, they are widely regarded as authoritative but they are sometimes challenged by proponents, by the competent authority or by environmental groups. The Evaluation Committee on the Environment Acts (ECW, 1990, 1996) saw the role of EIAC's published review of the EIS as critical to maintaining and improving the quality of EISs.

Canada

The situation relating to the review of environmental assessment (EA) reports in Canada varies with the type of report. In the self-directed assessment procedure there are differences between the screening and comprehensive study tracks. Public participation is discretionary in screening but, should the responsible authority (RA) determine that the public should be given the opportunity to comment on the screening report, it must provide an opportunity to do so. In practice, this happens in a minority of cases, where there is public demand or opposition. In some instances, however, extensive public participation in the review of screening reports has taken place. The RA has the discretion to seek comments by the relevant expert federal departments, and by other levels of government, on the screening report, and these are regularly sought.

The Canadian Environmental Assessment Act 1992 provides for central control in the review of comprehensive study reports. Once the RA has reviewed and finalised a comprehensive study report, the report must be submitted to the Canadian Environmental Assessment Agency (CEAA) for review and public comment. The Agency is supposed to explain when and where the report can be obtained and give a deadline for public comment. It should then file both the comments received and the summary of these that it is obliged to prepare in the 'public registry' (CEAA, 1997b). While the Agency's review of the report is mainly procedural, it has to then consult various federal departments to ensure that the content of the report is scientifically and technically accurate. The Agency thus asks, for example:

- Was the comprehensive study undertaken and conducted in accordance with the procedural requirements and intent of the Act?
- Were all the relevant expert federal authorities consulted and all concerns adequately resolved?

- Were there appropriate and sufficient opportunities for public involvement in the comprehensive study and are there any outstanding public concerns? (CEAA, 1994, p. 97)

The involvement of the Agency is intended to ensure that the quality of comprehensive study reports is adequate. It often results in the proponent being asked to produce further information.

Public involvement in EIS review has always been an important element in the independent panel review procedure. Over time, panels have organised the preparation of draft EISs and review of the actual EIS by the public as well as by expert federal departments and, where necessary, have demanded that the proponent provide further information where public or government agency concerns justified this. The panel guidelines, produced following scoping, are used as review criteria. This step (which is not shown in Figure 5.3) precedes the public hearings which panels hold into proposals. Since the EIS is the central document at these hearings, there has been adequate, if sometimes lengthy, public comment on the EIS. There is, necessarily, little EA report review in the mediation procedure, since no further documents are prepared and the negotiations (which should normally involve representatives of the public) take place in private.

There was considerable early Canadian interest in, and original research on, the quality of EISs and other EA reports (which are sometimes voluminous, running to several hundred pages) (see Beanlands and Duinker, 1983; Ross, 1987; Elkin and Smith, 1988). The quality of EA reports depends heavily on the professionalism of the RA and has varied considerably in the past, though it has improved over time as greater experience has been gained. While the quality of comprehensive study reports and EISs bears comparison with the best international practice, the quality of screening reports has left much to be desired. Thus, the Commissioner of the Environment and Sustainable Development (1998, pp. 6–20) reported that:

[a] majority of the 197 files we examined had screening reports that did not meet the minimum criteria set out in the Agency's [RA] guide. In particular, the description of the project or of the environment was often incomplete.

David Redmond and Associates (1999) confirmed that the quality of many screening reports was inadequate. The Minister of the Environment (2001) recognised that quality assurance by the Agency was needed and proposed powers to request information from RAs to assist in this (Government of Canada, 2001).

Commonwealth of Australia

The requirements relating to public and agency review of preliminary documentation, public environment reports (PERs) and environmental impact statements are specified in considerable detail in the Environment Protection and Biodiversity Conservation Act 1999) (EPBC Act). As mentioned in Chapter 11, the requirement relating to the approval of the draft EIS or PER prior to publication usually provides an invaluable check preventing the premature release of an inadequate document. The project-specific guidelines provide consultees and the public with a useful checklist for reviewing the draft EIS or PER but no other review criteria are employed and there is no published guidance on EIA report review. However, it is clear that, under the previous legislation, the guidelines (and the adoption of scoping) helped to improve the quality of draft EISs over time, though weaknesses remained (see below). There was previously provision for round-table discussions to be held between Environment Australia, the proponent and the public (Commonwealth of Australia, 1995, para. 6.6) but this was never formally invoked (Kinhill Engineers, 1994). Exceptionally, an independent review of an EIS has been commissioned, as in the case of the Second Sydney Airport (see Chapter 8).

The proponent must publish the draft EIS (or PER) and the preliminary documentation to enable consultation and public participation to take place. For EISs and PERs, at least 20 business days must be allowed for public comment. Notice of the availability of each document must be published on the internet but there is no provision for the documents themselves to be placed on Environment Australia's web site. Beyond this, the EPBC Act is silent on the availability of copies of documents. Under the previous legislation the charge made for these EISs was usually about A$25, enough to deter casual requests but not interested parties. Further, the proponent had to make a copy of the final EIS available free of charge to any person who had commented in writing on the draft EIS (Commonwealth of Australia, 1995, para. 8.2(c)). (Over 15,000 such submissions were received for the Second Sydney Airport.)

As in the United States, the proponent is required to produce a final EIS, following consultation and participation. In reality, this final document is usually a listing of the various comments received on the draft EIS, together with the proponent's response to them. This practice of preparing a supplement to the draft EIS, rather than a 'true' final EIS, has been criticised and can lead to over-long and confusing documentation. It is, however, usual for further modifications to the project design to be made at this stage by the proponent, and this two-stage EIS mechanism provides a record of these.

There was no provision in the previous Commonwealth EIA regime for the preparation of both a draft and a final public environment report. In this sense, the PER was similar to the UK environmental statement for which no draft is required. In practice, the difference between the information contained in Commonwealth PERs and EISs proved to be small. Perhaps for this reason, the EPBC Act requires the preparation of a final PER and indeed of revised preliminary documentation. The Environment Minister may refuse to accept the finalised EIS, PER or preliminary documentation if it is believed to be inadequate for the purpose of making an informed decision as to whether or not to approve the taking of the controlled action (see, for example, section 104(3)).

Notwithstanding the various checks and balances on the preparation and review of EIA reports, including the role of Environment Australia as an independent (supposedly impartial) arbiter, criticisms have been levelled in the past by environmental groups and other commentators (see, for example, James, 1995; Hughes, 1999) at the frequently lengthy EISs (which have usually been accompanied by detailed appendices). It has been claimed that the documents have been difficult for the public to understand, that social and cumulative impacts have been neglected (see Chapter 7), that alternatives have not been treated adequately and that, in particular, the no-action alternative has frequently been neglected (Chapter 8). The Commonwealth Environment Protection Agency (CEPA), in its agenda for reform, tacitly acknowledged many of these weaknesses by proposing that EISs be fully referenced, that judgements be justified, that quantifiable impact predictions be tabulated and that EISs be released only when they 'meet the requirements of the guidelines developed through the public scoping process' (CEPA, 1994, p. 40). Now that provisions for checking that EISs (and PERs) meet the requirements of the guidelines (and for the referencing of information sources) are in place, it will ultimately be the decisions of Environment Australia that determine the quality of EIA reports.

New Zealand

The review stage in the New Zealand EIA process serves two purposes. The first is to review the information provided by the developer to determine its adequacy and to permit the decision on the action to be taken. The second is to act as a second screening stage, not unlike the Canadian EIA review process. If the activity is thought likely to have major environmental effects, further information is usually requested from the developer. In effect, this provides for a two-tier EIA system in which the initial EIA report may be used to determine whether the project is likely to have significant effects. If the developer can mitigate the effects of the project so that residuals are only minor, and to the satisfaction of those affected, the local authorities need not notify the proposal and the further EIA information requirements will not be triggered (Chapter 9).

The first step is to review the submitted assessment of environmental effects (AEE) report to determine whether more information is required (see Figure 5.5). If it is, the developer must provide it. The Resource Management Act 1991 imposes time limits upon councils to process applications once they have sufficient information, but the clock is stopped if further information is required.

Having reviewed the initial AEE report and any further information, the local authority then determines whether the development is likely to have significant effects or not (in consultation with the people affected by the proposal). If the effects are judged to be significant, the application will be notified publicly and certain consultees will be informed. In the approximately 3,000 cases per annum where this occurs (see Chapter 9), the local authority may request further information on alternatives and on the consultation undertaken by the applicant (section 92(2)(a)), if this has not already been provided by the applicant (see Chapter 11), and/or commission a report 'on any matters raised in relation to the application, including a review of any information provided in an application' (section 92(2)(c)).

AEE materials are made available to the public for inspection. All submissions by organisations and by members of the public must be copied to the developer and the council has the power to ask the applicant to explain how these are to be dealt with. The Act provides that, if submissions requesting a council hearing at which opinions and objections can be expressed are received, one should be held by the council before any decision is reached. Such hearings are held for more than 50 per cent of notified activities. (There is also provision for pre-hearing meetings to resolve conflicts.) The Parliamentary Commissioner for the Environment (PCE) has maintained an interest in EIA cases, but does not possess the resources for intervention to be other than exceptional. Hughes (1996) described the use of independent panels set up at the Commissioner's instigation for three AEEs.

All the costs incurred by the local authority (LA) in dealing with the application (including staff and hearing costs) can be recouped from the applicant (though cost-recovery practice is variable (Ministry for the Environment – MfE, 2000a)). The ability to recruit consultants to review AEE reports without time or financial costs to the LA is a powerful tool for ensuring that the review is adequate. In practice, however, this provision has met great resistance from developers and, while consultants are retained to process more than 10 per cent of applications, few of these are specifically requested to assist in the environmental assessment of the proposal. Very few independent reviews of entire AEE reports have been commissioned (Morgan, 1995). This power to retain consultancy advice is supported by a provision in the Act for further information to be required of the applicant at any time prior to the hearing (section 92(1)). Further information is requested by local authorities in less than 33 per cent of cases (MfE, 2001c).

Belatedly, a good practice guide on local authority review procedures has been published (MfE, 1999b). This describes the procedures to be followed in dealing with an assessment of environmental effects accompanying an application for resource consent. While this does not recommend the use of review criteria, it states that the local authority should, for each effect, summarise any assessment techniques used and ask:

(1) Has the effect been adequately assessed?
(2) Is the assessment accurate? (Is it based on sound predictions or just guesswork?)
(3) Has the consent of all affected parties been obtained?
(4) In the particular environment, is the effect likely to combine with any other identified effect (be cumulative)?
(5) Are there any mitigation measures proposed?
(6) What is the effect's significance judged to be?
(7) If they are not minor, are the remaining effects acceptable?
(8) Do you need any more information before the above questions can be answered?
(MfE, 1999b, p. 43)

Further guidance on review is included in a template for the processing of resource consents (MfE, 2001a).[2]

Morgan and Memon (1993, pp. 61–2) suggested that, rather than relying solely on the checklist provided by the Fourth Schedule to the Resource Management Act, local authorities should be asked a set of questions when evaluating the proponent's AEE report. Morgan (2000a) developed these further to produce a straightforward review sheet for local authority use (see Box 12.1).

In practice, it appears that local authorities are suffering from a distinct lack of expertise in reviewing AEE reports and that, unsurprisingly, practice is very variable (Morgan, 1995). It appears that regional councils have relied 'on the judgement of the particular staff member to whom the application is passed' (Morgan, 2000b, p. 101). Nevertheless, 64 per cent of local authorities claimed to have followed a set process to check AEE reports (MfE, 2001c, p. 28). There have been some problems in meeting deadlines in the more complex cases (i.e. those dealt with at regional council level), but over 80 per cent of applications are processed on time (MfE, 2001c). Some district councils appear to be depending on the professionalism of the applicant's consultants for quality assurance, while others may be relying on council hearings for evaluation. There appears to be growing use of pre-hearing meetings and combined hearings between regional and district councils into all the consents required. There is, however, insufficient feedback to applicants and consultants following council review of AEE reports (Morgan, 2001) to improve practice (Dixon, 1993; PCE, 1995; Morgan, 2000b). It is to be hoped that the recently produced review guidance (MfE, 1999b, 2001a) will lead to an improvement in practice.

South Africa

The South African EIA Regulations are silent about EIA report review beyond demanding that the relevant authority consider the application after it has 'received an environmental impact report [EIR] that complies with regulation 8' (Republic of South Africa, 1997, Regulation 9) (see Chapter 11). The Regulations contain no formal requirements for review generally, for checks on objectivity (beyond those provided by the review of the plan of study), for an independent review body, for the publication of the review findings, for the provision of further information, for the preparation of a draft EIR, for consultation and participation in review, or for appeal against review decisions. However, the Environment Conservation Act 1989 makes provision for a board of investigation to review any matter pertaining to the Act, a power which was used to scrutinise the Saldanha Steelworks EIA.

On the other hand, the EIA guidelines (Department of Environmental Affairs and Tourism – DEAT, 1998c) contain considerable detail about EIA report review. In this they reflect the integrated environmental management (IEM) guidelines (Department of Environmental Affairs – DEA, 1992) which devoted a document (volume 4) solely to review. Some of its content has been carried forward to the new EIA guidance.

The review section of the EIA guidelines is divided into three: purpose, steps involved and evaluation of impacts. The purpose of the review is, *inter alia*:

- to determine whether the EIR presents an adequate assessment of the environmental impacts of the proposed activity and is of sufficient relevance and quality for decision-making;
- to collect and collate stakeholder opinion about the acceptability of the proposal and the quality of the EIA process undertaken; and
- to ensure that the EIR and the EIA process comply with the plan of study. (DEAT, 1998c, p. 29)

The section on the steps involved in reviewing the EIR states that 'the EIR should be reviewed by the relevant authority, with the assistance of the other authorities involved, specialists, all interested parties and the public' (DEAT, 1998c, p. 29). The guidelines go on to suggest that procedural review should be based upon:

- legal requirements;
- quality of scoping;
- quality of impact prediction;
- quality of determining impact significance;
- assessment of alternatives;
- quality of mitigation proposed; and
- public participation process. (DEAT, 1998c, p. 29)

The guidelines further suggest that a 'comments and response report' be prepared but provide little information about what this should contain.

The guidelines present helpful significance criteria for the evaluation of impacts before and after mitigation:

- effects on public health and risk of life;
- scale of the negative environmental impacts;
- geographical extent of impacts;
- duration and frequency of negative impacts;
- size of the affected community;
- degree to which negative impacts are reversible or irreversible;
- ecological context;
- international, national and provincial importance;
- degree and likelihood of uncertainty of negative environmental impacts. (DEAT, 1998c, p. 30)

The EIA guidelines generally reflect best previous practice, but although they indicate that EIRs can be amended (DEAT, 1998c, p. 18), they fall short of formally suggesting the preparation of a draft EIR.

Review practice under the voluntary IEM procedures was extensive and frequently followed the popular and useful published guidance (DEA, 1992, vol. 4). This practice often involved the production of draft and final EIA reports, the use of independent expert reviewers (see Chapter 11), ample (and sometimes novel) consultation and participation, and the use of review guidelines and criteria. Reviews of EIA reports revealed a number of weaknesses in practice, though significant impacts appear not to have been completely overlooked (see also Chapter 11).

It is clear that the emphasis during the review stage in South Africa is on feedback to those preparing the EIR, rather than to the relevant authority (Weaver *et al.*, 2000). Thus, the EIA guidelines do not suggest that the public's comments on the EIR should be considered by the relevant authority but rather by the proponent's consultants. This may be realistic, since the relevant authorities are often unprepared and understaffed. However, the poor quality of some EIA reports has exacerbated the problem by draining provincial authority resources (Duthie, 2001).

The National Environment Management Act 1998 (NEMA) specifies that procedures for independent review must be implemented (see Chapter 10). The regulations, expected to implement NEMA's EIA provisions in 2002, provide an opportunity to specify that the results of the independent review should be made available direct to the relevant authorities. The proposed Southern African Institute for Environmental Assessment (see Chapter 5) could potentially contribute as an independent reviewer.

Summary

Only the United States, the Netherlands and the Commonwealth of Australia fully meet the EIA report review criterion, as shown in Table 12.1. In South Africa, the emphasis is on feedback to the proponent rather than to the competent authority. In the UK it is not the public review of EIA reports which is missing but the duty on the proponent to respond formally to the points raised. In practice, the proponent usually provides further information if it is requested by the local planning authority (and such information must be made available for a further period of consultation and participation). Canada and New Zealand do not meet the criterion in all cases.

The two longest-established EIA systems (those in the United States and Australia) require the preparation of both draft and final EIA reports. In practice, the situation in three of the other five jurisdictions does not differ greatly since in the Netherlands, Canada and New Zealand, the proponent can be formally asked for further information. This supplementary information, which may consist of additional material, an elaboration of existing information or a response to comments, must be provided in all three countries. The form of this additional material may be disparate and may consist of several different documents. One advantage of final EIA reports is that they bring all the further information together within a single document.

Generally, most EIA reports in all the jurisdictions appear to require supplementing with additional data following formal or informal review. Apart from the obvious increase in quality of the information provided between the initial EIA report and the final documentation, there seems to be a general trend towards gradual improvement in the quality of EIA reports over time. There is, however, still much room for further increases in EIA report quality.

Table 12.1　The treatment of EIA report review in the EIA systems

Criterion 7: Must EIA reports be publicly reviewed and the proponent respond to the points raised?

Jurisdiction	Criterion met?	Comment
United States	Yes	Lead agency must respond to agency and public comments on published draft EIS in final EIS
UK	Partially	Local authority can request further information and, though not mandatory, proponents usually provide it
The Netherlands	Yes	EIA Commission reviews EIS and, where necessary, supplementary information requested by competent authority
Canada	Partially	Occasional discretionary review by public of screening reports and mandatory review of comprehensive study reports and EISs, with proponent response
Commonwealth of Australia	Yes	Proponent responds in published final EIS and public environment report (and in revised preliminary documentation) to comments on published draft EIA reports
New Zealand	Partially	Not all assessment of environmental effects (AEE) reports publicly reviewed. Notified AEE reports publicly reviewed: power to demand more information frequently used
South Africa	No	No requirement for scoping report or environmental impact report review but review guidance exists. Previous good practice augurs well for adequate review under new Act

Notes

1. These percentages are based upon unpublished figures produced by the Office of Federal Activities, Environmental Protection Agency, Washington, DC.
2. This document is available on the Ministry for the Environment web site at www.mfe.govt.nz/management/consents.htm

Chapter 13

Decision making

Decision making takes place throughout the EIA process. Many decisions are made by the proponent (for example, choices between various alternatives – see Chapter 8). Others may be made jointly by the proponent and the decision-making and environmental authorities (for example, screening and scoping decisions). However, the main decision in the EIA process, whether or not to allow the proposal to proceed (or, less frequently, which alternative to implement) is always taken in the public domain. While the decision-making body may have given previous indications of the likely outcome of this decision, it is normally taken by a central, state/provincial or local government agency, following consultation and public participation. The typical decision taken at this stage in the EIA process is not usually a choice between alternatives, but a seemingly simpler choice between authorisation and refusal.

This chapter presents a discussion of the decision-making stage in the EIA process. It presents a set of criteria for the evaluation of decision making in EIA systems. These criteria are then used to assist in the review of decision-making procedures and practice in the EIA systems of the United States, UK, the Netherlands, Canada, Commonwealth of Australia, New Zealand and South Africa.

Making decisions about actions

As Sadler (1996, p. 13) has stressed, the immediate aim of EIA: 'is to facilitate sound, integrated decision making in which environmental considerations are explicitly included.' It is therefore important to see EIA as a tool to aid the taking of decisions on the authorisation of projects and other actions, rather than as a method of decision-making in its own right. Authorisation decisions are frequently taken as part of procedures to which EIA contributes (for example, the town and country planning system in the UK). EIA only occasionally forms the basis of the authorisation decision in the way it does, for example, in Western Australia (Wood and Bailey, 1994).

There is a considerable literature on decision-making methods and their use in EIA[1] but little on how decisions are made (Bartlett and Kurian, 1999; Glasson, 1999a; Weston, 2000). While the use of quantified decision-making methods has been roundly criticised on a number of occasions (see, for example, Bisset, 1988), much effort continues to be devoted to them. One reason for this is that such methods are often devised and used by engineers and others whose training emphasises the use of quantified methods. Another 'is the desire of many decision makers to be faced with an easy decision, especially when comparing a complex variety of impacts from a number of alternatives' (Bisset, 1988, p. 60).

Much of the literature on quantified decision making (and especially about subjective weighting[2]) has, as Bisset implied, focused on the choice between alternative proposals, rather than on the more usual choice about whether to approve or refuse a proposal.[3]

In practice therefore, authorisation decisions on proposals subject to EIA, whether they are yes/no decisions or involve a choice between alternatives, are frequently made incrementally and often in the cyclical manner characteristic of the EIA process (Lawrence, 1994). Thus, it may become apparent at the design stage of the process that certain impacts are likely to be unacceptable, leading to withdrawal or to redesign. Equally, the weight and force of objections raised to the proposal during the review of the EIA report may lead to the proponent modifying the action further. Decision making on the frequently complex projects which necessitate EIA is seldom straightforward. For example, it may well involve a wide range of stakeholders and more than one decision-making body. Also, the decision is usually dependent not only upon the environmental merits of the proposal but also on political circumstances (see Chapter 1). It is apparent that the environmental impact of the proposal will usually be only one of the factors to be considered by the decision-makers.

The making of any decision will involve a large number of trade-offs in the information base: between simplification and the complexity of reality; between the urgency of the decision and the need for further information; between facts and values; between forecasts and evaluation; and between certainty and uncertainty. The people making a decision on a proposal involving EIA will frequently be elected central, state/provincial or local government politicians. They will seldom have time to read the EIA report and other EIA documentation and will therefore be dependent upon their officials for some form of summary evaluation of the earlier stages of the EIA process. (The existence of an objective, and well-presented, non-technical summary of the EIA report clearly makes this evaluation easier (see Chapter 11).)

This evaluation, which will often be very brief, will typically summarise the objectives of the proposal, any alternative to it, its principal positive and negative impacts and their significance, the mitigation measures proposed, the principal representations about the proposal and how objections have been met. Compatibility with relevant policies is usually discussed but formal decision-making methods are seldom employed. The evaluation is bound to subsume some element of decision making and may well contain recommendations.

However, while the use of quantified decision-making methods may be helpful in reaching a consensus among similarly trained professionals, their use in public decision making is regarded with suspicion and is discouraged by politicians. The furthest most politically accountable decision-makers would wish formal evaluation to proceed is the preparation of a summary set of quantitative forecast impacts, together with a separate set of adjectival indications of their significance. The presentation of detailed technical data on environmental impacts is therefore not appropriate for effective decision making. As Sadler (1996) suggested, the acceptance of EIA inputs by decision-makers is more likely if EIA is focused on providing information that is relevant to the decision they have to make.

Other, non-environmental, objectives and political factors may well outweigh the findings of the 'technical' evaluation in the interactions between elected or appointed representatives from which the decision emerges. There are likely to be value-laden trade-offs between environmental and socio-economic factors. Sadler (1996) observed that, in reality, many authorisation decisions involving EIA are usually taken behind closed doors, and therefore only a partial understanding of the process is available (Glasson, 1999a). Nevertheless, given the positive benefits that most proposals confer (for example, employment), decision-makers will frequently seek to approve the action, unless there are politically overwhelming reasons to refuse it, but to negotiate increases in benefits and further mitigation of its negative impacts.

The original intention of EIA was that environmental considerations should be given greater weight in the design of proposals and in the decisions taken upon them. EIA was intended to constrain, but not to control, discussions. Accordingly, most jurisdictions forbid the taking of a decision on the action until an EIA report has been prepared and subjected to review. This is a fundamental requirement of any EIA system.

If the decision on the action is to reflect the EIA process meaningfully, it should be possible to require modifications or, in the last analysis, to refuse permission for the proposal to proceed. Clearly, the power of refusal is important in ensuring that the aims of EIA are met. Refusal is the ultimate sanction on the proponent. This discretionary decision (i.e. where the decision-maker possesses the discretion to refuse, rather than the power only to set performance conditions which must be met) usually falls to the government authority responsible for land-use control (Wood, 1989). Other bodies will obviously also be involved in decision making, but they do not usually possess the power of refusal (for example, air pollution control authorities must grant permits provided emission and/or ambient air quality standards are met).

The EIA report, and the various stakeholder comments on it, provide excellent bases for the framing of detailed conditions and, subsequently, for monitoring (see Chapter 14). This has been a neglected aspect of EIA until recently. The conditions imposed upon the approval should be phrased to take account of:

1. the forecasts made in the EIA report (to help to ensure that the commitments made in that document are implemented); and
2. the uncertainty in the forecasts upon which the decision is based (to ensure that the conditions are realistic).

For the decision on the proposal to be seen to be fair it is obviously preferable that it should, in general, be made by a body other than the proponent.

Further, any summary evaluation prepared for the decision-makers by their advisers (for example, the Australian assessment report, the Canadian panel report, the Dutch EIA Commission report or the British report to the local authority planning committee) should be made public.

Without such a check, it would be too easy to meet procedural EIA requirements relating to preparation of the EIA report and review without meeting the substantive obligation to consider the outcome of the EIA procedures in reaching a decision. An additional, and even more important, check is that the decision, the reasons for it and the conditions attached to it, should be published. It is obviously desirable, if the extent to which the EIA process is actually taken into account in decision making is to be clear, that the reasons for making the decision include a statement confirming that EIA inputs have been fully considered in the decision (as required by the European Directive on EIA). In addition, an explanation of how the EIA report and its review influenced the decision is necessary. This latter is provided for in many EIA systems (for example, in the US and Dutch EIA procedures). California has perhaps gone furthest in trying to ensure that decisions are influenced by EIA by requiring 'a statement of overriding considerations' where significant impacts would result from the approval of an action (Bass *et al.*, 1999).

As at other stages of the EIA process, advice on the factors to be considered by decision-makers in reaching their decision is a valuable way of ensuring that procedures are complied with and that environmental factors are given appropriate weight. Such advice could provide guidance on whether or not certain impacts were likely to be environmentally acceptable. For example, a jurisdiction might well indicate that noise levels above a certain threshold, or landscape impacts within a national park, would not be tolerated and that the proposal should either be modified or face refusal.

Some jurisdictions allow for consultation and participation once the evaluation has been prepared for the decision-makers but before any decision has been reached. This is clearly not possible where no separation of the steps in decision making takes place. Morrison-Saunders and Bailey (2000) emphasised the importance of balancing transparency in decision making involving EIA, achieved through meaningful consultation and participation, with the need to maintain a flexible and discretionary approach to authorisation decisions. However, the provision of a public right of appeal against the decision can increase public confidence in the EIA process. Such an appeal could either be administrative (as in the Netherlands) or to the courts, as in the United States.

The possibility of a proponent and/or decision-makers eluding the various checks aimed at increasing the level of transparency within the decision-making process exists (though the more there are, the more difficult evasion becomes). Thus, in practice, it is still possible for decision-makers to ignore the EIA in California, despite the required statement of overriding considerations (see above). Dresner and Gilbert (1999, p. 128) found that selected European EIAs appeared to have little effect on authorisation decisions and suggested that the reason was an: 'inadequate level of transparency in the decision-

---

Box 13.1   Evaluation criteria for decision making

Must the findings of the EIA report and the review be a central determinant of the decision on the action?

- Must the decision be postponed until the EIA report has been prepared and reviewed?
- Can permission be refused, conditions be imposed or modifications be demanded at the decision stage?
- Is the decision made by a body other than the proponent?
- Is any summary evaluation prepared prior to decision making made public?
- Must the EIA report, and comments upon it, be used to frame the conditions attached to any consent?
- Are the decision, the reasons for it, and the conditions attached published?
- Must these reasons include an explanation of how the EIA report and review influenced the decision?
- Does published guidance on the factors to be considered in the decision exist?
- Is consultation and participation required in decision making?
- Is there a right of appeal against decisions?
- Does decision making function effectively and efficiently?

---

making process, breeding lack of trust in its outcome and ultimately exacerbating the dispute.'

It is, therefore, important that practice in decision making reflects the results of the EIA report and review in practice, i.e. that it is effective. If practice does not appear to be influenced by the EIA process, the implication must be that additional checks and balances are required, even though this will probably conflict with the criterion of efficiency (i.e. that the decision should be taken without undue expenditure or delay). The various criteria for the evaluation of decision making are summarised in Box 13.1. These criteria are employed in the review of decision making in each of the seven EIA systems below.

United States

No federal decision on a proposed action can be made until 30 days after the Environmental Protection Agency (EPA) has published a notice that the final environmental impact statement (EIS) has been filed. This is an important procedural provision but perhaps symptomatic of the National Environmental Policy Act 1969 (NEPA), which contains no substantive, enforceable requirement to protect the environment by, for example, refusal of consent. NEPA requires federal agencies to disclose the environmental effects of their actions and to identify alternatives and mitigation measures. Agencies may not select an alternative unless it has been discussed and evaluated in the final EIS, but they are not required to adopt the environmentally preferred alternative. The lead agency has the power to refuse consent on the basis of the EIS (i.e. to choose the no-action alternative) but it is likely only to refuse consent in responding to an application for a permit or for funding. It is very unlikely to turn down its own proposal and no other agency can override it.

While the lead agency, in its role as proponent, will clearly wish to implement the proposed action, other agencies may, and do, voice objections. These may occasionally lead to the abandonment of the proposal. One example of EIA resulting in the cancellation of a potentially damaging project is given in Box 2.2. In rare cases, objections may result in a referral to the Council on Environmental Quality (CEQ) in order to resolve interagency disagreements. Although a CEQ decision is not binding, agencies usually abide by its recommendations and disagreements have generally been effectively resolved by the referral process (Ortolano, 1997). Of the 26 referrals made between 1974 and 2001, 16 were originated by EPA.[4] The threat of referral is often sufficient to wring environmental concessions from the lead agency, since CEQ may, if necessary, submit its recommendation to the President for action, elevating the issue to a level that would embarrass a lead agency.

At the time of its decision, following the preparation of the final EIS, the federal agency must prepare a record of decision (ROD). This is a written public record which was introduced, like scoping, with the intention of ensuring that the EIS actually influences agency decisions. It must contain:

- a statement explaining the decision
- an explanation of the alternatives considered and those that are environmentally preferable
- the economic, technical and environmental factors considered by the agency in making its decision
- an explanation of the mitigation measures adopted and, if all practicable, mitigation measures were not adopted, an explanation of why not
- a summary of the monitoring and enforcement programme which must be adopted to ensure that any mitigation measures are implemented. (Regulations, section 1505.2)

Some agencies publish their RODs in the Federal Register though this is not a formal requirement.

Agencies, in making their decisions, must balance the relevant environmental and other factors, as in other jurisdictions. In practice, the EIS frequently influences the decision, though often not as a direct result of the information it contains. As Hyman and Stiftel (1988, pp. 426, 427) put it:

> while decision makers generally have not turned to EISs in the expected ways – for specific data, evidence or policy implications – . . . [EISs have] succeeded in jarring the consciousness of many decision makers and administrators.

However, it is unfortunately still true that 'senior government officials often do not make effective use of NEPA in evaluating alternatives in reaching decisions' (Clark, 1997, p. 22).

NEPA, the CEQ Regulations and the agencies' own regulations provide the necessary review guidance, and the environmental review procedure leading to the decision is open to appeal in the courts. However, provided the appropriate procedures have been followed to ensure that the agency has genuinely considered the environmental consequences of the action, including the making public of the final EIS and the ROD, the courts have not attempted to contradict agency decisions. Thus, notwithstanding the existence of the CEQ referral process (which is based upon the impacts, or merits, of the action), the Supreme Court has held that NEPA is in essence procedural and that agencies have no substantive duty, under the EIA provisions of NEPA, to protect the environment (Holland, 1996; Mandelker, 2000).

UK

The local planning authority (LPA) is required to have regard to the 'environmental information' (the environmental statement (ES) and the various submissions by statutory consultees and the public) before making its decision. As with any other planning application, the LPA may refuse permission or grant it with or without conditions. In reaching this decision, the LPA attempts to weigh all the planning advantages and disadvantages: the environmental impacts of the proposal are only one factor in the decision.

Planning permission is the only genuinely discretionary consent in the UK; other consents relating to, for example, pollution control are virtually always granted, provided the appropriate conditions are met (Wood, 1989, 1999a). LPAs are permitted to determine their own applications involving EIA (a very small proportion of the total) but must use the same procedures as for ESs prepared by external proponents.

It is normal practice for planning officers to prepare a report (to the planning committee) on applications to be determined by the LPA. This report should summarise the salient parts of the EIA process, including the ES, the comments received during consultation and participation, any other relevant matters and the officers' own evaluation and recommendation (Department of the Environment – DoE, 1994a, para. 8.8). This 'evaluation report' to committee is a public document.

It is mandatory that the LPA state in writing that the environmental information has been taken into account in reaching the decision (Regulation 3(2)). The planning decision must be publicised in a local newspaper. In addition, the decision, the conditions attached to it, and the reasons for the decision, whether the application is refused or approved, must be published in the planning register that every LPA is required to maintain. Where permission is granted, the main mitigation measures must be listed. While there is no formal specification of what areas 'the main reasons and considerations on which the decision is based' should cover, the Department for the Environment, Transport and the Regions (DETR) suggested that 'this requirement is met by the relevant planning officer's report to the Planning Committee' (DETR, 1999b, p. 27).

Guidance on decision making involving EIA has been issued to LPAs in the UK (DoE, 1994a). Basically, LPAs are required to make development control decisions 'in accordance with the development plan unless material considerations indicate otherwise' (DoE, 1994a, para. 7.5). In addition, the LPA:

> must take into consideration the information contained in the Environmental Statement (ES) (including any further information), any comments made by the consultation bodies, and any representations from members of the public about environmental issues. (DETR, 1999b, para. 119)

LPAs must first balance the relative merits of different environmental topics (DoE, 1994a, para. 7.10). Thus, the adherence of new sources to pollution standards is a relevant matter for LPAs to consider in reaching decisions (DoE, 1994c), as are other criteria of environmental acceptability. Although failure to meet such criteria would be a valid reason for refusal, there is no prohibition on the grant of planning permission if the EIA process reveals that the proposal does not meet them. Second, LPAs must draw together environmental, economic and social factors in reaching their decisions (DoE, 1994a, para. 7.10). Advice to LPAs has not been intended to match the specific methodological guidance available on incorporating EIA into decision making on major roads (DETR, 1998a).

There is no consultation and participation requirement during the decision-making process, although members of the public are often permitted to address the elected representatives who constitute the LPA while the decision is being discussed (usually by a committee). The public has no right of administrative appeal against the decision, unlike the developer. Planning appeals are normally determined by the Planning Inspectorate following written representation, hearing or inquiry procedures. Inspectors often make recommendations about decisions to the Secretary of State for Transport, Local Government and the Regions, usually based on similar criteria to those employed by the LPA. While such appeal decisions provide important precedents, they do not have legal force. There are restricted rights of further appeal to the courts.

In a study of 40 cases where an ES had been submitted, nearly two-thirds of the planning officers involved had found the ES, and especially the consultations on it, to be useful in helping them to reach their recommendations, which were nearly always accepted by the planning committee. Planning committees were believed to give considerable weight to the contents of the ES in reaching their decisions in about one-third of cases. However, the EIA was thought to have led to a reversal of the final decision in only one case (Wood and Jones, 1997). As Read (1997, p. 91) has stated:

> So often the fate of EIA cases depends on judgements that do not have a technical basis. It is the political process that has to complete a planning authority's review.

Environmental conditions were imposed by LPAs in all the cases where planning permission was granted, and many of these arose from the EIA process (Wood and Jones, 1997).

These results are reflected in a study of ten public inquiries which involved EIA. Jones and Wood (1995) found that the importance of the ES to inspectors in evaluating the information about a case varied, but was generally less than the additional evidence presented and examined at the inquiry. Weston (1997, p. 124), in a study of 54 public inquiry decision letters, confirmed this finding:

> other evidence is of more relevance than the ES and this may help to explain why in 53% of those cases examined the ES was not specifically referred to ... in ... decision letters.

Over half the inquiries studied by Jones and Wood (1995) resulted in refusal of permission. Where permission was granted, the ES was utilised in the framing of conditions.

It appears that EIA has had a gradual, rather than a revolutionary, effect on planning decisions, enhancing the provision of environmental information to decision-makers and, to a lesser extent, providing assistance in setting conditions (Glasson, 1999a). It must be concluded that, while the environmental information is an increasingly important material consideration in planning decisions, it is not yet a central determinant in many of them.

The Netherlands

The competent authority (which may be the same as the proponent for plan EIAs) publishes its draft decision with the EIS in licence application cases, and often does the same in other instances. However, this draft decision is frequently modified as a result of the intervention of the public or of the recommendations of the EIA Commission (EIAC) in relation to the EIS (it is not supposed to comment on the decision itself) so that the final decision is often significantly different.

The competent authority is obliged by the Environmental Management Act 1994, section 7.37, to incorporate the findings of the EIS, the comments of the consultees and the public, and the results of EIAC's review in its deliberations on the decision. This must include a discussion of the alternatives considered and a statement (if relevant) as to why the environmentally preferable alternative was not selected. The competent authority must explain fully the reasons for its decision in writing. These reasons must not only explain the weight given to EIA but also indicate the weight which has been attached to environmental parameters in comparison to other factors (Ministry of Housing, Spatial Planning and the Environment – VROM, 1991).

Irrespective of what sectoral legislation may say on the subject, all the environmental impacts must be considered when a decision requiring EIA is taken (section 7.35). The competent authority taking a decision requiring EIA prescribes the conditions, regulations or restrictions necessary to preclude or limit the impacts on the environment, including requirements as to how monitoring and auditing is to be carried out (see Chapter 14).

Furthermore, the competent authority may not accept a permit application unless an EIS is provided (section 7.28) or when the information in the EIS is out of date because the circumstances under which the statement was prepared have changed considerably (section 7.27). These substantial and comprehensive decision-making requirements together place a considerable duty, which it is difficult to avoid, on the competent authority to take the EIS into account. A copy of the decision, and the reasons for it, must be sent to those commenting on the EIS, including the statutory consultees and EIAC (section 7.38). The decision must also be made public through the media.

There is a general third-party right of appeal against decisions, and this is exercised in over 50 per cent of EIA cases. These appeals involve hearings, which often last a few hours, and take place under the relevant sectoral legislation. A poor EIS may be a reason for instituting an appeal against a decision for which the EIS was drawn up (VROM, 1991). The contents of an EIS may well be cited in support of environmental objectives to a proposal.

It is generally accepted that EIA is actually affecting decisions in the Netherlands, even if some decision-makers still decline to read EISs or to consider their findings, and many decisions are still determined mainly by economic factors. On the other hand, many EISs are used to phrase the conditions attached to permits. It is rare for applications to be refused. Of the 467 decisions made before the end of 2000, only three or four were refused because of environmental impacts and these all occurred in the early years of EIA (Scholten and van Eck, 1994; van Eck and Scholten, 1996; EIAC, 2001a). However, a significant number of proposals are not pursued or are modified as a result of negotiation between the proponent and the competent authority.

The Evaluation Committee on the Environmental Protection Act (ECW, 1990) found that the openness of decisions was leading to improvements in the way environmental considerations were taken into account. Ten Heuvelhof and Nauta (1997) reported, on the basis of a study of 100 EIAs for ECW, that 79 per cent had a 'direct' (if sometimes marginal) impact on the decision, either by causing the proponent to change the project early in the design process, or by influencing the competent authority's decision, or both. In 14 of the 21 cases where EIA had no direct impact, indirect impacts (such as changes in attitudes) were observed. Only in 7 per cent of cases did EIA appear to have no effect whatever (ECW, 1996).

Canada

It is a requirement of the Canadian Environmental Assessment Act 1992 that no responsible authority (RA) may 'exercise any power or perform any duty' in relation to a project until the environment assessment (EA) is complete (section 11(2)). In the self-directed assessment the RA can take action to allow the project to proceed if it is not likely to cause significant adverse environmental effects, after taking into account the implementation of appropriate mitigation measures. In other, very rare, circumstances a public review (see Figure 5.3) becomes necessary. The various decisions that can currently be taken by the RA and the Minister of the Environment (MoE), following consideration of a comprehensive study report, are shown in Table 13.1. The situation for screening reports is similar, save that the RA has to request the Minister to refer

Table 13.1 Decisions in Canadian comprehensive studies

Findings of comprehensive study report and public comments received	The Minister, taking into account appropriate mitigation measures	The responsible authority (RA)
1. The project is not likely to cause significant adverse environmental effects	Refers project back to RA	May provide federal support to the project
2. The project is likely to cause significant adverse environmental effects that cannot be justified	Refers project back to RA	May not provide any federal support to the project
3. It is uncertain whether the project is likely to cause significant adverse environmental effects		
4. The project is likely to cause significant adverse environmental effects and a determination must be made whether the effects are justified in the circumstances	Refers project to mediator or review panel	May not provide federal support to the project until the public review is completed
5. Public concerns warrant the referral		

Source: adapted from CEAA, 1994, pp. 100–1.

the project to public review rather than the Minister being required to make the decision in comprehensive study cases. Public notice of the RA's decision, and any conditions or follow-up measures, must be provided but no reasons need be stated.

Practice in Canada, as elsewhere, is not perfect. David Redmond and Associates (1999) found that 36 per cent of a sample of 191 screenings were started late in the planning process. The Commissioner of the Environment and Sustainable Development (CESD, 1998, pp. 6–12, 13) identified some instances where screenings were 'conducted either after the project is approved or so late in the planning phase that it cannot influence the project'. As he stated:

> Last minute environmental assessment can put pressure on officials conducting the screening to deliver a product that is less detailed than required and has little likelihood of reducing a project's potential environmental impact. (CESD, 1998, pp. 6–15)

The Commissioner recommended that action be taken to remedy this situation and, in response, it was proposed that the Canadian Environmental Assessment Agency (CEAA) be given:

power to request information from responsible authorities in support of a quality assurance program ... [which] would include an on-going mechanism to monitor compliance with the Act. (MoE, 2001, p. 24)

A project can be referred to public review at any time during the self-assessment procedure, not just at the end. This has sometimes been done to facilitate coordination with provincial EA procedures. Indeed, as mentioned in Chapter 9, it has been proposed that referral to public review at the end of the comprehensive study procedure will no longer be possible (Government of Canada, 2001; MoE, 2001).

For public review projects the panel or mediator must prepare a report and make it available to the public. For panel reviews, this must set out:

- the rationale, conclusions, and recommendations of the panel, including any mitigation measures and follow-up program;
- a summary of public comments. (CEAA, 1994, p. 121)

The report is submitted to the Minister and to the RA. Formally, the Governor in Council (the cabinet) must decide whether the RA can permit the project to proceed (with appropriate mitigation measures). The Act decrees that, if the cabinet concludes that 'the project is likely to cause significant adverse environmental effects that cannot be justified in the circumstances' (section 37(1)(b)), the RA must not permit the project to proceed. In practice, the RA makes the decision. If the RA is the proponent, its decision will determine the fate of the project. In some other cases, the proponent may be able to proceed without federal approval but, if federal land, money or permits are required, the project will be abandoned (CEAA, 1994, p. 121).

The RA may draw different conclusions from the panel or mediator but must explain its reasons in the lengthy public notice of decision issued on behalf of the cabinet. (There is no appeal against this decision except, on procedural grounds, through the courts.) Any disagreements about panel recommendations are resolved between ministries in cabinet. This (often lengthy) process also applies (in principle) to mediation reports. In the early years of the federal Environmental Assessment and Review Process (EARP), panel recommendations were accepted by initiating departments. More recently, as EARP was applied on a less voluntary basis and the Act was implemented, they have tended frequently to deviate from panel recommendations.

In one controversial case (the Oldman River Dam), one of the panel's alternative main recommendations was to 'decommission the dam by opening the low level diversion tunnels to allow unimpeded flow of the river' (Federal Environmental Assessment Review Office, 1992, p. i). Unsurprisingly, this particular recommendation (which arose as a result of retrospective imposition of a panel review by the courts) was not accepted. While this case remains unusual, decisions were often made so late in the design/construction process (largely because of uncertainties over EARP application) that the 'no-go' option was virtually precluded (Fenge and Smith, 1986).

The Voisey's Bay Mine and Mill Project panel report was over 200 pages in length. In recommending approval, the panel made over 100 broad-ranging environmental, economic and social impact recommendations, one of which was that the proponent:

> revise existing [proponent] employment assistance programs – including, but not limited to the women's employment plan and harassment policy – to address women's concerns. (CEAA, 1999d, p. 123)

The cabinet agreed that this project could proceed in 1999, more than three years after initiation.

The Act retains the same basic approach to EA as EARP but, while some uncertainties over its application remain, it seems unlikely that such exceptions from the federal EA process as the Oldman Dam will recur. The Act lacks any power to ensure that EA is a central determinant in the decision to approve or reject a project (Gibson, 1993). Even under the Act, therefore, decision making cannot be said always to have been either effective or efficient, especially in relation to screenings, though there have been successes involving panel reviews. The proposed amendments to the Act (Government of Canada, 2001) still do not provide for a documented final decision which could provide the basis for enforcing conditions of approval.

Commonwealth of Australia

Under the repealed Commonwealth legislation, virtually every proposal was modified during the EIA process to reduce impacts, thus ensuring that one aim of EIA was met. Some unsatisfactory proposals undoubtedly dropped out of the EIA process before the final decision was reached and it is possible that conditions placed upon some consents were so onerous that the proponents declined to progress them. However, only three or four of the 215 proposals that went through the full Commonwealth EIS or public environment report (PER) procedure were refused consent to proceed.

The very low number of refusals was undoubtedly partly testimony to the negotiating skills of the staff of Environment Australia and partly due to the low number of proposals assessed (Fowler, 1996). However, it was also attributable to a fundamental weakness of the Commonwealth EIA system, namely the absence of an environmental decision at the end of the EIA process. Carbon (1998, p. 7) illustrated this graphically:

> the new runway for Sydney, despite the $10 million EIS, . . . went ahead without a decision being made, without any binding environmental standards, and without any binding environmental conditions.

See also Harvey (1998, pp. 170–9).

The Commonwealth Environment Protection Agency reform proposals (CEPA, 1994) suggested a number of solutions to this problem, including the setting of legally binding environmental conditions. The Minister's consultation paper grasped the nettle firmly by proposing that 'the Minister will be responsible for deciding whether to grant consent' and for specifying conditions (Hill, 1998, p. 16). This was one of the fundamental improvements brought in by the Environment Protection and Biodiversity Conservation Act 1999 (EPBC Act).

The EPBC Act requires the preparation of an 'assessment report' on the proposal for the Environment Minister, as the previous legislation did. Environment Australia must prepare the report within 20 business days of accepting revised preliminary documentation or a final public environment report (PER) or within 30 business days of receiving a final EIS (section 105(1)). In accredited EIA process cases, the relevant state or Commonwealth agency provides the final assessment report to the Environment Minister for consideration in the approval decision.

The assessment report is intended to stand alone, to be read in isolation from the proponent's EIA report. Its aim is to provide an analytical summary overview of the environmental effects of the proposal based upon the proponent's documentation and the responses to it. (Assessment reports may be lengthy – one ran to over 200 pages.) The availability of a Commonwealth assessment report must be notified on the internet and, on request, made available (either free or at a reasonable charge).

The Environment Minister must decide whether or not to grant approval within 30 business days of receiving the assessment report. Before the decision is finalised and any conditions determined, the Minister must consult other relevant Commonwealth ministers who may comment on economic and social matters relating to the action. In practice, if there is any disagreement between the Environment Minister and the other ministers, liaison takes place, sometimes in cabinet, to resolve those differences, before the Minister makes a decision consistent with the principles of ecologically sustainable development (section 136). Such an arrangement is not dissimilar to that in the Netherlands. The Minister cannot make a decision until a notice from the relevant state is received confirming that the environmental impacts on 'matters other than those of national environmental significance' have been assessed.

The Minister can attach conditions to an approval if it is necessary to protect, or mitigate an impact on, a matter of national environmental significance. Conditions requiring the proponent to comply with any conditions imposed under other regulatory regimes are intended to avoid duplication (section 134(3)(c)). Other conditions relating to: the use of financial guarantees; insurance against remedial liabilities; environmental auditing; monitoring; impact management plans; and compliance with specified codes of practice, can be imposed (section 134(3)).

When deciding whether to approve the taking of an action, and what conditions to impose, the Environment Minister must consider :

- social and economic matters; and
- relevant environmental impacts.

In considering these matters, the Minister must take into account:

- the principles of ecologically sustainable development (i.e. long-term and short-term economic environmental, social and equitable considerations);

- the assessment report;
- the precautionary principle;
- any other information about the impacts of the action;
- relevant comments from other Ministers;
- the proponent's environmental history.

The Minister's decision is thus separate from, but normally coincides with, the recommendations in the assessment report and must be made public. The approval decision must be notified on the internet, and a copy must be made available on request, but reasons do not have to be provided. These arrangements are far superior to those under the previous legislation and it is clearly intended that the proponent's EIA report and the assessment report should be central determinants in the decision. The extent to which this proves to be the case in practice remains to be seen.

New Zealand

Environmental impact assessment should be central to decision making in New Zealand because the Resource Management Act 1991 requires policy statements, plans and consent decisions to avoid, remedy or mitigate any adverse effects of activities on the environment (see Box 5.1). Section 104 of the Act specifically states that the consent authority must have regard to the provisions of relevant policy statements and plans and of any assessment of environmental effects (AEE) report when considering an application for a resource consent. The AEE report (and the various submissions made) is intended to be central to the decisions since 'the consent authority shall have regard to any actual and potential effects of allowing the activity' (section 104(1)).

It is customary for the elected members of local councils to make the decision, on the basis of an evaluation report and recommendations prepared by officers (Ministry for the Environment – MfE, 2001a). The decision must be taken within 15 working days of the council hearing, which is held in more than half of the cases notified (see Chapter 12). Panel hearings are usually conducted by three elected members. In complex cases, or cases where the local authority is the applicant, one or more independent commissioners may become members of the panel. The Ministry for the Environment can submit its opinion but has done so in less that 10 cases to date. Panels almost always accept their officers' advice, and less then 1 per cent of all resource consent applications are refused (MfE, 2001c). Permission can be granted with conditions, about which drafting guidance has been issued (MfE, 2001b). There is a requirement that councils record the reasons for their decision on a resource consent application which may in some cases refer to the environmental impact assessment. A copy of the decision must be sent to the applicant and to all those who made a submission (MfE, 2001a).

The applicant or any person who has made a written submission about the application (including the minister) can appeal to the Environment Court against the consent authority's decision. The Court hears cases *de novo* (see Chapter 5), and has not hesitated to comment on the deficiencies of the documentation furnished in the past. Because it can award costs, the Court continues to provide a potential check on the quality of AEE reports and council evaluations.

The results of appeal decisions by the Environment Court are proving influential in the development of practice, as a result of specific guidance furnished by them. Rather as in the UK, there are powers for the Minister for the Environment to call in applications of national significance for decision in New Zealand. However, unlike in the UK, it was intended that these powers would be employed very exceptionally. There has been only one instance to date: in 1993 the air discharge permit for a large power station was called in because of its nationally significant carbon dioxide emissions.

In practice, decision making does not always appear to have met the aims of the Act. Dixon (1993) and Morgan (2000b) have pointed out that few checks exist on whether the EIA is actually considered when the decision is made. It appears that the parochialism of past practices continued under the far more sophisticated Resource Management Act (Parliamentary Commissioner for the Environment – PCE, 1995; McShane, 1998). There is, in particular, insufficient involvement by the public in the EIA process (Morgan, 2001). Morgan (1995) and PCE (1995) stated that some elected politicians in regional and (especially) territorial authorities tended not to be aware of (or sympathetic to) the basic principles of the Act and thus not to appreciate that AEE is intended to influence decision making. Nevertheless, this has changed with time and the integration of EIA into the resource consents process is widely believed to be resulting in improved environmental outcomes in an increasing number of cases as practice has developed.

South Africa

Implementation of certain activities cannot proceed without an authorisation under the Environment Conservation Act 1989 and this will only be forthcoming once either a scoping report or an environmental impact report (EIR) has been accepted. Once the relevant authority has decided that the EIR complies with the regulations (or that the information contained in the scoping report is sufficient) it must consider the application and the EIR or scoping report and may issue an authorisation, with or without conditions, or may refuse the application (Republic of South Africa, 1997, Regulation 9). Sections 2 and 3 of the Environment Conservation Act 1989 require this decision, *inter alia*, substantively to promote sustainability (Peckham, 1997).

In 1998, the High Court upheld a challenge by environmental groups against a large company's permit to mine coal under the Minerals Act 1991 on the grounds than an EIA was not carried out (and that there was no consultation with interested and affected parties) (Dutton and Dutton, 1998). This successful action must have encouraged environmental groups to consider bringing similar suits under the Environment Conservation Act where they feel they can argue that insufficient account has been taken of the EIA in decision making.

The EIA guidelines (Department of Environmental Affairs and Tourism – DEAT, 1998c) provide no detail about the factors which ought to be considered in reaching the decision or about the weight to be given to different factors. Should they follow best previous practice, many relevant authorities may provide 'an explanation of how environmental considerations were taken into account and weighed against other considerations' (Department of Environmental Affairs, 1992, p. 8) when explaining the key factors leading to their decisions.

Once it has taken its decision, the relevant authority must issue a record of decision (ROD) to the applicant and to any other interested party who requests it. The ROD must include, *inter alia*:

- a brief description of the proposed activity;
- the decision of the relevant authority;
- the conditions of the authorisation (if relevant) including measures to mitigate, control or manage environmental impacts or to rehabilitate the environment;
- the key factors that led to the decision. (Republic of South Africa, 1997, Regulation 10)

There is provision for an appeal against the decision to be made to the Minister or provincial authority by the developer, any interested party or any member of the public, within 30 days of the issue of the ROD. No provision is made for the result of the appeal to be made public, save where the original decision is not upheld, in which case the guidelines suggest that a revised ROD should be issued (DEAT, 1998c, p. 31). Generally, a report is prepared for the provincial minister by the officers of the relevant authority, summarising the main details of the case and offering a recommendation which includes a set of proposed conditions on any authorisation. This report can be accessed by members of the public who request it.

The writing of such reports has not been assisted in the past by the way in which EIA reports prepared under the integrated environmental management procedure have been presented. As well as being overly technical, EIA reports have rarely been decision-oriented and have seldom specified the developer's environmental management commitments (Joughin, 1997). However, there can be no doubt that EIAs prepared under the IEM regime were often a central determinant of the decision on the activity and in the setting of conditions (Hill, 2000).

While practice has developed under the EIA Regulations, it is apparent that the decision to grant authorisation has sometimes been made by overwhelmed provincial staff on narrow nature conservation or other grounds, rather than on the full range of factors normally considered in internationally recognised good EIA practice (Granger, 1998). In other instances, the EIA function has been located within an agency charged with economic development, for which there has been overwhelming pressure (see Chapter 5). As a consequence of these factors, the number of refusals has been very small (probably less than 2 per cent of applications).

The National Environmental Management Act 1998 (NEMA) requires that the findings of the EIA (and of the general objectives of IEM) are taken into account in decisions (section 24(7)(h)). The NEMA EIA regulations (expected in 2002) need to specify precisely how this is to be achieved under the very real constraints within which South Africa EIA operates.

Table 13.2   The treatment of decision making in the EIA systems

Criterion 8: Must the findings of the EIA report and the review be a central determinant of the decision on the action?

Jurisdiction	Criterion met?	Comment
United States	Partially	Consideration in, and explanation of, decision and disclosure of environmental effects mandatory. In practice, EIS often influences decision
UK	Partially	Environmental information is a material consideration but not necessarily a central determinant. Practice varies
The Netherlands	Partially	Explanation of way environmental impacts considered in decision is mandatory. In practice, EIA generally influences decision
Canada	Partially	Findings of self-directed assessment influence RA's or Minister's decision: reasons must be given by RA when cabinet disagrees with recommendations of public review report
Commonwealth of Australia	Partially	Environment Australia's assessment report based on EIA report must be taken into account in determining Environment Minister's decision on approval
New Zealand	Partially	Act makes EIA central to decision but, in practice, EIA is sometimes not given appropriate weight. Practice improving
South Africa	No	Environmental authorisation must be based on scoping report or environmental impact report (EIR) (and any review) but decision sometimes narrowly based on nature conservation matters, not on full range of EIR issues. Refusals rare

Summary

Appropriately, the decision about whether or not the EIA systems meet the decision-making criterion is a delicate one. The criterion states that the EIA report and the comments upon it must be a (not the) central determinant of the decision. EIA was never intended to provide the sole basis for decision making (see Chapter 1). However, to meet the criterion, an EIA system needs to demonstrate not only that the decision should be influenced by the EIA (all seven do so) but that, in practice, the EIA report actually influences each decision and is not just 'boiler-plate' paper. None of the seven EIA systems meets this interpretation of the criterion (Table 13.2).

The United States, UK, the Netherlands, Canada, Australia, New Zealand and South Africa use differing mechanisms for ensuring that the EIA is considered. In practice, however, it is still common for decision-makers to circumvent these EIA mechanisms where this is convenient. In the Netherlands the recommendations of the influential EIA Commission relating to the EIA are published and the competent authorities are in effect obliged to accept them. However, here as elsewhere it is possible to circumvent the findings of the EIA process in making decisions. The use of the EIA process to assist in the drawing up of conditions on approvals is, however, increasing. The challenge in the seven jurisdictions is to ensure that EIA becomes a more central determinant of decisions in practice.

Notes

1. See, for example, Bisset (1988), Hart *et al.* (1984), Canter (1996), Glasson (1999a).
2. The use of scoring methods (for example, of intervals, of ranking, of a binary yes/no approach or of normalisation) in the comparison of alternative values of the same type of impact (for example, noise) should be clearly distinguished from the use of weighting to compare different types of impact (for example, noise and air pollution).
3. Checklists and matrices and other approaches to the identification and communication of impacts have been modified to become decision-making methods by the integration of scales and weights, as in the Leopold *et al.* matrix (Leopold *et al.* 1971, in Canter, 1996), and in the environmental evaluation system advanced by Dee *et al.* (1973, in Canter, 1996). Here, and in simulation modelling, value judgements are implicit in the weights or scales used. There are various means of making these value judgements less opaque. One is to list the impact forecasts clearly and then to explain the basis on which scaling and weighting is undertaken. Another is to stop short of overall aggregation and to quantify only groups of impacts (for example, different types of air pollutants). Another is the use of sensitivity analysis in which the weights are varied to determine their effects on the outcome. Yet another is the opening up of the attribution of weights to consultees to make the judgements more representative.
4. See, for a listing of the 26 cases, Bass *et al.* (2001, pp. 167–8).

Chapter 14

Monitoring and auditing of impacts

Environmental monitoring and auditing in EIA, sometimes also referred to as EIA follow-up, has a number of purposes. Some of these are scientific while others are management related (Wilson, 1998). They include ensuring that the terms and conditions of project approval are met, verifying environmental compliance and performance, and indicating where the adjustment of mitigation and management plans may be necessary (Sadler, 1996 p. 127).

There are numerous definitions of monitoring and auditing in EIA. Thus Bisset and Tomlinson (1988) identified a total of seven different types of EIA audit. The first distinction to make is that between the monitoring of individual actions and of the EIA system as a whole (the latter is dealt with in Chapter 17). The second distinction is between the three main types of action monitoring and auditing with which this chapter is concerned: implementation monitoring, impact monitoring and impact auditing. The chapter presents a set of criteria for analysing the treatment of action monitoring in EIA systems. These criteria are then used to assist in the review of the monitoring of actions in the EIA systems in the United States, UK, the Netherlands, Canada, Commonwealth of Australia, New Zealand and South Africa.

Monitoring and auditing of action impacts

As mentioned in Chapter 11, the EIA report is no more than a record of the forecast impacts of the action as it is designed at a particular point in time. Further design work will take place once approval of the action has been granted (again on the basis of information available at a particular point in time) and this may well lead to modifications. Further, since even the best design may need to be altered to meet unexpected problems encountered during construction, further modifications may well take place during the process of implementing the action. It is inevitable, therefore, that the implemented action may well differ from that envisaged when the EIA report was prepared. Further, if the proponent's intentions fall short of full incorporation of the mitigation measures proposed in the EIA report, a significant 'implementation gap' may occur. Implementation and impact monitoring and impact auditing ensure that this gap is kept to a minimum and confer wider benefits.

The necessity of maintaining surveillance of, and control over, the implementation of actions has tended to be a neglected area in EIA. Sadler

(1996) cited monitoring (or follow-up) as one of the worst performed activities in the EIA process. As Sheate (1996, p. 111) rather graphically put it: '[p]ost project monitoring has tended to be the runt of the EIA litter of activities.' Inadequate checks have been applied in some jurisdictions to ensure that proponents negotiate any necessary changes with the appropriate authorities and thus that project construction, operation and decommissioning do not result in significant unanticipated impacts (Sadler, 1988, 1996). For example, a provision requiring monitoring where mitigation measures are agreed was added to the Californian Environmental Quality Act to close an obvious loophole but even its observance has been found to be wanting and there is no clear linkage back to the EIA report (Bass *et al.*, 1999).

The focus of EIA has traditionally been on the pre-decision stages leading up to project authorisation (Glasson, 1994; Sadler, 1996; Dipper *et al.*, 1998; Wood, G., 2000). The importance of monitoring and auditing in broadening the role of EIA into a broader, longer-term, environmental management tool has been repeatedly stressed (for example, by Holling, 1978; Beanlands and Duinker, 1983; Bisset and Tomlinson, 1988; Canter, 1993; Glasson, 1994; Sadler, 1996; Arts, 1998; Arts and Nooteboom, 1999; Glasson *et al.*, 1999). Shepherd (1998, p. 164) noted that monitoring 'transforms EIA from a one time pre-project document to a continual assessment of impacts.'

The absence of legislated monitoring requirements in most EIA systems is a major weakness, principally because the opportunity to improve practice by learning from experience is seriously compromised. As a result, there is widespread support for the introduction of mandatory monitoring requirements for certain types of project (Canter, 1993; Sheate, 1996; Dipper *et al.*, 1998). Glasson (1994) observed that resistance from industry was likely to limit the extensive application of monitoring and auditing, and practice would therefore be confined to 'enlightened developers'. Sheate (1996, p. 113) concurred:

> The experience in the EC and member states has clearly been that if post project monitoring or analysis is not contained in legislation, it is unlikely to happen or be effective.

Implementation monitoring involves checking that the action (normally a project) has been implemented (constructed) in accordance with the approval, that mitigation measures (for example, sound-proofing or habitat compensation) correspond with those specified or agreed and that conditions imposed upon the action (for example, process air or noise emission limits) have been met. Such checking may involve physical inspection (for example, of building location, wall construction or habitat establishment) or measurement (for example, of air or noise emissions) using various types of instrument, together with the application of professional judgement. This type of monitoring can be carried out either by the decision-making or environmental authorities or by the proponent (with appropriate checks and balances) or, as is frequently the case, may be divided between them.

Implementation monitoring frequently takes place under the provision of more than one set of legislative requirements (for example, land-use planning, building approval or pollution control procedures). Any necessary action to enforce compliance with the terms of the approval may be taken under non-EIA powers. Implementation monitoring is therefore essentially reactive, its principal purpose being to ensure that the action adheres to the conditions of its approval (Hollick, 1981; Tomlinson and Atkinson, 1987a; Sadler, 1988, 1996).

Impact monitoring involves measurement of the environmental impacts (for example, on ambient noise levels or upon a species of bird) that have occurred as a result of implementing the action. A variety of measurement techniques are likely to be needed, coupled with the exercise of expert opinion. This type of monitoring serves three purposes:

1. Impact monitoring provides a check on the extent to which vicinity authorisation conditions have been met (for example, ambient levels of sulphur dioxide, or visual impacts, at specified locations).
2. Where monitoring of the environment reveals unexpected or unacceptable impacts (for example, elevated noise levels at night) further design changes (baffling) or management measures (ensuring closure of doors and other openings) may be necessary. The monitoring results may indicate that the site approval conditions (for example, on noise emissions) have been breached. Even where this is not the case, voluntary action by the proponent may take place or action may be required under the provisions of other legislation.
3. Impact monitoring can provide useful feedback for the assessment of other similar actions by helping to ensure that relevant areas of concern are identified. It can also assist in indicating where existing environmental knowledge is deficient and thus where further research may be needed to improve environmental management practice.

In most EIA systems some impact monitoring is carried out by the proponent and/or the environmental authorities, though this is increasingly becoming a responsibility of the proponent. As with implementation monitoring, impact monitoring may be required under several legal provisions and there may be little coordination between, or compilation of, measurements. In some EIA systems, the proponent is required to specify proposed implementation and impact monitoring proposals in the EIA report (for example, in Western Australia where an environmental management programme may be required (Wood and Bailey, 1994)). Figure 1.1 shows the role of monitoring in the EIA process.

Impact auditing involves comparison between the results of implementation and impact monitoring and the forecasts and commitments made earlier in the EIA process (and especially in the EIA report). It is also frequently referred to as post-auditing. The principal purpose of impact auditing is to enable the effectiveness of particular forecasting techniques to be tested and thus to improve future EIA practice by reducing the uncertainty in impact prediction (Bisset and Tomlinson, 1988). If the reliability and accuracy of forecasts generated by specific impact techniques is to be improved auditing is essential

(Wood, G., 1999, 2000). A secondary purpose is in the management of the impacts of the action concerned. Since the conditions applied at the decision-making stage should reflect the outcome of the earlier stages of the EIA process (Chapter 13), implementation and impact monitoring should be designed to permit auditing of the proponent's mitigation commitments in the EIA report and any subsequent documentation submitted.

Auditing may be carried out by the decision-making or environmental authorities, by the proponent (possibly as part of internal auditing procedures) or independently by researchers. As a result of the orientation of most EIA systems to project authorisation, rather than to the ongoing management of impacts from projects, few jurisdictions require impact auditing, as opposed to implementation or impact monitoring under non-EIA legislative requirements, for particular actions (Tomlinson and Atkinson, 1987b; Sadler, 1988, 1996; Arts, 1998; Arts and Nooteboom, 1999). Wood *et al.* (2000, p. 46) highlighted a number of obstacles to effective impact auditing, the principal ones being a lack of mandatory requirements and proponent and public indifference.

In granting approvals, it is not always possible to impose conditions to cover every eventuality, and environmental impacts may arise which are unanticipated and therefore uncontrolled. If forecasts about these impacts have been included in the EIA report (for example, forecasts about the general appearance, landscaping and tidiness of a project), the proponent may be obliged to fulfil the undertakings made, for fear that a public commitment will be seen not to have been honoured. A requirement upon the proponent to produce impact auditing reports after a certain number of years (say, two and five years), subject to checks by the environmental authorities, would be invaluable in meeting both the primary and secondary purposes of such auditing.

Primarily, impact auditing, in which the EIA report and subsequent proponent documentation provide the basic point of reference, would generate results which would not only assist in the development of EIA forecasting and monitoring methodology but would provide public reassurance about impact management. As Buckley (1991a, p. 21) has stated: 'impact audits provide a means for both industry and government to demonstrate their competence in environmental management to the public'. Secondarily, the achievement of satisfactory auditing results could provide the basis of agreement between the environmental authorities and the proponent to terminate impact monitoring programmes (for example, on certain air or water quality parameters) though implementation monitoring would need to continue.

By far the largest proportion of the literature on EIA monitoring and auditing concerns impact auditing.[1] The main conclusions may be summarised as follows:

- There appear to be no standardised audit methodologies.
- Monitoring needs to be considered and designed very early in the EIA process.
- Monitoring requires coordination, information management and resources.

- Many EIA reports contain very few forecasts.
- Many EIA report forecasts are vague and qualitative.
- Little impact monitoring (especially at construction and decommissioning phases) takes place.
- Impact monitoring data (especially for socio-economic impacts) are often inadequate for auditing purposes.
- Routine impact auditing is rarely carried out.
- Only a minority of EIA forecasts have proved accurate or almost accurate.
- Few EIA report forecasts have proved totally inaccurate.
- Post-EIA-report design changes invalidate some forecasts but do not appear to affect accuracy significantly.
- Few unanticipated impacts have been detected.
- There is little evidence of systematic bias in forecasts.

Canter (1993, p.81) suggested that monitoring and auditing approaches should be planned early in the EIA process and set down in a 'monitoring programme'. Although the main elements would differ for each individual monitoring programme, there were several generally applicable points to consider:

1. An abundance of data is already available and should be utilised.
2. As monitoring programmes are expensive and difficult, extant monitoring programmes should be used or modified.
3. Coordination between agencies and companies is necessary to avoid overlap.

Many ecologists and other scientists have argued that EIA report forecasts should be framed as falsifiable hypotheses to facilitate auditing (see, for example, Beanlands and Duinker, 1983; Caldwell et al., 1983). However, Wilson (1998) suggested that a scientific approach to monitoring, centred on quantitative baselines, prediction and statistical comparisons, was unrealistic for reasons of complexity, uncertainty and lack of capacity of prediction methods. He proposed a more practical audit procedure.

Other commentators have suggested that, while precision is desirable where it is feasible and appropriate, the value of EIA report impact forecasts does not depend of their strict auditability (see, for example, Culhane et al., 1987; Bartlett and Kurian, 1999). As Culhane (1987, p. 236) stated, EIA reports are not intended 'simply to gratify theorists trained in the scientific method', but to force consultative and participative consideration of the consequences of proposals. Given that one of the principal purposes of EIA is to ensure that appropriate mitigation measures are utilised to minimise the impacts of approved actions, Bailey and Hobbs (1990, p. 165) believed that the crucial question was: 'did [EIA] result in appropriate management action being taken?'

The publication of monitoring results is clearly a necessary check on the operation of monitoring procedures. Such information is frequently currently available from a variety of sources but is often difficult to obtain and is seldom collated. The availability of all the relevant EIA monitoring data at a single location would be a significant advance on the situation in most current EIA systems. As mentioned above, the publication of the results of audits (and the environmental authorities' response to them) would be a considerable step forward. Sadler (1988, p. 141) supported publication of auditing results because public participation 'drives many innovations in EIA practice'. Public scrutiny would be even more effective if it were supported by a right of appeal to ensure that the environmental authorities take remedial action where impact monitoring results were demonstrably unsatisfactory.

It is important that, as at the other stages of the EIA process, monitoring should function effectively (i.e. that it should provide relevant information about implementation and impacts, linked to the earlier stages of the EIA process). The funding of monitoring programmes normally rests with the proponent but in some circumstances authority/proponent funding may be appropriate. Efficiency demands that needless monitoring is not undertaken. A targeted approach to monitoring and auditing, varying according to the environmental significance of the proposal and/or the uncertainty associated with the predicted impacts, is necessary (Sadler, 1996, p. 19). If nothing else, efficiency in EIA suggests that impact monitoring and auditing should not usually involve the time and volume of results usually associated with the writing of PhD theses.

Finally, the availability of clear guidance on the procedures and techniques of action implementation monitoring, impact monitoring and impact auditing will be helpful to proponents, the decision-making and environmental

---

Box 14.1   Evaluation criteria for the monitoring and auditing of action impacts

Must monitoring of action impacts be undertaken and is it linked to the earlier stages of the EIA process?

- Must monitoring of the implementation of the action take place?
- Must the monitoring of action impacts take place?
- Is such monitoring linked to the earlier stages of the EIA process?
- Must an action impact monitoring programme be specified in the EIA report?
- Can the proponent be required to take ameliorative action if monitoring demonstrates the need for it?
- Must the results of such monitoring be compared with the predictions in the EIA report?
- Does published guidance on the monitoring and auditing of action implementation and impacts exist?
- Must monitoring and auditing results be published?
- Is there a public right of appeal if monitoring and auditing results are unsatisfactory?
- Does action monitoring function effectively and efficiently?

authorities, consultees and the public. This and the other requirements for action impact monitoring and auditing are summarised in Box 14.1 in the form of evaluation criteria. These criteria are now used to assist in the analysis of the monitoring of actions in each of the seven EIA systems.

United States

The National Environmental Policy Act 1969 (NEPA) is silent on the issue of monitoring. However, as stated in Chapter 13, the Regulations (section 1505.2 (c)) require that a 'monitoring and enforcement program shall be adopted and summarized [in the record of decision] where applicable for any mitigation'. The Council on Environmental Quality (CEQ) has made it clear that:

> the terms of a Record of Decision are enforceable by agencies and private parties. A Record of Decision can be used to compel compliance with or execution of the mitigation measures identified therein. (CEQ, 1981a, question 34d)

This is significant since, provided the lead agency commits itself to mitigation measures, it must 'include appropriate conditions in grants, permits or other approvals' and 'condition funding of actions on mitigation' (Regulations, section 1505.3(a)(b)). The same is true of mitigation measures specified in findings of no significant impact (Clark and Richards, 1999).

Monitoring is, however, essentially discretionary. As the Regulations (section 1505.3) state: '[a]gencies may provide for monitoring to assure that their decisions are carried out and should do so in important cases'. The Regulations require some implementation monitoring, since they specify that the lead agency must, upon request, inform other agencies on progress in carrying out certain mitigation measures adopted (section 1505.3(c)). Further, it must (again upon request) make available to the public the results of relevant monitoring. If the monitoring requirements are specified in the record of decision, the lead agency is obliged to implement them.

In practice, despite these requirements, monitoring is generally perceived as a weak link in the US EIA system (Canter and Clark, 1997). For example, Blaug (1993) found that monitoring of conditions in findings of no significant impact was required by only half the federal agencies, and Clark (1993) has stated that, in practice, many agencies fail to monitor the environmental impacts arising from projects. In general, monitoring is not given high priority and some monitoring commitments are not honoured because of budgetary constraints or communication lapses. Canter (1993) suggested that several steps needed be taken to improve monitoring practice in the United States. However, monitoring practice varies substantially. Some agencies have instituted impact monitoring programmes (for example, for water pollution). In addition, a few agencies (for example, the Tennessee Valley Authority (Broili, 1993)) have undertaken auditing of selected projects.

The auditing research by Culhane et al. (1987) remains the only major US study comparing the accuracy of predictions with monitored results. They found that:

Despite some general cynicism about the veracity of government promises, agency managers prove to be quite responsible in carrying out promised mitigations. (p. 254)

Dickerson and Montgomery (1993), in surveying other post-auditing studies and their own work, concluded that the NEPA process is reasonably effective in producing useful predictions of impact but that there was scope for considerable improvement (Chapter 20). Wilson (1998) demonstrated that the minor errors he identified in environmental assessments could be remedied by better scoping and those in environmental impact statements (EISs) by improving the discussion of mitigation measures.

UK

Like the European Directive on EIA, the Planning Regulations and other regulations implementing its provisions in the UK are silent on the question of monitoring. This is not to say that monitoring, especially implementation monitoring, does not take place. It is customary for local planning authorities (LPAs) to impose planning conditions on permissions and for compliance with these to be monitored by LPAs as the need arises (generally when complaints are received) (Jones *et al.*, 1998). Such conditions may include emission standards (especially relating to noise), and may sometimes require the proponent to monitor these. Similarly, the monitoring of conditions on air pollution, water pollution or solid waste disposal permits granted under the pollution control legislation tend to be carried out separately, by environmental authorities administering separate legislation.

LPAs and the pollution control authorities may also impose conditions requiring impact monitoring (especially in relation to ambient pollution levels) which may be carried out by the LPA/environmental authorities or by the proponent. (LPAs are discouraged from imposing conditions which may overlap with the requirements of other environmental control authorities (see Chapter 15).) However, impact monitoring is by no means a general requirement for projects approved under the Planning Regulations, being confined to major developments like power stations or waste disposal operations.

The requirements relating to the making public of monitoring results vary in the UK. However, EIA does not provide any regulatory mechanism for bringing together the monitoring results arising from different legislative requirements (Glasson, 1994). There is no formal right of appeal if monitoring reveals unsatisfactory emissions or impacts, but monitoring results can be used to bring pressure to bear on proponents and/or environmental authorities. Where the proponent is shown to be in breach of the conditions on the planning approval or a pollution control permit, enforcement action can be taken. However, the enforcement of both planning and pollution control conditions has left much to be desired in the past.

The existing guidance on EIA in the UK makes no reference to the monitoring of implemented project impacts. There is, however, a mention of monitoring in the Circular: 'developers may adopt environmental management systems ... to demonstrate implementation of mitigation measures and to monitor their effectiveness' (Department of the Environment, Transport and the Regions, 1999b, para. 124).

There is no formal linkage between either implementation or impact monitoring and the earlier stages of the EIA process (save, of course, for the making of the decision). Audits are not a requirement, though voluntary proponent auditing is beginning to increase in importance.

An analysis of monitoring intentions in a sample of almost 700 UK environmental statements (ESs) prepared prior to 1993 indicated that about one-third contained some reference to impact monitoring but that, even in these cases, monitoring proposals were far from comprehensive. A study of 17 implemented projects revealed that, in practice, the monitoring indicated in these ESs was carried out and that the ESs understated by about 30 per cent the amount of monitoring actually undertaken (Frost, 1997).

In a study of 28 projects granted planning permission, it was confirmed that very little monitoring of environmental impacts actually occurs. Just over half the 865 predictions in the relevant ESs were found to be auditable; the others lacked data or were too vague or ambiguous for auditing to take place. There were only six unpredicted impacts. Eighty per cent of auditable predictions were deemed to be reasonably accurate and there was no evidence of systematic bias or underprediction of impacts (Wood *et al.*, 2000). Dipper *et al.* (1998) reported broadly similar findings, as did Chadwick and Glasson (1999) from their audit of the socio-economic impacts of the construction of a nuclear power station. Graham Wood's (1999, 2000) research revealed a tendency to overestimate the magnitude of impacts. Despite these studies, it is apparent that there is still ample scope for improvement in monitoring and auditing practice in the UK.

The Netherlands

The Dutch Environmental Management Act 1994 does not require any information on monitoring to be submitted as part of the EIS. However, the EIA Commission (EIAC) advises that the proponent covers monitoring and auditing in the EIS by specifying, in its recommended scoping guidelines, that specific impact measurement and possible corrective arrangements should be described. Most guidelines issued by the competent authorities make this advice a requirement. The Ministry of Transport has issued guidance to ensure that all its EISs contain a section of at least two pages on how monitoring will be undertaken.

As mentioned in Chapter 13, the Act requires the competent authority to specify how it intends to carry out the mandatory post-decision evaluation (section 7.34(2)). The Act contains five further sections devoted to evaluation, i.e. to auditing. In summary, these are:

1. The competent authority must monitor and evaluate the consequences of the implemented action (section 7.39).
2. The proponent must provide the competent authority with monitoring information (section 7.40).

3. The competent authority must prepare an evaluation report, publish it and send it to EIAC and to the statutory consultees (section 7.41).
4. The competent authority must take such remedial action as it sees fit (for example, tightening licence conditions) if impacts are much more severe than anticipated when the decision was taken (section 7.42).
5. Detailed regulations relating to monitoring can be made (section 7.43: none has yet been issued).

Three of the Dutch guidance documents relate to auditing, and further guidance has been published by the Ministry of Transport and by some of the provinces. There is, therefore, more than adequate provision for monitoring and auditing in the Dutch EIA system. However, this provision is not proving to be universally effective.

No mention has been made of monitoring in about half of the decisions reached to date, and evaluation has been specified in less than 30 per cent of the decisions made (Arts, 1998). Although monitoring takes place under pollution control and other legislation, few evaluation reports have been published under the EIA requirements (see Chapter 5). By 1998, an evaluation study had been begun in 16 per cent of cases for which a decision had been taken and an evaluation report had been published in 6 per cent of cases (Arts, 1998, p. 539). These proportions still applied at the end of 2000 (EIAC, 2001a). These evaluation studies were concerned mainly with implementation monitoring and with comparisons with conditions and standards rather than with impact auditing (comparison with EIS predictions).

The Evaluation Committee on the Environmental Protection Act (ECW, 1990, p. 13) stressed the importance of post-project evaluation in relation both to minimising the project impacts and to improving EIA generally, and later commissioned a study of EIA monitoring. The main reasons that evaluation practice, though increasing, was still limited were found to be:

> low policy priority, lack of external pressure, lack of surveillance and sanctions, and lack of insight in the usefulness and added value of EIA evaluation in relation to the effort needed. In addition, financial, personnel and time constraints hamper EIA evaluations. (Arts, 1998, p. 539)

As a result of these findings, ECW recognised that few authorities were pressing proponents for the necessary information and most were not preparing the monitoring reports for publication and transmittal to EIAC. Accordingly, in its second report, ECW (1996) proposed that the Act be altered to allow the competent authority to determine whether or not an ex-post evaluation should be undertaken. While this recommendation was not adopted, and proponents remain reluctant to undertake evaluation, practice is slowly improving (Arts, 1998).

Canada

Monitoring (or follow-up) is emphasised strongly in the Canadian environmental assessment (EA) legislation. Section 14(c) of the Canadian

Environmental Assessment Act 1992 states that the EA process includes, where applicable, 'the design and implementation of a follow-up program', and section 16(2)(c) requires that EA reports other than for screenings must include a section on 'the need for, and the requirements of, any follow-up program in respect of the project'. 'Follow-up program' is defined in section 2 of the Act as being a programme for:

(a) verifying the accuracy of the environmental assessment of a project, and
(b) determining the effectiveness of any measures taken to mitigate the adverse environmental effects of the project.

However, the Act contains no means of ensuring that monitoring is implemented.

While the responsible authority's (RA's) guide states that 'the need for and requirements of a follow-up program need not be considered during preparation of a screening report' (Canadian Environmental Assessment Agency – CEAA, 1994, p. 102), the RA has the discretion to require 'details on monitoring programs to evaluate the effectiveness of mitigation measures as well as to determine the accuracy of the EA' (p. 94). Further, for screenings, the RA 'must make a decision about whether a follow-up program is appropriate. If so, it must ensure that one is designed and implemented' (p. 102).

For projects subject to comprehensive study or to public review the RA must decide whether to implement a follow-up programme. In the case of panel and mediation reports, the guide (CEAA, 1994, p. 124) revealingly states:

> The RA is not obliged to follow the recommendation for a follow-up program, but if it does not, it must justify the decision publicly. RAs should keep in mind that the report from a mediator or panel is the product of an open, fair and rigorous review involving all key interests, and that the report's recommendations cannot be treated lightly.

It goes on to state that the critical question in determining the need for a follow-up programme is one of uncertainty or unfamiliarity (Box 14.2). In practice, however, the cost of follow-up provides a powerful disincentive to the RA.

The implementation of appropriate mitigation measures is discussed in Chapter 15. Together with the follow-up programme, these measures provide for the monitoring of both the accuracy of predictions of the environmental effects of the project and the effectiveness of mitigation measures applied to projects subject to comprehensive study or public review. Although the Act contains no absolute means of ensuring either compliance or monitoring by RAs, the extensive requirements for the provision of information to the public about follow-up and monitoring (section 38) – but not for public appeal if the results fail to meet expectations – provide a valuable check.

The monitoring programme is clearly linked to the earlier stages of the EA process in Canada, but the Act provides no mechanism for ameliorative actions to be taken where monitoring reveals the need for them: this is left to such means as the enforcement of permit conditions under other legislation.

Box 14.2   Canadian guidance on monitoring predicted effects

When a follow-up program may be appropriate

The RA [responsible authority] should develop a follow-up program for a project when the circumstances warrant. Examples include situations where

- the project involves a new or unproven technology
- the project involves new or unproven mitigation measures
- an otherwise familiar or routine project is proposed for a new or unfamiliar environmental setting
- the assessment's analysis was based on a new assessment technique or model, or there is otherwise some uncertainty about the conclusions
- project scheduling is subject to change such that environmental effects could result.

Source: CEAA, 1994, p. 124.

The Canadian Environmental Assessment Research Council commissioned various early studies on EA monitoring (see, for example, Munro *et al.*, 1986; Krawetz *et al.*, 1987). More recently, David Redmond and Associates (1999) reported that follow-up requirements were included in about half the 191 screening reports they examined. The Commissioner of the Environment and Sustainable Development (CESD, 1998) found that, of the 48 screenings for which follow-up of environmental effects was appropriate, RAs planned to follow up about three-quarters. He was 'concerned that one quarter of required follow-ups are not requested as a condition of approval' (CESD, 1998, pp. 6–24). He recommended that monitoring be strengthened and that the results be made widely available. Following the five-year review of the Act, the Minister of the Environment (MoE, 2001, p. 24) proposed that it be strengthened to require RAs:

> to ensure that a follow-up program ... is conducted for projects that have undergone a comprehensive study or panel review ... [and] to consider whether follow-up programs are appropriate for screenings they are conducting.

These provisions, and others relating to the use of monitoring information for both adaptive project management measures and for improving the quality of future EAs, were contained in the bill published by the Government of Canada (2001). CEAA is to publish guidance on follow-up (MoE, 2001, p. 25).

Commonwealth of Australia

Monitoring provisions existed in the Australian EIA system under the repealed Impact of Proposals Act but they were in essence discretionary. Perhaps unsurprisingly, monitoring of action implementation and impacts by Environment Australia was rarely employed in practice. Organisational arrangements, lack of staff and lack of political determination rendered the Commonwealth EIA monitoring provisions virtually inoperative. (Industrial monitoring and monitoring by other agencies took place, but was not related to the EIA process.) Ongoing control over implementation and performance was frequently lacking and formal comparison with impacts predicted in EISs did not take place (Jambrich et al., 1992).

Various recommendations to overcome the shortcomings in the Commonwealth system were suggested over the years, most involving the use of enforceable and auditable conditions and an active monitoring programme (see, for example, Australian and New Zealand Environment and Conservation Council, 1991). The Commonwealth Environment Protection Agency (1994, p. 45), in its agenda for reform, proposed 'the introduction of post-assessment monitoring as a standard element of the Commonwealth environmental impact assessment process'. All the results from, and reports on, monitoring and auditing were to be made public (p. 46). The ministerial consultation paper proposed that the Minister 'will have the power to monitor compliance with conditions, enforce conditions and take action in the event of default' (Hill, 1998, p. 16). Some of these proposals were given legal force in the Environment Protection and Biodiversity Conservation Act 1999 (EPBC Act), which represents a considerable improvement on the previous situation.

Public environment reports and EISs must contain a section on the monitoring arrangements for ensuring that mitigation is effective. This is achieved by requiring:

> an outline of an environmental management plan that sets out the framework for continuing management, mitigation and monitoring programs for the relevant impacts of the action, including any provisions for independent environmental auditing. (EPBC Regulations 2000, Schedule 4, para. 4(d))

In addition, the information which the proponent must provide about other approvals and conditions includes 'a description of the monitoring, enforcement and review procedures that apply, or are proposed to apply, to the action' (EPBC Regulations, Schedule 4, para. 5(d)).

The EPBC Act contains additional powers. As mentioned in Chapter 13, the Environment Minister, in granting approval for an action, can impose conditions relating to auditing and monitoring. These are specified in section 134(3) of the EPBC Act:

(a) conditions requiring an environmental audit of the action to be carried out periodically by a person who can be regarded as being independent from the person whose taking of the action is approved; and
(b) conditions requiring specified environmental monitoring or testing to be carried out . . .

There is, however, no provision for monitoring results to be compared with predictions, though other powers can be triggered on the basis of auditing. The Environment Minister can direct that an 'environmental audit' be carried out if it is believed that a person has contravened an environmental approval issued under the EPBC Act. This power applies not only to the contravention of conditions but can be utilised where:

the impacts that the action authorised ... are significantly greater than was indicated in the information available to the Minister when the authority was granted. (section 458(1)(b))

In certain circumstances, civil or criminal penalties can apply to executive officers of a corporation that contravenes the requirements for environmental approvals under the EPBC Act, including the provision of false or misleading information (in EIA reports) to obtain approval (Part 17, Division 15). Further, there is a power to require payment for the remediation of environmental damage arising as a result of contravention of the Act (Part 18) (Environment Australia, 1999a). Finally, the Minister may revoke an approval if a serious contravention occurs or if the proponents did not accurately identify the impacts either deliberately or due to negligence. These powers are substantial and, though Hughes (1999) considers them to be inadequate, the threat of utilising them should provide Environment Australia with a valuable weapon for enforcing both EIA report quality and compliance.

Despite the absence of monitoring administered by Environment Australia under the previous legislation some empirical results are available from academic research. Buckley (1991a) found few verifiable predictions in an audit of several Commonwealth and state EISs. There was little obvious bias in the limited number of forecasts he was able to check: 57 per cent of impacts were less severe than predicted, 43 per cent more severe. These results do not differ greatly from those reported elsewhere (Culhane *et al.*, 1987; Bisset and Tomlinson, 1988; Wood *et al.*, 2000). Morrison-Saunders (1996) demonstrated that monitoring plays an important role in the strong relationship between EIA and environmental management that exists for some (Western) Australian projects.

New Zealand

The Resource Management Act 1991 makes both general and specific reference to monitoring. Every local authority in New Zealand is required to 'gather such information, and undertake or commission such research, as is necessary to carry out effectively its functions under this Act' (section 35(1)). Each local authority must also monitor the exercise of the resource consents that have effect in its area (section 35(2)(d)). The Fourth Schedule (see Box 7.3) specifies that assessment of environmental effects (AEE) reports should contain proposals for impact monitoring where the scale and significance of impacts require it. This provides the opportunity for a strong linkage between assessment and monitoring. There is a public right to ask the Environment Court to issue an enforcement order if conditions are not met.

It appears that the principal aim of the monitoring provisions is the checking of compliance or enforcement rather than EIA system enhancement, for which predictions need to be checked against actual impacts to test their accuracy (see above). No guidance on monitoring was initially issued in New Zealand and the practical guide (Morgan and Memon, 1993) made very little reference to monitoring. The recent Ministry for the Environment guidance to applicants (MfE, 1999a) was silent about monitoring the implementation of current conditions. The fullest advice on monitoring is that published by the Parliamentary Commissioner for the Environment (PCE, 1996a) in the form of a good practice guide following a study of compliance with resource consents.

However, the Ministry for the Environment guide on review ('auditing') stressed that councils needed to consider whether any effects needed to be monitored, on the basis of the proponent's AEE: 'how much monitoring or review of conditions is required should reflect the level of risk of adverse effect' (MfE, 1999b, p. 67). It appears that emphasis on monitoring in New Zealand is increasing as the strongest monitoring advice to date was issued in 2001:

> Monitoring provides information on consent compliance or non-compliance and is useful when reviewing or renewing consents. It also provides feedback to the consent authority's consent procedures, and plans and policies. The proactive monitoring of consent conditions is strongly recommended. ... If monitoring is not undertaken ... any effects of the activity on the environment are unable to be determined and managed. (MfE, 2001a, para. 10.3.2)

Where monitoring does take place, the results are normally made public.

Montz and Dixon (1993) suggested that because monitoring falls to regional authorities, there were likely to be variations in the extent to which the monitoring provisions were implemented and so it has proved, though the situation has improved over the years. It appeared that few authorities had developed monitoring systems or appropriate methods by the mid-1990s, not withstanding the legal requirement to monitor (PCE, 1995, 1996a; Smith 1996).

However, by 2000 all regional and unitary authorities allocated some resources to consent and action monitoring and 96 per cent of territorial authorities carried out some monitoring (MfE, 2001c). Some 57 per cent of resource consents administered by local authorities responding to the Ministry for the Environment were monitored during 1998/99 and compliance was nearly 70 per cent. However, a lack of local authority resources appeared to be the most common reason why more local authorities were not monitoring consents for compliance with conditions (MfE, 2000a, p. 39). Most councils mainly utilise public complaint rather than monitoring to indicate problems, despite the difficulties of relying solely as such an approach (PCE, 1995).

It appears that, despite recent progress, there remains considerable scope for improvement in monitoring practice in New Zealand.

South Africa

Monitoring has long been recognised as a crucial component of environmental management in South Africa. Thus, the Environment Conservation Act 1989, section 26(c), empowered the Minister of Environmental Affairs to make regulations concerning:

> the procedure to be followed in the course of and after the performance of the activity in question or the alternative activities in order to substantiate the

estimations of the environmental impact report and to provide for preventative or additional actions if deemed necessary or desirable.

The well-known integrated environmental management (IEM) guidelines (Department of Environmental Affairs – DEA, 1992) were very specific:

> A monitoring programme should be required for all approved proposals. . . . This programme should include clear guidelines as to what should be done, who should do it, and who should finance it. Aspects to be covered . . . include verification of impact predictions, appraisal of mitigatory measures, adherence to approved plans, and compliance with conditions of approval. (DEA, 1992, vol. 1, pp. 8–9)

The IEM guidelines went on to suggest that 'periodic assessments of the positive and negative impacts of proposals should be undertaken' (DEA, 1992, vol. 1, p. 9). Auditing was intended to provide feedback on, *inter alia*, the accuracy of predictions. Volume 4, on report requirements, devoted a helpful section to management plans, monitoring and environmental contracts.

It is therefore surprising that neither the EIA regulations (Republic of South Africa, 1997) nor the EIA guidelines (Department of Environmental Affairs and Tourism, 1998c) refer to monitoring. It is at this stage that the differences between the legislated EIA system and the voluntary IEM procedure are most apparent. Hill (2000, p. 52) felt that 'the lack of regulations on EIA follow-up constitutes a retrograde step for environmental management in South Africa.'

As in other jurisdictions, some monitoring has been carried out under non-EIA legislation (for example, the air pollution control and water pollution control Acts). It was common for monitoring conditions to be included in approvals under the voluntary IEM procedure. These often included requirements for an environmental management plan to be implemented and independently audited, for an on-site environmental control officer to be appointed during construction (Barker, 1996; Hill, 2000) and for subcontractors to be penalised if environmental safeguards were violated.

Best IEM practice was continued in the use of conditions attached to authorisations issued under the EIA Regulations. These have included the use of audits, the keeping of records of compliance, the notification of non-compliance (and their inspection by the relevant authority on demand), the provision of copies of the approval and the associated conditions to all interested and affected parties, and the right of the relevant authority to review and amend the conditions periodically. This last condition, with others, allows the relevant authority to require the proponent to take ameliorative action if necessary.

Despite the IEM guideline recommendations and the emphasis by many consultancies on environmental management, it is generally accepted that monitoring has often been inadequate. There has been insufficient verification of the implementation of mitigation measures (Weaver, 1996) and monitoring and auditing has rarely been undertaken (Hill, 2000). Thus, Duthie (2001, p.221) stated:

> with the exception of Gauteng and Western Cape, most provinces do not undertake follow-up auditing of projects to assess compliance with conditions of approval.

The problem of underfunding and understaffing of provincial and local authorities has meant that the relevant authorities must rely on the complaints of neighbours and the integrity of developers and their consultants for information about non-compliance (Hill, 2000). The capacity of relevant authorities to take enforcement action if admonition proves ineffective has obviously been severely limited.

The National Environment Management Act 1998 (NEMA) requires the:

> investigation and formulation of arrangements for the monitoring and management of impacts, and the assessment of the effectiveness of such arrangements and their implementation. (section 24(7)(f))

This is clearly intended to permit a considerable strengthening of the 1997 EIA Regulations when the Department of Environmental Affairs and Tourism develops EIA regulations under NEMA (see Chapter 5).

Summary

Table 14.1 demonstrates that none of the EIA systems fully meets the impact monitoring evaluation criterion. Monitoring is an acknowledged weakness of the US EIA system, and there is no provision for monitoring in the UK or South African EIA systems (though some monitoring can be accomplished under other legislative or voluntary means). The New Zealand Resource Management Act 1991 imposes a general duty upon local authorities to monitor project impacts but this is infrequently undertaken and there is no linkage to the earlier stages of the EIA process in these essentially discretionary monitoring requirements.

The Dutch EIA system contains several impact monitoring provisions but, in practice, these are not always implemented. The same is true in Canada, where the Canadian Environmental Assessment Act also contains extensive impact monitoring (follow-up) requirements which are often not observed. This largely applies to the Australian EIA system, where the monitoring provisions are quite clearly discretionary rather than mandatory. Because of these weaknesses, these three EIA systems can be adjudged only partially to meet the evaluation criterion. It is quite clear that monitoring and auditing is one of the weakest areas of EIA activity and that EIA systems need either to be strengthened or better coordinated with ongoing project environmental management requirements and with other environmental monitoring programmes.

Note

1. See, for example, Holling (1978), Bisset (1981), Caldwell *et al.* (1983), Beanlands and Duinker (1983), Sewell and Korrick (1984), Culhane (1987), Culhane *et al.* (1987), Tomlinson and Atkinson (1987a,b), Bisset and Tomlinson (1988), Sadler (1988, 1996), Bailey and Hobbs (1990), Buckley (1991a), Bailey *et al.* (1992), Canter (1993), Glasson (1994), Arts (1998), Dipper *et al.* (1998), Arts and Nooteboom (1999), Chadwick and Glasson (1999), Glasson *et al.* (1999), Wood, G. (1999, 2000), Wood *et al.* (2000).

Table 14.1   The treatment of impact monitoring and auditing in the EIA systems

Criterion 9: Must monitoring of action impacts be undertaken and is it linked to the earlier stages of the EIA process?

Jurisdiction	Criterion met?	Comment
United States	No	Monitoring essentially discretionary but some requirements where mitigation measures specified in record of decision. Practice often weak
UK	No	No provision for monitoring. Uncoordinated implementation monitoring takes place under planning and other legislation unrelated to earlier stages in EIA process
The Netherlands	Partially	Specific requirements relating to monitoring and comparison with EIS. Often not observed in practice
Canada	Partially	Extensive provision in Act for follow-up of mitigation measures involves monitoring but no mechanism for ensuring full compliance
Commonwealth of Australia	Partially	Extensive discretionary provisions for monitoring and enforcement measures exist, including power for Environment Australia to demand an environmental audit
New Zealand	No	Duty of local authority to monitor impacts of project resource consents in Act often not complied with. Seldom linked to earlier EIA stages
South Africa	No	No formal requirements for monitoring exist but use of monitoring (not auditing) conditions common

Chapter 15

Mitigation of impacts

If the consideration of alternatives lies at the heart of the environmental impact statement (see Chapter 8) then the mitigation of environmental impacts lies at the heart of the EIA process. In practice, the consideration of alternatives is intertwined with the consideration of mitigation measures. In the Netherlands, for example, mitigation measures are called 'alternatives'. One twin purpose of EIA is, in essence, to persuade the decision-making authority to authorise the action by reducing the adverse impacts of a proposed project to an acceptable level while maximising the beneficial impacts. The second purpose of EIA is to prevent unsuitable development by demonstrating that certain impacts cannot be mitigated to the point of acceptability. This chapter explains why the mitigation of environmental impacts is important and advances several evaluation criteria to assist in the review of this element of the EIA process. These criteria are then employed in the analysis and comparison of the EIA systems in the United States, UK, the Netherlands, Canada, Commonwealth of Australia, New Zealand and South Africa.

Mitigation of impacts within the EIA process

Mitigation, or amelioration, of the severity of impacts arising from an action can take a variety of forms. Canter (1996, p. 47) favoured an approach in which sequential consideration of mitigation measures was undertaken. Similarly, the Department of the Environment, Transport and the Regions (DETR, 1997d) ranged possible mitigation measures in a mitigation hierarchy:

- Avoidance at source.
- Minimise at source.
- Abatement on site.
- Abatement at receptor.
- Repair.
- Compensation in kind.
- Other compensation and enhancement.

The mitigation measures higher in the hierarchy were recommended as preferable in most circumstances.

Examples of mitigation measures may include process alterations to reduce emissions, altering pollution control equipment to render it more effective, adjusting the hours of operation of a plant, changing site layout to reduce visual, noise or air pollution impacts, the requirement of fencing and walls, the amendment of vehicular access arrangements, the provision of mounding and planting, the creation of replacement habitat, and many others.[1] Indeed the whole range of land-use planning, pollution and other controls and measures should be considered during the EIA process to ensure that suitable mitigation measures are adopted.

As Figure 1.1 demonstrates, mitigation is iterative: different measures may be proposed at the various stages of the EIA process. Glasson *et al.* (1999) confirmed that mitigation is inherent in all aspects of the EIA process. For example, the results of the review of the EIA report, and of the consultees' and public's comments upon it, may yield proposals for mitigation additional to (or different from) those proposed in the EIA report itself.

The adoption of some mitigation measures may involve considerable costs, though other effective measures may cost very little (for example, alteration of road access, alternative material storage arrangements). Canter (1996) suggested that their cost-effectiveness should be assessed prior to the final choice of mitigation measures. There will normally come a point at which the developer may withdraw a proposed development because the additional costs associated with mitigation measures are deemed to be too high (Wood, 1989). On the other hand, the decision-making authority may ask whether mitigation of the impacts of a proposal will contribute towards the achievement of sustainable development. In other words, it may be that the authorities will seek either no deterioration of environmental resources, or even a net improvement of environmental resources through offsets or compensation (as in US air pollution non-attainment areas), rather than merely a reduction of (or a remedy for) impacts. In these circumstances, it may be necessary to consider radical alternatives to the proposal.

Mitigation measures can themselves have impacts which need to be identified, predicted and evaluated. For example, the use of earth mounding to provide noise baffles or to screen development from public view can create unnatural landforms which are themselves visually obtrusive. The use of interaction matrices, as part of 'a schedule of mitigation' to be incorporated in the EIA report, has been suggested as a way of identifying the knock-on effects of mitigation measures (DETR, 1997d). It was recommended that the potential effect of each measure on significant impacts, the commitment to implementing the measure, and the reasons for not mitigating certain impacts be outlined. The schedule was to be amended and updated as the EIA process proceeded (DETR, 1997d). It is therefore clear that careful planning is an important element of any mitigation strategy.

Not only may there be a comparison of the benefits of mitigation but there may also need to be a trade-off between the mitigation of different impacts. Where these relate to pollution, the best practicable environmental option needs to be sought (Royal Commission on Environmental Pollution, 1988). As Glasson *et al.* (1999, p. 155) stressed:

Mitigation measures must be planned in an integrated and coherent fashion to ensure that they are effective, that they do not conflict with each other, and that they do not merely shift a problem from one medium to another.

However, it will frequently be necessary to trade off pollution and other types of impact in addition. These trade-offs can be complex. It is for this reason, among others, that consultation and participation are so important in the EIA process. Consultees and the public can provide invaluable assistance not only in suggesting mitigation measures but also in determining which residual impacts are tolerable and which cannot be countenanced. As well as determining the acceptability of impacts, Canter (1996) noted that consultation and public participation could also give an indication of whether the proposed mitigation measures are themselves acceptable.

Mitigation measures, to be effective, must be implemented. This is particularly important because the significance of potential impacts detailed in EIA reports is often dependent on the implementation of certain mitigation measures. Several EIA systems (including those in the United States) initially possessed no effective means of ensuring that the mitigated measures proposed in EIA reports were actually carried through (Hollick, 1981). Many of these shortcomings have now been corrected (see, for example, in relation to California, Bass *et al.*, 1999, p. 121). Perhaps as a consequence of the former lacuna in California, it has been suggested that the criteria shown in Box 15.1 should be employed in drafting effective mitigation measures. The preparation of a mitigation schedule can increase the commitment of a proponent to implementing mitigation measures, particularly if it is tied to a monitoring programme (see Chapter 14). Indeed, Glasson *et al.* (1999) have described the

---

Box 15.1   Guidelines for drafting effective mitigation measures

WHY: State the objective of the mitigation measure and why it is recommended.

WHAT: Explain the specifics of the mitigation measure and how it will be designed and implemented:
- Identify measurable performance standards by which the success of the mitigation can be determined.
- Provide for contingent mitigation if monitoring reveals that the success standards are not satisfied.

WHO: Identify the agency, organisation, or individual responsible for implementing the measure.

WHERE: Identify the specific location of the mitigation measure.

WHEN: Develop a schedule for implementation.

Source: Bass *et al.*, 1999, p. 114.

incorporation of a monitoring programme as one of the most important of all mitigation measures. The use of a negotiated environmental management plan with appropriate monitoring and opportunities for modification of monitoring arrangements, as in Western Australia (Wood and Bailey, 1994) is an example of such an approach.

Clearly, financial compensation (or remuneration in kind) has a role to play in gaining public acceptance of unmitigated environmental impacts. Though such payments may be controversial, and though there may be considerable problems in determining who should receive payments, the use of compensatory measures can help to resolve disputes and achieve the aims of EIA. Several instances of the use of financial compensation exist in siting decisions involving EIA in the United States, New Zealand and Australia. In the UK, compensation is often achieved through the use of planning gain, which may involve the provision of off-site landscaping or of community facilities.

As mentioned in Chapter 8, preliminary documentation produced during the early stages of the EIA process should show clear evidence of the mitigation/avoidance of environmental impacts in the initial action design. Similarly, documentation prepared for screening and scoping purposes should also address the question of mitigation of impacts. Clearly, details of mitigation should be set down in the EIA report, which should provide a record of all the mitigation measures and modifications suggested or accepted by the proponent, preferably in the form of a schedule. Lee and Colley (1992, in Lee *et al.*, 1999, p. 43) suggested that EIA reports should deal fully with the:

> Scope and effectiveness of mitigation measures: All significant adverse impacts should be considered for mitigation. Evidence should be presented to show that proposed mitigation measures will be effective when implemented.

Mitigation should therefore continue to be considered during the review and revision of EIA reports, during decision making and during the monitoring stages of the EIA process. As indicated in Chapter 13, carefully worded conditions to any approval are frequently used to codify the set of designs and mitigation measures approved at the decision-making stage. These conditions need to be monitored and enforced to ensure that mitigation remains effective (see Chapter 14). However, some flexibility is needed to ensure that, on the one hand, unexpected impacts are mitigated and that, on the other hand, unnecessarily expensive mitigated measures can be modified.

As at other stages of the EIA process, the existence of published guidance on the mitigation and modification of actions to render them environmentally more acceptable is helpful to developers, consultants, environmental and decision-making bodies and the public. In practice, it appears that relatively little specific advice on mitigation in the EIA process, and especially at stages other than EIA report preparation, has been published in many jurisdictions. The UK research report on mitigation measures, which contains some guidance (DETR, 1997d), is an exception.

Finally, the mitigation of action impacts should take place effectively (i.e. mitigated measures should actually ameliorate impacts) and efficiently (i.e.

> ## Box 15.2   Evaluation criteria for the mitigation of impacts
>
> **Must the mitigation of action impacts be considered at the various stages of the EIA process?**
>
> - Must clear evidence of the mitigation/avoidance of environmental impacts be apparent in the initial action design described in preliminary EIA documentation?
> - Must a schedule of mitigation measures and their implementation be set down in the EIA report?
> - Must evidence of the consideration of mitigation be presented during screening, during scoping, during EIA report review and revision, during decision making and during monitoring?
> - Does published guidance on mitigation and modification exist?
> - Does the mitigation of action impacts take place effectively and efficiently?

they should not involve the expenditure of unnecessary time, manpower or financial resources). Clearly, the earlier in the EIA process that the mitigation proposals are made, the more effective and efficient they are likely to be, since they will be progressively refined during the consideration of the proposal. This and the other criteria for reviewing the treatment of mitigation measures in EIA systems are summarised in Box 15.2. These criteria are now employed to help to analyse the procedures for mitigation in each of the seven EIA systems.

United States

Although the treatment of alternatives is the 'heart of the environmental impact statement' (EIS) in the United States (see Chapter 8), mitigation of environmental impacts is probably the most important outcome of the American EIA process. As indicated in Chapter 13, the abandonment of unsatisfactory proposals is rare. Mitigation certainly pervades the US EIA system and most proposals are heavily modified by the end of the EIA process. As Clark (1997, p. 22) has stated:

> [the US] EIA process has caused the modification of thousands of proposed actions in ways that have reduced or avoided impacts and in many cases saved money associated with the project.

Mitigation is, of course, the central determinant of one outcome of screening, the mitigated finding of no significant impact (FONSI). Following the preparation of an environmental assessment (EA) a mitigated FONSI may be (and almost always is) prepared where the potentially significant environmental effects of a proposal can be rendered acceptable by the adoption of appropriate mitigation measures. The courts have ruled that mitigated FONSIs are legally adequate where, *inter alia*, the agency can convincingly show that its mitigation measures will reduce all the significant environmental impacts identified in the EA to less-than-significant levels (Bass *et al.*, 2001, p. 60). In practice, there is heavy reliance on mitigation measures to justify EAs and FONSIs, showing either that environmental considerations are increasingly being integrated early in the decision-making process or that agencies are desperate to avoid preparing EISs (or both).

Mitigation measures must be considered during scoping and should be summarised in the recommended report of scoping. Both draft and final EISs must contain a discussion of appropriate mitigation measures. The Council on Environmental Quality (CEQ) Regulations (section 1508.20) specify that mitigation must involve avoiding, minimising, rectifying, reducing or compensating for significant environmental effects. The implication is that mitigation measures, which are not specific and tangible (for example, proposals to consult, to conduct further studies or to monitor) will not generally solve the environmental problems identified.

All relevant, reasonable mitigation measures that could improve the action must be identified in the EIS, even if they are outside the jurisdiction of the lead agency (CEQ, 1981a, questions 19(a), (b)). However, the lead agency is not obliged to commit itself to implementing the mitigation measures identified in the EIS unless its own EIA regulations require their adoption. This is a crucial lacuna.

Those mitigation measures adopted by the lead agency must be specified in the record of decision (ROD), together with a monitoring and enforcement programme for each measure (see Chapter 13). As stated in Chapter 14, a ROD can be used to compel compliance with, or execution of, the mitigation measures contained in it. While these powers are valuable, the lead agency can easily circumvent them by failing to adopt relevant mitigation measures. Indeed, the implementation of mitigation measures, like monitoring, is a significant weakness of the US EIA system.

The Environmental Protection Agency (EPA) conducted a study in 1987 which showed that, using the definition of adequacy in the Regulations (see above), the effectiveness of the mitigation measures in about 20 per cent of the EISs that it reviewed was questionable (EPA, quoted in Bass *et al.*, 2001, p. 118; see also Wilson, 1998). Even where mitigation measures are specified in EISs and RODs, implementation may be unsatisfactory because of budgetary constraints or failure to inform relevant personnel or to incorporate measures in construction contracts. However, the effective implementation of mitigation measures is gradually improving as public concern and knowledge grow.

UK

The UK Planning Regulations require that the environmental statement (ES) contain a description of the measures envisaged in order to avoid, reduce and, if possible, remedy significant adverse effects' (Schedule 4, Part II). The Circular makes it clear that local planning authorities are expected to impose conditions designed to mitigate impacts, or to require a legal planning agreement for this purpose, when granting planning permission:

Mitigation measures proposed in an ES are designed to limit the environmental effects of the development. Planning authorities will need to consider carefully how such measures are secured.... Conditions attached to a planning permission may include mitigation measures.... A planning condition may require a scheme of mitigation ... to be submitted.... However, planning authorities should not seek to substitute their own judgement on pollution control issues for that of the bodies with the relevant expertise and the statutory responsibility for that control. (Department of the Environment, Transport and the Regions – DETR, 1999b, p. 26)

There are no formal requirements for the treatment of mitigation measures at earlier stages in the EIA process.

In addition to the Circular (DETR, 1999b) there is other published guidance on mitigation and modification, most of which focuses on avoidance, reduction and remedy (Department of the Environment, 1995; DETR, 2000a). Examples of a 'mitigation hierarchy' are shown in Box 15.3 (DETR, 1997d).

There have been several studies of practice relating to modifications and mitigation. In a sample of 40 projects, Jones *et al.* (1998) found that 70 per cent were modified as a result of EIA. Modifications took place prior to ES submission in 50 per cent of cases, both before and after submission in 30 per cent of cases, and following the ES in 20 per cent of cases. Most modifications arose as a result of formal consultations. These results tended to confirm the findings of earlier studies. Barker and Wood (1999) found that EIA resulted in modifications in all the cases they examined and that 75 per cent of these were

---

Box 15.3   Different types of UK mitigation measures

Avoidance or minimisation at source
Investing in public transport rather than building a new road.
Demand management rather than building a new reservoir.

Abatement on site
Installing air pollution control or effluent treatment plant.
Enclosing plan to contain noise or odours.

Abatement at the receptor
Planting trees to hide the site at the nearest residential properties.
Installing noise insulation at the receptor property.

Repairing or remedying the impact
Installing clean-up equipment for spills and training staff in its use.
Establishing a procedure for handling and responding to complaints and training staff.

Compensation (in kind or other) including enhancement
Replacing homes and other property.
Creating new habitats.

Source: adapted from DETR, 1997c, pp. 17–18.

regarded as being of major significance. Most developers incorporated measures to mitigate adverse impacts into project design.

This finding was confirmed in a study of the treatment of mitigation in the EIA process (DETR, 1997d). Other notable findings were that:

- most ESs addressed mitigation;
- the range of mitigation options was restricted;
- little attention was paid to construction phase impact mitigation;
- mitigation descriptions were often imprecise;
- commitment to implement mitigation measures was unclear;
- residual impacts following mitigation were poorly specified. (DETR, 1997d, p. 2)

It is apparent that while there is scope for further improvement in the mitigation of project impacts throughout the UK EIA process, practice (particularly in the treatment of mitigation in ESs) has developed as experience has been gained.

The Netherlands

Mitigation is not referred to by name in the Dutch Environmental Management Act 1994. Indeed, there is little specific provision for mitigation measures to be considered (as demanded by the European Directive on EIA) in the Dutch legislation. It is not a requirement that mitigation be mentioned in the notice of intention and, frequently, mitigation measures are not specifically included in the scoping guidelines. EISs frequently (but not invariably) contain specific reference to mitigation measures. Perhaps surprisingly, the suite of Ministry of Housing, Spatial Planning and the Environment EIA guidance documents does not include a volume on mitigation, although the Ministry of Transport has produced such guidance.

Mitigation is largely subsumed in the treatment of alternatives in the Dutch EIA system (see Chapter 8). The environmentally preferable alternative is required not only to incorporate all feasible mitigation measures but also to reduce effects on the environment 'as far as possible using the best means available of protecting the environment' (section 7.10(3)) but may also be required to include a description of any measures necessary 'to compensate for the remaining adverse effects' (section 7.10(4)). For example, the compensation of habitat loss by the planting of trees is often regarded as an alternative to the proposed action rather than as mitigation. The Dutch would claim that strong mitigation measures thus provide the starting point for consideration of a project rather than being elicited from the proponent later. Although there is no requirement to select the environmentally preferable alternative, information about this alternative provided in the EIS has influenced the final decision in about 75 per cent of cases.[2] The competent authority generally implements the mitigation of the impacts of the proposed action by imposing enforceable conditions on its approvals.

There is no doubt that mitigation of environmental impacts is taking place as a result of the EIA process. Although not universal, changes to improve the proposal's environmental compatibility usually take place by iteration during the EIA process. Thus there have been elements of mitigation in nearly all the 25 per cent of cases where the alternative most favourable to the environment did not influence the final decision. Changes in location, in design and in technical controls are all common. There are numerous examples of good mitigation practice (see, for example, Morel, 1996; Scholten, 1998). Box 15.4 summarises the approach to mitigation of the effects of gas exploitation in the very sensitive Wadden Sea.

Box 15.4   Wadden Sea natural gas extraction impact mitigation and compensation measures

Oil companies want to exploit natural gas reserves under the Wadden Sea. Though the oil companies have concession rights for exploitation dating back to 1960, exploration and exploitation was postponed through adoption of a moratorium.

The Dutch government has [now] decided that, in the public interest, there is a need to prospect for and extract natural gas reserves in the North Sea coastal zone and the adjacent Wadden Sea. As highly important nature values are at stake, environmental assessments at both the strategic and the project level would have contributed to a balanced decision-making process. However, the assessment was restricted to the project level; the disadvantage of this was that discussions on strategic topics were not resolved at the strategic level and complicated decision-making at the project level.

The government has determined that exploration for natural gas (and later its exploitation) must satisfy the most stringent environmental conditions. Amongst others, this involved that production should take place from outside the Wadden Sea, through lateral approaching the geological targets. Remaining impacts should be prevented and mitigated by application of best available technological means, to be identified in project-EIAs.

Research indicates that despite adoption of all possible preventive and mitigatory measures there will remain impacts. The most important is supposed to be soil subsidence due to natural gas extraction. This would result in a loss of tidal sand flats. Those sand flats are of evident importance in the ecosystem. One of the preconditions in the agreement on lifting the moratorium was the obligation to quantify this impact exactly before starting exploitation. The oil companies are responsible to compensate this impact by supplying huge amounts of sand at a determined distance from the Wadden Sea. The sand has to be carried in by tidal movement to compensate for the soil subsidence.

The main purpose of EIA, therefore, was to identify the 'alternative most favourable to the environment' (AMFE). Although complicated by the absence of [strategic environmental assessment], project EIA still proved to be a strong tool for guiding the development of the initiative in a more sustainable direction. Key elements were:

(continued)

(continued)

- EIA stimulated a *proactive approach* – the vulnerability of the area determined the project formulation.
- The obligation to *develop and compare alternatives* in the EIS enabled the selection of the AMFE, which was essential in this process.
- The *involvement of the public and the independent Commission for EIA* in reviewing the EIS led to the formulation of additional mitigation measures in the AMFE.
- The need to *identify gaps in knowledge* in the EIA process resulted in a recommendation by the Commission to further study the natural values in the area.
- The competent decision-making authority had to *substantiate in the decision the significance assigned to the environmental information*. This showed clearly that the decision did not fully comply with the elements of the AMFE identified in the EIS, which was one of the reasons the decision was challenged successfully in a court case.
- In the EIA process it became clear that *protection criteria for sensitive areas* such as the North Sea coastal zone had not been specified in sufficient detail to allow clear conditions to be placed on the proposed activities. As a result of the EIA, a start has been made in drawing up these further specifications.

Sources: adapted from Morel, 1996, p. 53; 1998, p. 43.

Canada

No Canadian EA, from screening through to panel review, can be undertaken without consideration of 'measures that are technically and economically feasible and that would mitigate any significant adverse environmental effects of the project' (Canadian Environmental Assessment Act 1992, section 16(1)(d)). The responsible authority (RA) is enjoined in section 20(1)(a) of the Act to 'ensure that any mitigation measures that the responsible authority considers appropriate are implemented'. Section 2(1) of the Act defines mitigation as:

> the elimination, reduction or control of the adverse environmental effects of the project, and includes restitution for any damage to the environment caused by such effects through replacement, restoration, compensation or any other means.

As with monitoring (see Chapter 14), however, mechanisms for ensuring the implementation of mitigation measures are weak.

The responsible authority's (RA's) guide to the Act provides brief advice on mitigation measures, emphasising that they should be considered first during the design phase of the project and be gradually refined as the EA progresses (Canadian Environmental Assessment Agency – CEAA, 1994). The use of specialists and public consultation is recommended where circumstances demand. The guide suggests that mitigation measures can be implemented through, for example, 'the issuance of conditional approvals, the holdback provisions of funding arrangements, and contractual arrangements' (CEAA, 1994, p. 103). For example, the scope of a permit issued under the navigable waters legislation can be expanded to cover a wide range of mitigation measures. It also proposes the placing of performance bonds by proponents to ensure implementation. Conditions on leases and covenants can also be employed.

To a very real extent, mitigation of adverse environmental effects, rather than the decision to permit or refuse a project, is the principal objective of EA in Canada. As mentioned previously, few projects have been subject to panel review under either the Environmental Assessment and Review Process or under the Act. In theory, this means that environmental effects have been mitigated to the point where they are not significant and that considerable expertise must have been developed in achieving this. In practice, while expertise has developed, it has been the express aim of virtually every department to avoid a panel review because of the expense, time and potential embarrassment involved. Hence, the conclusion that effects are no longer significant following mitigation must be doubtful in many cases but has seldom been tested (in the courts or elsewhere). Where panels have been set up, mitigation measures have been an important element of their reports.

In principle, public access to the documents in the EA process and the reference of comprehensive study reports to the Minister of the Environment should lead to effective mitigation. However, the political position of a relatively weak government department informing powerful departments that, in effect, their mitigation measures leave significant residual adverse effects is a difficult one. The emphasis on mitigation in the Act derives from a considerable research programme funded by the Canadian Environmental Assessment Research Council (CEARC) (see, for example, CEARC, 1988).

David Redmond and Associates (1999) reported that steps were taken to ensure that mitigation measures were implemented in about half the medium and large project screenings they examined. The Commissioner of the Environment and Sustainable Development (CESD) reported, on the basis of his study of 187 screenings, that:

> Responsible authorities have generally been including mitigation measures in the terms and conditions of their approval, or building them into related contract documents. (CESD, 1998, pp. 6–23)

However, it was less obvious that RAs were verifying 'whether mitigation measures were actually implemented by project proponents'. He recommended that 'mitigation measures are included where required and that project proponents implement them' (CESD, 1998, pp. 6–24). The bill to amend the Act, following the five-year review of its operation, contained provisions to strengthen the implementation of mitigation measures (Government of Canada, 2001).

Commonwealth of Australia

As indicated in Chapter 13, the main aim of the Australian EIA system is the mitigation of impacts, and numerous references to mitigation measures are made in the Environment Protection and Biodiversity Conservation Act 1999 (EPBC Act) and in the accompanying EPBC Regulations. Thus, the preliminary information must include a description of the mitigation measures to be taken to reduce the potential environmental impacts of the proposal. Unless the action is expected to be subject to an environmental impact statement or public environment report (PER), this should furnish:

> details of measures that will be implemented to avoid or manage any relevant impacts of the action including, if appropriate, evidence in the form of reports or technical advice on the feasibility and effectiveness of the proposed measures. (Regulations, Schedule 3, para. 7.01)

The mitigation information requirements for EISs and PERs are substantial, covering the expected effectiveness, the statutory basis and the cost of the measures. Details of the relevant environmental management plan and of the responsible agencies must also be provided. Finally:

> a consolidated list of mitigation measures proposed to be undertaken to prevent, minimise or compensate for the relevant impacts of the action, including mitigation measures proposed to be taken by State governments, local governments or the proponent must be provided. (Regulations, Schedule 4, para. 4(f))

These provisions considerably extend those employed under the repealed Impact of Proposals Act 1974. Under that legislation, the draft EIS (and the PER) always covered mitigation in some detail. In practice, many important mitigation measures tended to be introduced in the final EIS as a way of overcoming objections raised to the draft EIS. The assessment report produced by Environment Australia nearly always contained further suggestions for the mitigation of environmental impacts, but these were usually of less significance than the proponent's proposals. In order for the Environment Minister to impose mitigation conditions to any approval under the EPBC Act (Chapter 13), the assessment report must continue to deal with mitigation at some length.

It is generally accepted that mitigation has long been a genuine consideration in the Australian EIA system and virtually all proposals have been extensively modified (see Chapter 13). The main area of weakness in the treatment of mitigation related to implementation and action monitoring but these provisions have been strengthened considerably in the EPBC Act. Much depends on the way in which the Environment Minister exercises the new powers over the delivery of mitigation measures.

New Zealand

The duty to avoid, remedy or mitigate environmental impacts is one of the main aims of the Resource Management Act 1991 (see Box 5.2). The Act requires an application to contain an assessment of effects 'and the ways in which any adverse effects may be mitigated' (section 88(4)(b)). The Fourth Schedule (see Box 7.3) specifies that 'a description of the mitigation measures' to prevent or reduce impacts should be included in the assessment of environmental effects (AEE) report.

There are no other specific requirements relating to mitigation in the Act beyond the general criteria contained in section 108 relating to conditions, and no guidance relating to amelioration of impacts has been issued. This is somewhat surprising since the Environmental Protection and Enhancement Procedures (EPEP) lay great stress on mitigation measures (Ministry for the Environment – MfE, 1987) and considerable expertise exists in New Zealand (though EPEP mitigation recommendations were often ignored). Guidance to applicants explains and encourages (but does not emphasise the use of mitigation measures throughout the AEE process (MfE, 1999a).

The review guidance to local authorities makes it clear that conditions should be employed where, 'after analysing the AEE and the application as a whole, effects that need to be avoided, remedied or mitigated have been identified' (MfE, 1999b, p. 65). Conditions requiring management plans detailing how environmental effects are to be mitigated can be imposed (MfE, 2001b).

The principal method of ensuring that mitigation is actually working in practice (i.e. that local authority conditions on any permission specify mitigation measures derived from the EIA process) is public scrutiny of the implementation of conditions attached to resource consents and permits, complaint to the local authority, and the right to appeal to the Environment Court (see Chapter 15). Monitoring (see Chapter 14) is usually secondary to public vigilance in ensuring enforcement.

In practice, the inexperience of developers and local authorities has meant that the treatment of mitigation in EIA has varied considerably in New Zealand. The Parliamentary Commissioner for the Environment reported (PCE, 1995) that local authorities endeavoured to grant consents and to impose conditions to mitigate adverse impacts in some instances where this was inappropriate. Recent Ministry for the Environment guidance (see above) has strengthened the need to avoid this. There are, undoubtedly, examples of good practice by developers and local authorities, especially for more complex projects, but it is apparent that, too often, mitigation measures have been belatedly conceived, ill-considered or unimplemented.

South Africa

Mitigation has been a great strength of EIA in South Africa. Because there have been so few refusals of authorisation, almost the whole emphasis of integrated environmental management (IEM) has been on avoiding, reducing or remedying the negative environmental impacts of development and enhancing the positive impacts. As the IEM guidelines stated: 'the focus of opposition to, or approval of, developments frequently centres on the management of mitigation measures' (Department of Environmental Affairs – DEA, 1992, vol. 3, p.

17). These guidelines stressed the benefits of preparing a management plan, which may be required as a condition of approval, to 'describe how negative environmental impacts will be managed, rehabilitated or monitored and how positive impacts will be maximized' (p. 17). Further advice on the coverage of mitigation in EIA reports was presented in another of the IEM guideline documents, where it was stated that the EIA report should 'indicate whether the mitigatory measures have been agreed to by the proponent' (DEA, 1992, vol. 4, p. 11).

Given the emphasis on mitigation in IEM, it is surprising that the only mention of mitigation in the 1997 EIA Regulations is the requirement that the environmental impact report must include particulars about 'the possibility for mitigation of each identified impact' (Republic of South Africa, 1998, Regulation 8(a)(ii)) (see also Chapter 8). The EIA guidelines state that this potential should be explained 'in terms of its nature, its extent, its duration, its intensity, its probability and its significance' (Department of Environmental Affairs and Tourism – DEAT, 1998c, p. 28). They go on to state that:

Mitigation options include:

- alternative ways of meeting the need;
- changes in planning and design;
- improving monitoring and management;
- monetary compensation; and
- replacing of e.g. wetlands by constructing other wetlands, relocating villages or people displaced by projects and rehabilitating sites.

Often a combination of compensation, relocation and rehabilitation is needed. It is also important to specify when and how mitigating measures should be done, also including an indication of the effectiveness of these measures. (DEAT, 1998c, p. 28)

Previous IEM mitigation practice (good and bad) appeared to continue under the EIA Regulations. Mitigation measures were usually considered at each stage in the EIA process under the IEM procedure, with particular emphasis on the later stages. Many EIA reports presented an evaluation of the significance of environmental impacts with and without mitigation and discussed the practicability of the proposed mitigation measures. The summary of the Maputo Iron and Steel Project EIR (see Chapter 11) proposed the preparation of environmental management plans (EMPs) for the construction, operational and decommissioning phases and the appointment of both an environmental control officer and a social officer (Gibb Africa, 1998). Weaver *et al.* (2000, pp. 13–11) provided 'an example of guidelines to EIA specialists designed to ensure that there is a smooth transition between the EIA and the EMP' (Box 15.5).

Many relevant authorities have endeavoured to ensure that the mitigation measures proposed in the EIA report have been implemented by encapsulating them in authorisation conditions. Weaver (1996, p. 115) counted 'identifying appropriate mitigation measures and minimizing environmental damage' as one of the key improvements to South African EIA. However, despite the

---

Box 15.5   South African mitigation guidelines for an EIA

(a) *Mitigation objectives – What level of mitigation must be aimed for?* For each identified impact, the specialist shall provide mitigation objectives (*tolerance limits*) which would result in a measurable reduction in impact. Where limited knowledge of expertise exists on such tolerance limits, the specialist shall make a 'guestimate' based on experience.

(b) *Recommended mitigation/enhancement actions:* For each impact, the specialist shall recommend practically attainable mitigation actions, which can measurably affect the significance rating. They shall also identify management actions which could *enhance* the condition of the environment (i.e. potential *positive* impacts of the proposed project). Where no mitigation is considered feasible, this must be stated and reasons provided. Specialists shall record the significance rating both with and without mitigation/ enhancement actions.

(c) *Effectiveness of the mitigation actions:* The specialists shall provide quantifiable standards *(performance criteria)* for reviewing or tracking the effectiveness of the proposed mitigation actions.

(d) *Recommended monitoring and review programme:* The specialists shall recommend an appropriate monitoring and review **programme** which can track the achievement of the mitigation objectives.

Source: Weaver *et al.*, 2000, pp. 13–11.

---

emphasis on mitigation, practice in South Africa has not always been exemplary. Thus, the IEM guidelines stated that:

> Assessments have a tendency to indicate mitigating measures that should be done without specifying when or how they should be carried out and without indicating the effectiveness of the measures. (DEA, 1992, vol. 4, pp. 10–11)

Mafune *et al.* (1997, p. 208) in their survey of early EIAs, reported that 'where mitigation was recommended in the EAs, mitigation of social impacts was given far less attention than mitigation of biophysical impacts'. Barker (1996, p. 6) criticised the treatment of mitigation in EIA reports prepared under the IEM procedure: 'mitigatory actions that are recommended in EIA reports are rarely presented in sufficient detail to support their effective implementation'. She suggested that measurable mitigation targets be specified (for example, a quantitative boundary noise level) and that environmental control officers be appointed to supervise the implementation of measures to meet these targets at construction and operational stages of the development. Weaver (1996, p. 115) cited 'poor requirements for enforcement of compliance' and 'setting enforceable terms and conditions' as problem areas. Hill (2000, p. 54) believed that:

> Practice in South Africa falls short of the ideals of IEM, partly because authorities do not have access to adequate human and financial resources.

The National Environmental Management Act 1998 (NEMA) requires the 'investigation of mitigation measures to keep adverse environmental impacts to a minimum' (section 24(7)(c)). Hill (2000, p. 54) recommended that, in promulgating EIA regulations under the provisions of NEMA (see Chapter 5), the Department of Environmental Affairs and Tourism needed to ensure that 'requirements for post-decision implementation' were included.

Summary

Unsurprisingly, Table 15.1 shows that each of the seven EIA systems meets the mitigation criterion. 'Unsurprisingly' because mitigation is the principal aim of the EIA process. However, the length to which mitigation is taken in the EIA

Table 15.1   The treatment of mitigation in the EIA systems

Criterion 10: Must the mitigation of action impacts be considered at the various stages of the EIA process?

Jurisdiction	Criterion met?	Comment
United States	Yes	Formal requirement to incorporate mitigation measures in record of decision. Effectiveness of implementation varies, but generally improving
UK	Yes	ES must cover mitigation and local planning authorities impose conditions upon permissions to mitigate impacts. Practice improving at various stages in EIA process
The Netherlands	Yes	Mitigation subsumed in treatment of alternatives but not separately required by EIA Act. Practice often satisfactory
Canada	Yes	Mitigation and its implementation are central considerations in EA process. Practice satisfactory in panel reviews, comprehensive studies, less so in screening
Commonwealth of Australia	Yes	Mitigation measures are explicitly and extensively provided for at various stages in EIA process. Practice often satisfactory
New Zealand	Yes	Mitigation of environmental impacts is one of main purposes of Act. Practice varies at various stages in EIA process
South Africa	Yes	Mitigation is main focus of South African EIA. Mitigation measures proposed in previous EIA reports not always effectively implemented

systems varies and the emphasis on the implementation of mitigation measures also differs both between and within jurisdictions.

It is probable that, as concern over the sustainability of development grows, more emphasis will be placed on the avoidance of impacts by the consideration of alternative approaches, as in the Dutch EIA system. It is likely that increasing attention will be paid to the notion of 'no net deterioration' or 'net amelioration' of the environment. This will affect not only the choice of alternatives and mitigation measures (for example, the establishment of a new and larger recreational open space as compensation for the loss of, or damage to, the original space) but also the implementation of these measures.

It is apparent that there is considerable scope for improving the implementation of mitigation measures in all seven jurisdictions. This is bound up with impact and implementation monitoring (see Chapter 14), where practice also needs to be improved. It is probable that the greatest single contribution to sustainable development which EIA could deliver (beyond the prevention of environmentally unsatisfactory actions) is a major improvement in the nature of mitigation measures and their implementation.

Notes

1. See, for example, Morris and Therivel (2001) and Canter (1996) for valuable discussions of generic mitigation measures for different environmental impacts.
2. EIAC web site at www.eia.nl (Information/EIA/Products)

Chapter 16

Consultation and participation

As stated in Chapter 1, consultation and participation are integral to environmental impact assessment: EIA is not EIA without consultation and public participation. Almost all EIA systems provide for consultation and participation after publication of the EIA report, prior to the decision on the action, though a few (particularly in developing countries) do not (see Chapter 12). However, many jurisdictions either require or encourage consultation and participation at earlier stages of the EIA process, for example during scoping. Indeed, there appears to be a growing consensus that increased consultation and participation, using one or more of the large number of means of participation which exist, can produce significant benefits for the proponents of actions and for those affected.

This chapter presents an examination of the role of consultation and participation in the EIA process and advances a set of evaluation criteria for the treatment of consultation and participation within EIA. These criteria are then employed in the analysis of procedures for consultation and participation in the United States, UK, the Netherlands, Canada, Commonwealth of Australia, New Zealand and South Africa.

Consultation and participation within the EIA process

In recent years, there has been a general increase in demand for public involvement in environmental decision-making generally (Sheate, 1996). As Roberts (1995, p. 243) noted: 'People expect and demand to be involved.' Tilleman (1995) pointed out that EIA demands individual participation rather than representative democratic involvement. Petts (1999) suggested that public participation in EIA should be seen in the context of the public involvement requirements of the Rio Declaration, Agenda 21, the European Commission Fifth Environmental Action Programme, the Espoo and Aarhus Conventions and the sustainable development agenda.

The objective of consultation and participation in EIA is to improve the quality of environmental decisions by the identification of, assignment of significance to, and mitigation of, impacts and the prevention of environmentally unacceptable development. Sheate (1996, p. 83) suggested that the underlying objective of EIA, to evaluate the environmental impacts of a proposal fully, could not be achieved without first obtaining the views of people most likely to be affected by that proposal. Glasson *et al.* (1999, p. 160) observed that consultation and public participation in EIA:

> ... can help to ensure the quality, comprehensiveness and effectiveness of the EIA, as well as to ensure that the various groups' views are adequately taken into consideration in the decision-making process.

The principles for involvement by organisations and individual members of the public in EIA have been summarised by the Australian and New Zealand Environment Conservation Council (ANZECC) (Box 16.1).

Glasson *et al.* (1999, p. 163) noted that it was possible to classify the public into two main groups:

1. Voluntary groups, quasi-statutory bodies or issue-based pressure groups concerned with a specific aspect of the environment or environmental issues in general.
2. People living near a proposed development who may be directly affected by it.

Sinclair and Diduck (1995, p. 222) suggested that: 'the public is a constantly shifting multiplicity of organisations, individuals, interests and coalitions.' It must therefore be recognised that 'the public' consists of different groups, each having different objectives concerning the EIA process, and consequently wishing to participate in different ways.

Box 16.1   EIA public participation principles

(a) Participate in the evaluation of proposals through offering advice, expressing opinions, providing local knowledge, proposing alternatives and commenting on how a proposal might be changed to better protect the environment.

(b) Become involved in the early stages of the process as that is the most effective and efficient time to raise concerns. Participate in associated and earlier policy, planning and programme activities as appropriate, since these influence the development and evaluation of proposals.

(c) Become informed and involved in the administration and outcomes of the environmental impact assessment process, including:
- assessment reports of the assessing authority
- policies determined, approvals given and conditions set
- monitoring and compliance audit activities
- environmental advice and reasons for acceptance or rejection by decision-makers.

(d) Take a responsible approach to opportunities for public participation in the EIA process, including the seeking out of objective information about issues of concern.

Source: ANZECC, 1991, p. 7.

There are several different types of public participation. These can be distinguished by the nature of the relationship between the public and the decision-making body or the proponent (Arnstein, 1969; Sewell and Coppock, 1976; Canter, 1996; Petts, 1999). These relationships range from the provision of information, through a range of types of consultation, to direct public control. All three types of relationship may be identified in EIA systems at different times and in different circumstances. For example, Australian Aboriginal people may sometimes control whether a particular mining project affecting lands over which they have acknowledged rights (but not formal ownership) can proceed or not.

In addition to the various means of involving consultees and the public during the EIA process (see below), the use of mediation or environmental dispute resolution in certain circumstances has been suggested. Mediation involves the assistance of a mediator in negotiations between the parties in a dispute over a new development and requires a willingness to compromise and utilise environmental mitigation. Canter (1996, p. 609) observed that it is common for conflicts and disputes to arise over environmental impact issues in the United States and that a key feature of conflict resolution techniques was 'collaborative problem solving'. While it is not easy to state precisely when and if mediation will help negotiations towards completion, there appear to be four prerequisites to its success: a stalemate, or the recognition that stalemate is inevitable; voluntary participation; some room for flexibility; and a means of implementing agreements (Bingham, 1986). However, these prerequisites have tended to apply in only a small minority of siting decisions involving EIA. Despite this, Canter (1996, p. 615) envisaged a long-term role for dispute resolution in EIA in the United States:

> With increasing environmental awareness, additional environmental-protection and resource-management legislation, and greater societal pressure for responsible environmental management, there is every reason to believe that dispute resolution will remain a part of and even increase in importance to the EIA process.

It may be that environmental dispute resolution is particularly appropriate in the United States. Jeffery (1987) saw little role for mediation in EIA in Australia and Canada, except during scoping. In particular, he did not consider mediation to be appropriate where public hearings were used in arriving at decisions. Buckley (1991b) was also wary of the use of environmental dispute resolution approaches in Australia which, he felt, were no substitute for planning and impact assessment legislation, providing several opportunities for formal public information and participation or for third-party recourse to the courts if agencies failed to discharge their responsibilities adequately. In many jurisdictions, such as the UK, there is often only one major siting decision and stalemates seldom apply. In these circumstances, the local planning authority often acts as mediator and negotiations are used to reduce the adverse effects of proposed developments. Canada is one example of a jurisdiction outside the United States which has instituted a formal mediation procedure as one track in its environmental assessment system (see Chapter 5), but this has not yet been utilised (Sadler, 1994).

Figure 1.1 shows that consultation and participation can be employed at every stage in the EIA process. While the involvement of agencies and the public in the very early consideration of alternatives and of preliminary design of the proposed action is not usually feasible, consultation and participation in screening is possible and is usual in, for example, Western Australia (Wood and Bailey, 1994). Consultation and participation in scoping are common-place in many EIA systems, and are a requirement in, for example, the Netherlands (see Chapter 10). Similarly, the involvement of consultees and the public in EIA report preparation (as in Victoria (Wood, 1993)) should lead to improved quality, or at least to improved acceptability (see Chapter 11). As mentioned above, almost all jurisdictions provide for consultation and partici-pation during the review process. In a fully participative EIA system, these rights should also extend to the review of further information submitted by the proponent, to any evaluation report relating to the action prepared and to the making of the decision (see Chapter 13). Finally, the ability to comment upon monitoring results is a necessary check on the operation of monitoring pro-cedures (see Chapter 14).

Roberts (1995, p. 230) suggested that the appropriate level of consultation and public participation in EIA varied with the circumstances:

> The need to involve the public in the decision making process is directly related to the significance of the decision to the public, and the extent to which the decision under consideration is controversial.

Whatever level of consultation and participation takes place, it can be effec-tive only if copies of EIA documents are made public at each stage of the EIA process (for example, at the scoping stage as well as on completion of the EIA report). Such documents need to be readily available at a number of locations convenient to those most likely to be affected by the proposal concerned. The documents also need to be accessible in the sense of being clear and compre-hensible. This is especially true of the non-technical summary of the EIA report (Sheate, 1996). As well as making EIA documents available, it may sometimes be necessary to educate the public about the nature of the EIA process in order to enhance the effectiveness of public involvement. As Sinclair and Diduck (1995, p. 228) have noted: '[a]n effective transfer of power requires an effective transfer of knowledge.'

Equity demands that copies of EIA documents can be obtained and/or pur-chased at a reasonable price for detailed perusal. In the United States such doc-uments are generally free of charge, whereas a substantial charge may sometimes be made in other jurisdictions. In Australia it has been usual for a nominal charge for EIA reports to be made (say A$25) to discourage requests for large numbers of such documents from, for example, schools. The cost of EIA reports in the UK has varied from nothing to over £100. There is an increasing tendency to make EIA reports and other documents available through the internet.

Prior to the introduction of EIA, proponents frequently invoked confidentiality and secrecy as reasons for not making information about a proposed action available to the public. Many EIA systems expressly permit restrictions on the availability of information where the case for withholding it can be demonstrated. In general, these restrictions are seldom invoked in most jurisdictions, though instances of national security or commercial sensitivity do sometimes arise.

There are numerous methods of consultation and participation, each with its own advantages and disadvantages (see, for example, Canter, 1996; Petts, 1999). Examples include:

- questionnaires and surveys
- advertisements
- leafleting
- use of media
- displays
- exhibitions
- telephone hotlines
- open houses
- personal contact
- community liaison staff
- community advisory committees
- group presentations
- workshops
- public meetings
- public inquiries.

The choice of method depends on the stage of the EIA process at which it is employed, on the public being engaged, on the nature of the action and on the particular circumstances which apply. For example, while a public inquiry may well be appropriate immediately prior to decision making, it is very unlikely to be suitable at the screening stage (Glasson *et al.*, 1999). In practice, most jurisdictions leave considerable discretion to decision-makers and proponents in their choice of consultation and participation methods, and in most circumstances a variety of methods will be needed. EIA consultation and participation strategies need to be planned and managed with clear goals, and indicators to measure their achievement of these goals, but be flexible (Roberts, 1995; Canter, 1996). Petts (1999) warned that such strategy goals tend to be seen from the proponent's, rather than the public's, perspective.

Intervenor funding is often difficult to arrange and control, but access to such funding can beneficially affect the outcome of the EIA process by making participation more effective. Without financial assistance, local groups may feel at a great disadvantage relative to the proponent at all stages in the EIA process. However, Canadian (and other) experience has shown that when funding has been made available to help participants prepare for public hearings they have frequently made well conceived and constructive contributions which have led to greater consensus about the environmental consequences of the proposed action (Lynn and Wathern, 1991).

As well as the public, various consultees have valuable contributions to make at the different stages in the EIA process. While it is usual to consult the bodies which are thought likely to provide useful information on an ad hoc basis, there are advantages in specifying a list of consultees who must be consulted at the various stages of the EIA process by the proponent and/or the decision-maker. It is clearly equitable that neighbouring authorities, states and countries be consulted where proposals are made which could affect their environments. The European Directive, which was amended partly to incorporate the provisions of the 1991 Espoo Convention on Trans-boundary Impacts in EIA (and much other legislation), requires this. International environmental disputes may still arise, but their likelihood is considerably reduced by transboundary consultation. As at other stages of the EIA process, the availability of clear guidance on the procedures and techniques for consultation and participation will be helpful to proponents, the decision-making and environmental authorities, consultees and the public.

The publication of the results of consultation and participation is clearly a necessary check on their use in the EIA process. There should be a right to inspect both public and consultee submissions and the use made of them by the responsible agencies. Thus, an indication of the changes to the proposal made as a result of public involvement, and the reasons for not acting on other suggestions, should be furnished. It is clear that the role of public involvement in the success of the US National Environmental Policy Act 1969 in influencing decisions on actions owes much to two factors: the first is the right to participate and to gain access to relevant documentation; the second is the public right of appeal to the courts over EIA decisions (see Chapter 2). Other jurisdictions provide opportunities for appeal against the various decisions made during the EIA process. For example, an appeal against the decision to permit an action can be made to the Environment Court in New Zealand.

Clearly, while such appeal rights should make the EIA process more effective by influencing decisions taken at different stages in the process, they need to be tempered by the need to make the process efficient. A balance has to be struck between the positive benefits of consultation and participation in ameliorating the impacts of actions and in reaching consensus on environmental outcomes and the financial and time costs involved. Since the expenditure of time, rather than money, often appears to be the principal criticism of those EIA systems with extensive consultation and participation requirements, such a balance could imply limiting the amount of time taken to complete each stage in the EIA process while providing adequate information and the opportunity for appeals to be dealt with effectively. This need not conflict with fairness, competence and the achievement of the goals of the participation strategy employed (Petts, 1999).

The various criteria for the evaluation of consultation and participation provisions within the EIA process are listed in Box 16.2. These criteria are used to assist in the analysis of consultation and participation procedures in each of the seven EIA systems which now follows.

Box 16.2   Evaluation criteria for consultation and participation

Must consultation and participation take place prior to, and following, EIA report publication?

- Must consultation and participation take place prior to scoping, during scoping, during EIA report preparation, during review and following revision, during decision making and during monitoring?
- Must a public participation strategy be initiated for each EIA?
- Are copies of EIA documents made public at each stage of the EIA process?
- Can copies of EIA documents be accessed free of charge or purchased at a reasonable price?
- Do confidentiality/secrecy restrictions inhibit consultation and participation?
- Are consultation and participation methods appropriate to the stage of the EIA process at which they are employed?
- Is funding of public participants provided for?
- Are obligatory consultees specified at various stages in the EIA process?
- Must adjoining authorities/states/countries be consulted?
- Does published guidance on consultation and participation exist?
- Must the results of consultation and participation be published?
- Do rights of appeal exist at the various stages of the EIA process?
- Does consultation and participation function efficiently and effectively?

United States

Consultation and participation have been the driving force in the evolution of EIA in the United States. As Clark (1997, p. 22) has reported:

> NEPA [the National Environmental Policy Act 1969] has been and still is the principal avenue for public involvement in the planning and decision-making processes of federal agencies, and the public has thus played a major role in maintaining the viability and importance of NEPA.

NEPA requires that relevant federal agencies be consulted during the preparation of the environmental impact statement (EIS) and that the public be involved. The Council on Environmental Quality (CEQ) Regulations, section 1506.6, specify that agencies must:

> a) Make diligent efforts to involve the public in preparing and implementing their NEPA procedures.
> b) Provide public notice of NEPA-related hearings, public meetings, and the availability of environmental documents so as to inform those persons or agencies who may be interested or affected.

The Regulations make provision for agency consultation and public participation at the following stages of the EIA process:

- in screening (preparation of, and comment upon, the environmental assessment, and comment upon the finding of no significant impact);
- on publication of notice of intent;
- in scoping;
- in preparation of, and comment upon, the draft EIS;
- in preparation of the final EIS;
- on the record of decision;
- on monitoring results following implementation.

Participation takes a variety of forms, from the making public of documents, through the circulation of documents, to meetings and hearings (Tilleman, 1995). EIA documents must be notified in the Federal Register[1] and made available to the public either free of charge or for not more than the cost of photocopying. (Where changes for EISs are made, these vary from $25 to $75.) EISs are made widely accessible for inspection and many are published on the internet (where there is a web site devoted to information about NEPA (Jessee, 1998; see also Chapter 17)). In cases where there is substantial controversy or interest, or where another agency with jurisdiction over the action requests it, a public hearing on the draft EIS must be held (Kreske, 1996). Apart from this, no methods of public participation (Canter, 1996) are specified. However, CEQ (1997b) has suggested that conventional consultation and public participation methods may be inadequate to ensure the involvement of low-income, minority or tribal populations. The lead agency must consult other agencies with jurisdiction or special expertise.

NEPA specifies no enforcement mechanism and CEQ and the Environmental Protection Agency have no enforcement authority. This is why public access to the NEPA process is so crucial. There are opportunities for recourse to the courts at various stages in the EIA process, and interest groups, private citizens, state and local agencies and businesses have taken advantage of these by filing thousands of lawsuits, at great expense. Certainly, environmental non-governmental organisations use NEPA as a crucial means of access to the courts and, indeed, now see this as its principal asset. They argue that the threat of litigation is sufficient to improve agencies' environmental behaviour (Ortolano, 1997). Nevertheless, despite many successes in the lower courts, the Supreme Court has consistently supported the federal agency in each of the 12 cases it has heard (Bass *et al.*, 2001, pp. 172–3). The courts have, however, required good-faith efforts to comply with NEPA's full disclosure provisions.

There appear to have been very few difficulties relating to commercial confidentiality but a limited number of EISs, prepared for actions involving national security, has not been made available to the public (Fogleman, 1990, pp. 7, 8). The response of the public is variable: some actions result in vociferous objections and well-attended meetings, but others elicit little or no public interest (CEQ, 1997d). This may depend partially on the nature of the action, and partially on the participation methods employed. There is virtually no funding for participants in the EIA process.

In practice, as well as weaknesses in some EIS consultation procedures, there are significant weaknesses in participation in the preparation of environmental assessments (EAs). Blaug (1993) reported that there was public involvement in less than half of the EA cases examined, despite the requirements of the Regulations.

Agency consultation in EIA preparation is more effective. It is generally accepted that consultation results in informed comments on the draft EIS from the relevant federal, state, tribal and local agencies, which must be made public. These responses have helped to avoid or resolve many conflicts (CEQ, 1997d). In a positive development, some states, counties and tribal governments have recently shown greater interest in being involved in EIA as 'cooperating agencies'.

UK

The use of consultation and participation is officially encouraged throughout the EIA process in the UK. Consultation can, and sometimes does, take place at the screening stage and, if a formal opinion is requested of the local planning authority (LPA), the material provided by the developer is made public and representations can be made about the need for EIA (see Chapter 9). Consultation must take place (and the public may be invited to participate) at the scoping stage if the developer requests an 'opinion' (see Chapter 10). The provision of information by the statutory consultees frequently occurs during the preparation of the environmental statement (ES) (Chapter 11). However, it is only once the ES has been submitted that the LPA must consult the public. The LPA is required to forward, or arrange for the forwarding of, copies of the ES to the statutory consultees and to take their comments, together with those of the public, into account before reaching a decision.

Consultees must be allowed at least 14 days to comment, and the public has at least 21 days. Where additional information is provided by the developer, following a request by the LPA, this too must be circulated to consultees and the public. As mentioned in Chapter 13, there is no formal consultation and participation requirement during the LPA decision-making process, though lobbying and, sometimes, the opportunity to address LPA decision-makers while the decision is being discussed, are permitted. (There is, of course, a right of the public and consultees to be heard at public hearings and inquiries.) Similarly, there exists no public right to participate in the monitoring of implemented projects.

Advertisements and site notices must be placed where EIA is required. The environmental statement must not only be made readily accessible to the public, but available for purchase at a reasonable charge (see Chapter 12). On the whole, the purchase prices of ESs in the UK are indeed reasonable (many being free of charge) but a minority are expensive and some, as mentioned above, have been priced in excess of £100. Issues of confidentiality and secrecy have seldom arisen in relation to the EIA process in the UK.

There is some published guidance on consultation and participation. The Department of the Environment (DoE, 1995, p. 17) reiterated the advice in the procedural guide (DoE, 1989: now replaced by Department of the Environment, Transport and the Regions – DETR, 2000a) about involving the public at the scoping stage (see Chapter 10) but recognised that developers 'may be reluctant to make a public announcement about their proposals at an early stage.' The Circular is conspicuously circumspect about consulting the public during scoping:

> There is no obligation for the developer to consult anyone about the information to be included in a particular ES. However, there are good practical reasons to do so. Local planning authorities will often possess local and specialised information. ... It will normally also be helpful to a developer preparing an ES to obtain information from the consultation bodies. ... Non-statutory bodies also have a wide range of information and may be consulted by the developer. (DETR, 1999b, paras 87, 88).

No requirements as to consultation and participation methods are laid down in the Regulations, beyond those relating to the availability of environmental statements. DoE (1994a, p. 12) has suggested that it is helpful to the LPA, in gauging public opinion, if 'the developer has carried out a series of well-publicised public meetings and/or exhibitions, and has reported the response in the ES', but that 'a series of standard letters, or a petition' are likely to be less helpful. 'For this reason the planning authority may decide to hold its own meetings' (DoE, 1994a, p. 12). This amplifies other advice on gaining public views on the ES and the proposal:

> The authority and the developer may wish to consider the need for further publicity at this stage, for example, publication of further details of the project in a local newspaper, or an exhibition. (DETR, 2000a, para. 46: originally DoE, 1989)

Similarly, there is no provision for the funding of public participants in the EIA process. Apart from the usual statutory consultees for planning applications, the Countryside Agency, English Nature and, for certain developments, the Environment Agency, must be consulted where an ES is received in England. As required by the European Directive, adjoining member states must be notified by the British government where a project is likely to have significant effects on their environment though this has seldom occurred. Consultation of neighbouring local authorities is at the discretion of the LPA.

As mentioned in previous chapters, LPAs are required to keep planning registers and, while there is no separate publication of the results of consultation and participation, it is normally possible to inspect the responses, as well as the other documents prepared during the EIA process, at the offices of the LPA. Although the developer has rights of appeal against the LPA's screening and scoping decisions (see Chapter 9) and against its decision on the planning application (Chapter 13) no similar right of appeal by statutory consultees or by the public exists at these or any other stages in the EIA process.

In a sample of 40 ESs, consultation with statutory EIA consultees, major public interest and local action groups took place in 95–98 per cent of cases. The ES was supplied to statutory consultees in about 90 per cent of cases but to public interest and local action groups in only half the cases (Jones *et al.*, 1998). These results represent a significant improvement over those reported by DoE (1991a). McNab (1997) clearly illustrated the dilemma facing a developer over early public involvement in the EIA of a holiday village, despite its desirability (Box 16.3).

To summarise, consultation is not a requirement of the UK EIA system prior to the submission of the ES unless the developer requests a scoping opinion. It is, nevertheless, normal practice. Public participation during scoping is not mandatory, though it sometimes occurs informally as well as formally subsequent to submission. Practice in consultation and public participation appears to be improving but there is clearly scope for an increase in

---

Box 16.3   Scoping and public participation: Cotswold Water Park

The assessment process

CRC [a consultancy] was approached ... to provide planning advice to a company who were proposing a tourism and recreational facility in the Cotswold Water Park. ... A rapid initial review of planning and environmental factors suggested that there were strong arguments in favour of the development. ... However, there were a number of immediately apparent environmental problems. ...

This was a proposal by two major public companies to enter a new market and involved an option to purchase a large site. Lakewoods [the developer] required the project to remain confidential as long as possible. ...

CRC recommended, and Lakewoods agreed to, a programme of confidential officer discussions to establish the scope of the proposed assessment. ... The consultations were used both to gather information and to define the scope of the assessment. ... During this process, public relations consultants were appointed to deal with the non-technical aspects of the consultations, i.e. press, political and community relations. ... However, initial contracts almost immediately led to information about the proposed development coming to the attention of Parish Councillors.

Lakewoods subsequently agreed to their request to brief the Parish Council and to an open meeting. The meeting was acrimonious. The introduction by the Chairman was tendentious and opposed to the development. The developers and their consultants could not answer all the questions because, whilst the village had been designed, the environmental assessment was continuing and a number of issues, such as the means of sewerage disposal, remained to be agreed. ...

Shortly after the public meeting, an outline planning application was lodged with an ES. It was at this point that a substantial programme of consultations and public involvement was initiated. ... An exhibition, incor-

(continued)

(continued)

porating a large-scale model of the proposed development, was prepared and open for view weekly on-site, when representatives of Lakewoods and their consultants were available to answer questions. Presentations were made to the County and District Councils and to the five adjacent Parish Councils. Open days were held at the site with conducted tours and evening visits were specially arranged for individual groups such as the Council for the Protection of Rural England (CPRE). Copies of the ES and the non-technical summary were made available free to all. . . .

Subsequently, . . . a supplement to the ES was produced setting out further information on a number of issues. This was formally submitted to the local planning authority and, again, made freely available to all. This information was supplemented by a series of local newsletters, produced by the public relations consultants with inputs from the planning and environmental team, and distributed to all local households.

The decision

The proposal was eventually called in by the Secretary of State for the Environment and was the subject of a marathon public inquiry. . . . The opposition thus won the initial political battle, having the application called in, but Lakewoods won all the technical arguments. Following further delay, the application was finally approved. . . .

Reflections on the assessment process

For the reasons set out, it was decided not to involve the public in scoping. It was always intended to organise a programme of public consultation following the lodging of the planning application. However, despite the best efforts to keep the proposal confidential, it was leaked prior to this stage. The subsequent public meeting, from the developer's viewpoint, was a public relations disaster and stimulated the formation of an organised opposition to the proposal. The elaborate consultation process which followed patently failed to alter the views of this opposition group. However, a petition in favour of the development was successful in attracting substantially more signatures than the petition against the development. . . .

Planning decisions are political decisions and at the end of the day success or failure depends upon the attitude of individual Councillors. In addition to winning the technical argument, developers must win the political argument. In the Cotswold Water Park case much effort was spent in trying to change the views of vehement opponents of the scheme who were intransigent and not enough time on garnering the support of the often more reticent but nevertheless important supporters of the scheme.

Source: McNab, 1997, pp. 63–8.

effectiveness, especially in relation to the largely marginal role played by the general public in the EIA process. As Glasson *et al.* (1999, p. 170) have stated:

> it is clear that in the UK the requirements for public participation have been implemented half-heartedly at best, and developers and the competent authorities have in turn generally limited themselves to the minimal legal requirements.

The Netherlands

The Dutch Environmental Management Act 1994 specifies two occasions in the EIA process on which the statutory consultees and the public must be given the opportunity to comment. The first is public participation in relation to the establishment of the scoping guidelines (section 7.14(3)) which are often drawn up following a public meeting. The second is when the EIS is reviewed (section 7.23(1)). There is a four-week period for comments in both cases. There is usually a public hearing (generally lasting less than four hours) at the review stage. The notification of intent, the decision and the auditing report are also published. Confidentiality restrictions have rarely been invoked in the Dutch EIA process.

While copies of EISs are available for purchase, they are often expensive (more than €40). However, some are free of charge and some have been made available on internet sites or as inexpensive CD-Roms. EISs must be widely available for public inspection in the evenings as well as during the day, and proponents sometimes prepare leaflets and publicise the availability of EISs imaginatively. Unfortunately, some EISs are too lengthy and technical to encourage a lay public readership, notwithstanding the availability of the summary. Copies of some of the documentation (for example, the scoping guidelines) are made available at no charge to those participating in the process. The various comments of statutory consultees and the public must be made available for public inspection and taken into account in the decision.

The Environmental Management Act contains provisions for the consultation of neighbouring member states if significant adverse transboundary impacts are possible (section 7.38). Inception memoranda, scoping guidelines and EISs are sent to neighbouring countries for proposals thought likely to have transfrontier impacts, with translations where appropriate. This formal procedure has been followed in less than 20 cases to date. De Boer (1999) described a number of examples of cooperation on transboundary impacts between the Netherlands and its neighbours.

Some large environmental pressure groups like Stichting Natuur en Milieu receive public funding on a project-by-project basis and have sometimes become involved in selected EIA cases. The EIA guidance deals with public participation and some advice about methods and procedures exists (for example, from the Ministry of Transport). The right to participation is supported by the possibility of administrative appeal to the courts against the decision (exercised in over 200 cases to date). This right of appeal is important in ensuring that EIA procedures are properly followed.

Generally, consultation and public participation in the Netherlands work reasonably effectively. In some instances (for example, for roads and dike reinforcements) competent authorities have gone well beyond the legislative requirements in gaining public comments. Many members of the public and non-governmental organisations object to proposals at both the scoping and EIS review stages, but others make useful comments on such matters as alternatives, vulnerable people or receptors, potential damage to people and difficulties in predicting impacts. There have been instances where an environmentally and economically preferable alternative has been put forward as a result of public participation, and public comments have often helped to refine proposals (Mostert, 1995; van Eck and Scholten, 1996). More important to the decision, however, is the informal consultation between the proponent and the competent authority. (The EIA Commission's role is limited to commenting on the need for additional information prior to the competent authority taking the decision.)

The Evaluation Committee on the Environmental Protection Act (ECW, 1990) stressed the importance of public participation in EIA ensuring that the main participants, and especially the competent authorities, discharged their responsibilities properly. The open nature of the Dutch EIA process, with consequent minimisation of the possibility of abuse, was seen as one of its great strengths. In its second report, ECW (1996) recommended the improvement of screening decisions by greater public participation and by the involvement of EIA Commission (see Chapter 9).

Canada

One of the five purposes of the Canadian Environmental Assessment Act 1992, as amended, is 'to ensure that there be an opportunity for public participation in the environmental assessment process' (section 4(d)). Every EA report must contain a discussion of any comments received from the public (participation is discretionary in screenings but mandatory in other EAs). The provisions for public participation vary according to the type of assessment being undertaken. In screenings, notice and participation are at the discretion of the responsible authority (RA) which can provide for the screening report to be made available for comment prior to decision making. It is supposed, in any event, to be filed in the public registry for the project (see below), providing an opportunity for public scrutiny and post-decision protest and lobbying. In comprehensive studies, the public may, at the discretion of the RA, be involved in the preparation of the report and must be provided with the opportunity to comment on the report once it has been submitted to the Canadian Environmental Assessment Agency (CEAA).

In practice, relevant experts outside the RA were consulted in over 80 per cent of cases studied by David Redmond and Associates (1999) and the public was consulted in all the comprehensive studies analysed by Shillington and Burns Consultants (1999). Despite the fact that the public was involved in only 10–15 per cent of screenings (mostly for larger projects), RAs:

cited numerous examples of public participation contributing to project decisions in the form of project approval conditions, inclusion of specific mitigations and follow-up measures, and changes in project decision and siting. (Shillington and Burns Consultants, 1999, p. 5)

It is striking that, given the purpose of the Act, there is so little public participation in screenings, notwithstanding the improvements that have taken place. It is small wonder that there are concerns about 'limited opportunities for meaningful public involvement in screenings and comprehensive studies' (Minister of the Environment – MoE, 2001, p. 14). As a result, it was proposed that the Act be amended to clarify that an RA can establish opportunities for public participation at any stage in a screening (Government of Canada, 2001). Provisions for public participation in comprehensive studies were to be increased and additional guidance on participation was to be prepared (MoE, 2001, p. 29).

The level of public concern is one reason for the RA to refer a project to public review under the Act (section 25(b)). In fact, this had not happened by mid-2001, all the public reviews having been initiated under other provisions. The Minister, after consulting the RA, sets the terms of reference for a review panel (or mediation). Review panels normally consist of three independent members appointed by the Minister of the Environment for each referred proposal. Members are selected for their expertise, credibility and knowledge of factors associated with the proposed undertaking, and the Minister appoints one of the members to chair the panel (Ross, 2000). Legal representation at hearings is discouraged to facilitate participation (Mostert, 1995; CEAA, 1997c). The Minister must give public notice of the availability of reports of panels (or mediators) and of how copies may be obtained. In cases where a panel is convened, the public:

- provide input and comments throughout the panel process;
- participate in public hearings convened by the panel. (CEAA, 1994, p. 115)

This type of public involvement, at the scoping stage, in the preparation of the EIS (where appropriate), in review of the EIS, in hearings and in commenting on the panel's report, is well established in Canada. It has evolved over time, the principal early contributor being Justice Berger in the MacKenzie Valley Pipeline Inquiry (Sewell, 1981), who is credited (Smith, 1993) with:

- broadening the interpretation of impacts to social perspectives;
- utilising various hearing formats, including local hearings;
- intervenor funding;
- accepting individual concerns as valid information;
- using the media to publicise the process (and popularise EA).

As Clark and Richards (1999, p. 216) have stated:

These panels tend to be controversial and politically sensitive, and ... to concentrate large quantities of scientific, technical and social information into clear recommendations to the Minister. . . . Participants may not always like the conclusions of the panel, but generally agree they have had a fair hearing.

One of the most significant aspects of the Canadian situation is the public participation funding programme for panel review cases (Lynn and Wathern, 1991; CEAA, 1999b). Section 58(1)(i) of the Act states that the Minister shall 'establish a participant funding program to facilitate the participation of the public in mediations and assessments by review panels'. This type of funding was first formally established in 1990 under the federal Environmental Assessment and Review Process, after several years of informal practice. An annual budget of over C$2 million has been allotted and criteria worked out for disbursement. In practice, average expenditure for the three years 1997–2000 was less than C$200,000 per annum (CEAA, 2000). Some provincial intervenor funding is also made available.

The money has been used by aboriginal peoples to employ researchers to prepare and, sometimes, to argue a case. Participant funding is widely regarded as being a great success and is credited with increasing the quality of information, of debate and, subsequently, of decisions. The Minister proposed that it be extended to comprehensive study assessments to raise public participation standards (Government of Canada, 2001; MoE, 2001). Generally, public participation, often supported by expert department information provision and comment, has worked well in panel review cases. Box 16.4 sets down some of the perceived benefits of the Voisey's Bay Project panel review (CEAA, 1999d), achieved only after legal action was brought by the aboriginal peoples (Hazell, 1999).

A citizen's guide to public participation has been published which explains how members of the public can take part in the EA process (CEAA, 1995). Most documents, including many EISs, have been available free of charge in the past. There has been considerable research prior to the more recent studies on public participation (see, for example, Weston, 1991).

A second significant provision of the Act relates to the setting up and maintenance of the public registry for each project subject to EA by the RA. Section 55(3) of the Act specifies that the public registry must include:

(a) any report relating to the assessment;
(b) any comments filed by the public in relation to the assessment;
(c) any records prepared by the responsible authority for the purposes of [follow-up];
(d) any records produced as the result of the implementation of any follow-up program;
(e) any terms of reference for mediation or a panel review; and
(f) any documents requiring mitigation measures to be implemented.

This ambitious public registry is supposed to consist of a computerised index to all EA cases, a listing of all the documents for each EA and the documents themselves. Documents are intended to be easily accessible electronically or to be copied at reasonable prices. In effect, the aim of this public registry system is to extend the very widespread availability of information that has always been provided for panel review projects to all EAs. However, the Commissioner of the Environment and Sustainable Development (1998) found that the index was difficult to use and:

Box 16.4   Benefits of public participation in the Voisey's Bay Mine and Mill Project panel review

The Voisey's Bay review panel was the first of its kind in terms of the participation of Aboriginal groups throughout the EA process. It was established under a memorandum of understanding (MOU) with the federal government, the province of Newfoundland and Labrador, the Labrador Inuit Association and the Innu Nation as signatories.

This MOU provided a co-operative framework for the governments and Aboriginal groups to ensure a single and effective environmental assessment was conducted for the project. At the same time, the MOU met legal requirements of the federal and provincial governments while addressing the concerns of the Aboriginal groups.

As a result of the full participation of the Aboriginal groups, the quality and credibility of the environmental assessment process were strengthened. Local community participation was enhanced since they deemed the process was open and balanced. The outcome was a thorough and comprehensive review of all key issues by the panel.

The long-term impact of the public review process is still to be determined since the government only recently responded to the panel's recommendations and construction has not begun. However, the review process strongly influenced the design of the project and mitigation measures intended to address environmental and socio-economic issues. As a result of these measures, adverse impacts relating to the project will be reduced or eliminated. In addition, the review panel process identified and enhanced economic growth opportunities for local communities.

The end result was a review process that facilitated discussion among stakeholders. This is expected to improve the project design and lead to a more sustainable development in the region.

Source: CEAA, 1999c, p. 19.

that fewer than half of the 187 projects we reviewed had been registered in the Index before the assessment was completed and a decision made.... These delays may effectively preclude public input. (pp. 6–20)

This finding was confirmed in a study by Shillington and Burns Consultants (1999). The Minister has proposed amendments to the Act and various other actions to improve the operation of the public registry (Government of Canada, 2001; MoE, 2001).

Commonwealth of Australia

Despite the number of different stages in the Australian EIA process and the necessity to produce documentation at each of these, public participation provisions were often permissive rather than mandatory under the repealed Impact of Proposals Act 1974. In practice, participation was more pervasive than the requirements specified but tended to rely heavily on written representations rather than allowing sufficiently for oral communication. Consultation of government departments and relevant agencies took place throughout the EIA process. However, there was no public participation in the vast majority of cases assessed, i.e., those examined by Environment Australia that did not require the preparation of an EIS or public environment report (PER) (Kinhill Engineers, 1994).

Some funding of public interventions took place under the previous legislation but this was very limited. As mentioned in Chapter 6, the number of legal actions mounted by environmental groups or others against the EIA procedure used in particular cases has been very low, in contrast to the US situation. In practice, public participation in Australian EIA has been very influential but not so wide or effective in determining outcomes as in the United States.

There were criticisms of inadequate participation by indigenous peoples in the EIA process (Craig & Ehrlich *et al.*, 1996). There were suggestions for strengthening various aspects of participation and for selective intervenor funding to ensure fuller public participation at several stages in the EIA process (see, for example, Kinhill Engineers, 1994). Richardson and Boer (1995) believed that increased use of public inquiries (and the funding of participation) could provide a much needed means of increasing oral participation in EIA. The Commonwealth Environment Protection Agency (1994) proposed a number of measures to formalise and strengthen public participation, including the provision of selective intervenor funding and greater use of public inquiries, in its reform package. The Ministerial consultation paper proposed an enhanced inquiry process and participation before a decision not to require an EIS or a PER was taken (Hill, 1998). These proposals were implemented in the Environment Protection and Biodiversity Conservation Act 1999 (EPBC Act).

The range of EIA information which the Environment Secretary must publish on the internet is broad. Every week, notice of:

- the Minister's intention to develop a draft bilateral agreement;
- referrals received by the Minister;
- decisions that an action is a controlled action;
- decisions on assessment approaches;
- information and invitations relating to assessment on preliminary documentation;
- guidelines for PERs and EISs;
- public invitations to comment on draft guidelines;
- draft or finalised PERs and EISs; and
- availability of assessment reports

must be published (EPBC Act, section 170A). These requirements represent a number of advances over the previous situation.

In particular, there is now consultation and public participation on the first screening decision, that assessment is or is not required, and both the decision and the reasons for it may be published. There is provision for consultation with the relevant state or territory on the second screening decision (i.e. about which type of assessment approach is to be employed) but there is no right of public participation. The decision (but not the reasons for it) must be published, however, giving an opportunity for legal challenge. Unlike the previous situation, the EPBC Act provides for the preliminary information to be made public through the assessment approach selected, matching previous discretionary practice.

There is now a discretionary requirement for public participation during the preparation of scoping guidelines. (Under the previous legislation, discretionary public involvement became common (see Chapter 10)). There is no formal requirement for public participation in the preparation of the draft EIS or PER but the EPBC Regulations 2000 (Schedule 4(2)) require the proponent to furnish information about:

(h) any consultation on the action, including:
   i   any consultation that has already taken place;
   ii   proposed consultation about relevant impacts of the action;
   iii if there has been consultation about the proposed action – any documented response to, or result of, the consultation;
(i) identification of affected parties, including a statement mentioning any communities that may be affected and describing their views.

Such consultation can clearly be extended to the public though, unsurprisingly, this tended not to occur under the previous legislation (see Chapter 11). There is, of course, consultation and participation once the draft EIS or PER has been produced (though the period of time permitted, not less than 20 business days, has been criticised by environmental groups), and the proponent must respond to the comments received in the final EIS or PER. The availability of the assessment report prepared by Environment Australia must be notified on the internet and a copy must be made available on request (either free or at a reasonable price). As mentioned in Chapter 13, there is a requirement for the Minister's approval decision to be notified on the internet. As under previous legislation, there are provisions for the confidentiality of EIA documents but these were hardly ever invoked in the past and, despite fears about their usage (see, for example, Hughes, 1999), there is no reason to expect them to be overused under the EPBC Act.

Although there are no provisions for funding participation in EIA, the EPBC Act meets criticisms of inadequate participation by indigenous peoples by providing for an Indigenous Advisory Committee to be set up to advise the Environment Minister. This is intended to enable the significance of indigenous people's knowledge of land management and the conservation and sustainable use of biodiversity to be taken into account (section 505B(1)).

New Zealand

The Resource Management Act 1991 contains several provisions relating to public participation. The Fourth Schedule (see Box 7.3) specifies that the proponent should include a list of affected or interested persons, the consultation undertaken and any response to the views of those consulted, in the assessment of environmental effects (AEE) report that is submitted. As mentioned in Chapter 10, where an application is found to have major effects and is notified by the local council, the Resource Management Act empowers local authorities to ask for an explanation of 'the consultation undertaken by the applicant'. Consultation and negotiation with affected parties is not new in New Zealand but the Act requires the developer to include a description of pre-application discussions and their outcome in the AEE report, since the local council can, and does, insist on this. It is also necessary to gain the written agreement of adversely affected parties to ensure that notification, and a consequent council hearing, are to be avoided. Consultation and participation prior to submission of the AEE report is thus not mandatory but is strongly advised.

Where projects are notified, local authorities are required to make AEE documentation available to the general public, who may obtain photocopies at a reasonable charge (Ministry for the Environment – MfE, 2001a). Any member of the public can make a submission. Certain consultees must be informed of the existence of the AEE report. The applicant or a submitter (objector) can insist that a council hearing be held (section 100). However, public participation is much more limited where the application is not notified.

There have been several guides to public participation in AEE in New Zealand, commencing with general guidance on scoping and public review (MfE, 1992b). There is now a readable citizen's guide to the Act (MfE, 2001d) and a practice guide on public participation and consultation for applicants has been published (MfE, 1999c). The Ministry for the Environment guidance to applicants (1999a) and to local authorities (1999b), in keeping with earlier Ministry advice, laid substantial stress on early and sustained consultation between the applicant and both the local authority and the various affected parties. As Morgan (1998) has pointed out, European approaches may not be the best way of ensuring Maori participation in AEE (Box 16.5).

As in other EIA systems, the role of the public is crucial in ensuring that the EIA system functions effectively (Dixon, 1993). However, Morgan (1995, 2000b) and the Parliamentary Commissioner for the Environment (PCE, 1995, 1996b, 1998) discerned real shortcomings in the involvement of the public. They reported that councils rarely consulted the public early enough in the process and often failed to encourage applicants to consult the public. Local authority publicity for applications often neglected to mention that environmental information was available for examination. Many AEE reports were found to be overly technical. Morgan (1995, 2001). It was suggested that non-technical summaries needed to be made available and that the accessibility of AEE documentation needed to be improved (see also Smith, 1996).

Box 16.5   Community meetings in New Zealand

In New Zealand, the usual European-derived approach to community participation often involves holding meetings in a town or community hall. People are invited to attend if they are interested and, perhaps after hearing speeches from representatives of various interest groups, will then be invited to ask questions. Although a certain protocol is often followed, it is not unusual for such meetings to be marked by confrontation and acrimony if the topics are contentious. The Maori, the indigenous people of Aoteoroa-New Zealand, view such behaviour as culturally inappropriate. Not only is this type of public meeting very different from their customary practices, it can also represent a real barrier to Maori involvement in public debate on important issues.

Seeking Maori participation necessitates an understanding of their protocols for initiating and conducting community meetings. The local elders will call the meeting, which would be hosted by the local *iwi* (tribal group) on their *marae* (the cultural and spiritual centre of their community, with the meeting-house as its focus). Participants in the meeting would be greeted and escorted on to the *marae* according to long-established ritual. There are well-defined rules governing who has the right to speak, and in what order. Protocol requires all speakers to greet their hosts, the ancestors of the local community members, and other participates, in a formal and respectful manner. Once due regard has been given to the needs of protocol, matters of concern can be addressed in what is often a passionate, but also usually a polite, debate. Oratory is a much-valued feature of such meetings, but displays of bad manners are poorly regarded.

Source: Morgan, 1998, p. 155.

In practice, only about 5 per cent of applications are notified (MfE, 2001c) despite the fact that many applicants fail to discuss their proposals either with council staff or with those likely to be affected prior to submission. Overall, Morgan (2001) felt that public involvement in the AEE process was frequently very poor.

Non-governmental organisations in New Zealand generally do not have the resources to intervene frequently and, where they do, there may be insufficient consultancy expertise available to enable them to present a credible case. However, while a consent application is referred to the Environment Court, there is a recent provision for environment, community and Maori groups to receive environmental legal assistance to enable them to participate more effectively in the resource management process, up to a ceiling of NZ$20,000.

There is some evidence of 'greenmail' (i.e. payments to neighbours for written consents) and of the use of environmental objections to make business competitiveness arguments (McShane, 1998; PCE, 1998). The media have not played a crusading role in New Zealand as they have in many other countries. It is apparent that, in practice, the admirable public participation provisions of the Act have not been implemented enthusiastically by councils or by applicants. However, there is evidence that, as experience has been gained and as precedents have been created by the Environment Court, practice is improving.

South Africa

Consultation and participation have been emphasised more during the early stages of the EIA process in South Africa than in many other jurisdictions. Despite this, the 1997 EIA Regulations (Republic of South Africa, 1997), while providing several opportunities for participation in the preparation of both the scoping report and the environmental impact report, are strangely silent about their public review. They emphasise that the applicant may be required to advertise the application (Regulation 4(6)). In addition, the applicant:

> is responsible for the public participation process to ensure that all interested parties, including government departments that may have jurisdiction over any aspect of the activity, are given the opportunity to participate in all the relevant procedures contemplated in these regulations. (Regulation 3(1)(f))

Both the scoping report (see Chapter 10) and the environmental impact report (Chapter 11) must contain appendices on public participation.

As mentioned in Chapter 13, the record of decision must be issued to any interested party who requests it (Regulation 10(1)). There is a third-party right of administrative appeal against the decision (Regulation 11) together with the possibility of appeal to the courts. To ensure transparency, any report submitted for the purposes of the EIA Regulations becomes a public document once the record of decision has been issued (Regulation 12).

It is the proponent and not the relevant authority who is responsible for consultation and public participation. While this is the consequence of an understandable endeavour to reduce the workload of the relevant authorities, it results in a failure to ensure prior unfettered public scrutiny of the reports which form the basis of the authorisation decision (Granger, 1998). The impression that information, rather than opinion, is being sought from the public participation process is overwhelming.

Interestingly, the EIA guidelines (Department of Environmental Affairs and Tourism – DEAT, 1998c) go much further than the Regulations. They recommend that the characteristically named 'interested and affected parties' (which include members of the public) are involved in reviewing both the scoping report and the environmental impact report (EIR) and they give brief guidance on the conduct of these reviews.

The integrated environmental management (IEM) guidelines provided considerable guidance about the involvement of the interested and affected parties in scoping. It is widely accepted that, despite the existence of these guidelines, there was limited public participation in EIA under the IEM procedure in

South Africa (Ridl, 1994; Republic of South Africa, 1998). Mafune *et al.* (1997, p. 206) found that only 5 of the 28 early EIA reports that they studied documented some form of public involvement in the undertaking of the EIAs. Ridl (1994) and Weaver (1996) have also emphasised the difficulties in managing public involvement in EIA, which has sometimes made EIA report preparation a lengthy and controversial process.

In particular, there have been severe limitations upon the participation of disadvantaged sections of society in EIA. These have resulted from, *inter alia*, a primary concern with survival (rather than conservation), illiteracy, the legacy of apartheid, the use of technical language, the holding of formal public meetings in an unfamiliar language, and suspicion of consultants, relevant authorities and certain developers (Goudie and Kilian, 1996; Burger and McCallum, 1997; Khan, 1998). Nevertheless, a number of specialist (and often innovative) public participation consultants in South Africa continue to endeavour to rescue the public involvement of disadvantaged communities in EIA from disrepute.

Certainly, the advantages of incorporating consultation and public participation in scoping are widely recognised (see, for example, Weaver *et al.*, 1999). In practice, scoping reports are generally circulated in draft form (see Chapter 10) and most draft EIRs are publicly reviewed before being finalised, as recommended in the EIA guidelines (DEAT, 1998c; see also Chapter 12). For example, to ensure that the review of the draft EIR for the Maputo Iron and Steel Project (see Chapter 11) was as inclusive as possible, the summary was released in Portuguese, Tsonga, Siswati and Northern Sotho, as well as English (Gibb Africa, 1998).

There is no direct funding of participants in the EIA process. However, several of South Africa's environmental groups are well funded and influential. They intervened effectively in many EIAs undertaken under the IEM procedures, commenting, gaining publicity, calling meetings and applying pressure on developers and politicians. While they have only rarely taken legal action in the past (see Chapter 14), their opportunity to do so increased when EIA became a legal requirement in 1997. However, many environmental non-governmental organisations have experienced consultation fatigue as a result of the sheer pace of change, and their ability to engage in the numerous EIAs being undertaken, let alone take legal action, has consequently diminished (Weaver *et al.*, 2000).

One of the objectives of the EIA provisions of the National Environmental Management Act 1998 (NEMA) is to:

> ensure adequate and appropriate opportunity for public participation in decisions that may affect the environment. (section 23(2)(d))

NEMA somewhat ambitiously requires public information and participation and conflict resolution, in all phases of the environmental assessment process (section 24(7)(d) – see also Chapter 10). It is based on an important principle of environmental justice:

> adverse environmental impacts shall not be distributed in such a manner as to unfairly discriminate against any person, particularly vulnerable and disadvantaged persons. (section 2(4)(c))

Given the emphasis in the constitution on the citizen's rights to be heard and to a healthy environment (see Chapter 5), it will be surprising if the admirable provisions for public participation early in the existing EIA process are not matched by similar provisions for involvement later in the process when the 1997 EIA Regulations are repealed on the promulgation of the NEMA EIA regulations (Chapter 5). NEMA itself is to be modified to give better effect to the realisation of environmental rights in EIA (DEAT, 2000a). Already, the user guide to NEMA (DEAT, 1999) makes it clear that citizens can ask the courts to set aside permissions if the recommendations of an EIA are ignored.

Table 16.1   The treatment of consultation and participation in the EIA systems

Criterion 11: Must consultation and participation take place prior to, and following, EIA report publication?

Jurisdiction	Criterion met?	Comment
United States	Yes	Consultation and public participation take place at several EIS stages, limited in environmental assessment stages
UK	Partially	Consultation often takes place prior to environmental statement (ES) if scoping undertaken: public participation rare. Both must be undertaken following ES release
The Netherlands	Yes	Formal requirements for consultation and public participation in both scoping and review
Canada	Yes	Consultation and public participation mandatory throughout panel reviews, required following comprehensive studies and discretionary in screenings
Commonwealth of Australia	Yes	Formal requirement for public participation in screening and discretionary power for involvement in scoping expected to be used routinely. Agency consultation takes place throughout EIA process
New Zealand	Partially	Duty to consult public following EIA report publication and virtually compulsory to consult earlier for notified projects
South Africa	Partially	Provisions for public involvement in scoping report and environmental impact report (EIR) preparation not matched by formal rights to comment on completed scoping reports or EIRs

Summary

Six of the EIA systems meet the requirement that there must be consultation and participation following the release of the EIA report but two do not make consultation and participation prior to the EIA report mandatory (Table 16.1). While some consultation and public participation often takes place following the preparation of the EIA report in South Africa, this is not mandatory.

Table 16.1 masks some significant variations in consultation and participation between and within countries. The weakest requirements for pre-EIA report participation are those in the UK which, with New Zealand, is the only EIA system not formally to require scoping. While consultation often takes place in the UK prior to the ES where the proponent requests it, the involvement of the public is relatively rare. In New Zealand, the local authorities have the power to demand that consultation and participation take place prior to submission of EIA reports for notified projects.

In Australia the existence of a scoping stage effectively ensures public participation, even though this is not mandatory. Although full participation takes place during panel reviews in Canada, there is no mandatory requirement for involvement in the preparation of comprehensive studies though this is customary. Participation in the preparation of Canadian screening reports is entirely discretionary. In the United States there is full provision for early participation and consultation in the preparation of the EIS but provisions relating to public involvement in the preparation of environmental assessments are often not observed. Public participation and consultation are most strongly embedded in the EIA system in the Netherlands. Generally, however, participation provisions could beneficially be strengthened in most EIA systems.

Note

1. Details of EISs and EAs are also published on the EPA Office of Federal Activities web site at www.es.epa.gov/oeca/ofa

Chapter 17

Monitoring of EIA systems

In addition to the monitoring and auditing of impact actions (see Chapter 14), it is increasingly being recognised that some form of EIA system monitoring is needed (Canadian Environmental Assessment Research Council – CEARC, 1988; Sadler, 1996). The principal purposes of EIA system monitoring are the consideration of EIA system effectiveness, the amendment of the EIA system to incorporate feedback from experience and to remedy any weaknesses identified, and the diffusion of best EIA practice. As Sadler (1998, p. 37) has stated: 'there is an evident requirement to use effectiveness reviews as an integral strategy for building quality control and assurance throughout the EA [environmental assessment] process.'

EIA system monitoring requires records to be kept of the: numbers of EIA reports produced, types of project assessed, decisions reached, numbers of implemented projects, availability of documents, etc. This chapter outlines the issues involved in the monitoring of EIA systems and puts forward a set of evaluation criteria for EIA system monitoring. These criteria are then utilised in the comparative review of the EIA systems in the United States, UK, the Netherlands, Canada, Commonwealth of Australia, New Zealand and South Africa.

EIA system monitoring

Reviews of any EIA system need to be carried out from time to time and any necessary changes to the system implemented. The better the EIA system monitoring information available, the easier such a review will be. As with other elements of the EIA process, the role of consultation and participation in reviews of the EIA system is important and should be adequately provided for. The opinions of the various stakeholders in the EIA process: politicians, other decision-makers, government advisers, proponents, consultancies, environmental authorities, non-governmental organisations, selected members of the public, researchers, journalists, etc., should be sought. Where formal reviews of EIA reports are undertaken (for example, the formal assessment reports drawn up in the Australian EIA system), a record of these and the results obtained should be kept and made public. In addition, experience of specific EIAs may reveal that changes in practice or procedure within the EIA system more generally need to be made.

Most jurisdictions have learned from the experience of individual EIAs and have carried out reviews of their EIA systems from time to time. Provision for the feedback of findings, including recommendations regarding the improvement of EIA system effectiveness and for taking appropriate action, should therefore exist. Clearly such feedback tends to be most effective where only a limited number of responsible authorities are involved in the EIA process, and practice varies between jurisdictions. Appropriate action may take the form of legislation, regulations, circulars, practice advice notes, amendment of project-specific or generic guidelines, training, or other means. Most jurisdictions have amended their EIA systems at one time or another and most would probably wish to implement further modifications, whether major or minor, at any specific time.

There are numerous elements of any EIA system which can be monitored to establish the effectiveness of the whole EIA process and subsequently to amend the system to incorporate feedback from experience. Therefore, prior planning of the approach to system monitoring is necessary. System monitoring might be focused on particular dimensions of EIA effectiveness, perhaps by targeting certain aspects of EIA acknowledged to be weak. For example, Glasson *et al.* (1999) cited the consideration of cumulative impacts and public participation requirements as areas of EIA requiring improvement in many jurisdictions, while Sadler (1996) reported that scoping, review and monitoring were generally poorly handled (see Chapter 20). It is also necessary to develop indicators to judge the effectiveness of the EIA system. Only after such targets and indicators have been established should attention be focused on gathering the documentation and other information to undertake the EIA system monitoring programme. The diffusion of best EIA practice within and outside the EIA system relies not only upon EIA system monitoring but also upon the provision of published guidance, training and the undertaking of research.

In any EIA system, a definitive record of the number of EIA reports undertaken should be maintained and made public, to enable system monitoring research to be practised. This record should relate both to total numbers of EIA reports and to EIA reports for different types of action. Clearly, sufficient details relating to the precise title of each document, its length, its date, its price, where it may be accessed or obtained and any other relevant matters should be made available. Such records are available in many EIA systems, although their quality and level of accessibility is not always satisfactory (Sadler, 1996).

In addition, the existence of other EIA documents, such as scoping reports, should be recorded. Similar details to those listed for EIA reports should be maintained for this documentation generated within the EIA system. Practice in the maintenance of such records tends to vary from one EIA system to another. Ideally, all EIA reports and other EIA documents should be publicly available at one or more central locations during reasonable hours. Collections of EIA reports provide an invaluable source of information to those engaged in preparing such documents, to those responsible for reviewing them, to those likely to be consulted, to the public and to those undertaking research. In

practice, these documents may be consulted in many EIA systems, with varying degrees of difficulty.

There are considerable difficulties in obtaining accurate information about
the financial costs involved in undertaking EIA. However, while the costs of
EIA report preparation may be difficult to distinguish from those of other
activities associated with the action (see Chapter 18), some information about
expenditure incurred in preparing and processing EIA documents in every EIA
system should be obtained, perhaps on a sample basis, and centrally recorded.
Details about numbers of staff involved in EIA, as well as about consultancy
costs and any fee payments should be maintained and made public. As with
other aspects of EIA system monitoring, such information is easier to collect if
a single agency is responsible (as, for example, in the Netherlands), and practice varies accordingly.

Similar information should be collected and maintained in relation to the
time required to undertake EIA. Data obtained should include the amount of
time needed to process each EIA report once it has been received. As elsewhere
in the EIA process, there are numerous measurement difficulties (Hart, 1984),
but reasonably reliable records, possibly on a sample basis, can be kept if the
will to do so exists.

Finally, and again in common with other elements of the EIA system, the
monitoring of the EIA system should be effective (i.e. lead to achievement of
its goals) and efficient (i.e. not consume disproportionate financial, managerial
or time resources). This and the other criteria discussed above are summarised
in Box 17.1. The various evaluation criteria are used in the comparative review
of EIA system monitoring which now follows.

Box 17.1   Evaluation criteria for EIA system monitoring

**Must the EIA system be monitored and, if necessary, be amended to incor-
porate feedback from experience?**

- Is there a legal provision for periodic review of the EIA system?
- Have reviews of the EIA system been carried out and changes made?
- Is consultation and participation required in EIA system review?
- Is a record of EIA reports for various types of action kept and made public?
- Are records of other EIA documents kept and made public?
- Are EIA reports and other EIA documents publicly available at one or more
  locations?
- Are records of the financial costs of EIA kept and made public?
- Is information on the time required for EIA collected and made public?
- Are the lessons from specific EIAs fed back into the system?
- Does the monitoring of the EIA system function efficiently and effectively?

United States

The National Environmental Policy Act 1969 (NEPA) contains no requirement for periodic review of its provisions. However, the Council on
Environmental Quality (CEQ) was created by NEPA and given the responsibility for environmental policy development and the duty to review and
appraise federal agency compliance with NEPA. Part of this CEQ oversight
has involved the preparation of annual reports which summarise the trends in
the implementation of NEPA (numbers of statements, numbers of court cases,
significant developments, etc.). The annual reports of CEQ have provided an
invaluable picture of the operation of NEPA over the years (and of environmental trends in the United States generally) but are now no longer required
(that for 1997 was the last (CEQ, 1999)).[1]

Since the Environmental Protection Agency (EPA) must notify each draft
and final environmental impact statement (EIS) in the Federal Register, list-
ings of all EISs can readily be obtained from EPA, if necessary broken down
by type of action, by agency and by geographical location. Copies of EISs must
be filed by EPA but the Agency now only maintains a library of EISs in hard
copy form for two years. However, EISs are available for inspection or loan at
the library of Northwestern University, Illinois, and can be purchased from a
private company.[2]

A web site, NEPAnet,[3] was established in 1995 as a repository of baseline
environmental impact information and to give better public access to NEPA
information (Jessee, 1998). However, there is no central record of other NEPA
documents (for example, environmental assessments, findings of no significant
impact, records of decision) but each of the agencies maintains at least some
statistics on these documents. Copies of other NEPA documents are not kept
centrally or filed on a long-term basis by the relevant agencies. No regular
records of the financial costs or time requirements of EIA are maintained. This
makes quantification of the successes and failures of NEPA impossible (Welles,
1997).

A number of reviews of the operation of the whole EIA system has been
carried out, most notably those which led to the 1978 CEQ Regulations, that
initiated during the early Reagan years (see Chapter 2), the Environmental
Law Institute reappraisal (Environmental Law Institute, 1995b) and the 1997
effectiveness study (CEQ, 1997d; Welles, 1997). There have also been federally funded reviews of parts of the EIA system, for example, on agency compliance (Environmental Law Institute, 1981), on the scientific quality of EISs
(Caldwell *et al.*, 1983), on auditing (Culhane *et al.*, 1987) and on environmental assessments (Blaug, 1993). In addition, federal agencies have conducted reviews of their own NEPA procedures and fed lessons from specific EIAs
(perhaps as a result of court cases) back into their systems, for example, by substantially revising their original regulatory procedures. These reviews have generally involved extensive consultation. For example, the Department of Energy
commissioned the National Academy of Public Administration (NAPA) to
undertake a review of its EIA activities (NAPA, 1998).[4] Further, there have

been numerous academic reviews of the US EIA system, especially as it has come of age (see, for example, Blumm, 1990; Hildebrand and Cannon, 1993; Clark and Canter, 1997; Caldwell, 1998).

UK

There is no single official comprehensive listing of all the environmental statements (ESs) which have been prepared in the UK. The Planning Regulations for England and Wales and for Scotland require local planning authorities (LPAs) to send three copies of any ES to the appropriate government department when it is received. On the basis of this information, the Department of Transport, Local Government and the Regions (DTLR) prepares lists of ESs prepared under the Planning Regulations which were, until 1994, published. Information including the name of the LPA and about the nature of the development and the category of project within Schedule 1 or 2 to the Regulations is kept. Because of incomplete compliance with the requirement for LPAs to send copies of ESs to central government, these lists tend to be incomplete. Nevertheless, the monitoring situation with regard to planning ESs is generally better than for projects approved under other regulations.

There is a published list of all the ESs prepared for planning projects in Scotland (Scottish Executive, 1999), despite the absence of an equivalent list in England. Unofficial lists of ESs have been prepared by the EIA Centre at Manchester University, by the Institute of Environmental Assessment (1993) (now the Institute of Environmental Management and Assessment – IEMA) and by the Impacts Assessment Unit at Oxford Brookes University (see, for example, Wood and Bellanger, 1998). Summaries of opinions, notifications and directions were published prior to 1995. There is no monitoring of LPA EIA decisions or of decisions on planning appeals or call-in cases involving EIA.

There is no single repository of ESs for all types of project for the whole of the UK. English ESs prepared under the Planning Regulations may be inspected at the library of DTLR in London, but the collection is far from complete. Other collections, such as those at Manchester University, at IEMA and at Oxford Brookes University, are also incomplete (Glasson *et al.*, 1999, p. 223). No record of the monetary costs and time required for EIA are kept though some information on these topics is available (see Chapter 18).

Inevitably, as experience has been gained with EIA, practice has improved (see, for example, Department of the Environment – DoE, 1996; Leu *et al.*, 1996; Jones *et al.*, 1998), and modifications have been made to the operation of the EIA system. As mentioned in Chapter 4, a series of studies has been commissioned: on the early operation of the EIA system (DoE, 1991a); on the utilisation and preparation of ESs (DoE, 1994a,b, 1995); on the quality of ESs (DoE, 1996); and on the role of mitigation in EIA (Department of the Environment, Transport and the Regions – DETR, 1997d). Prior to the modifications necessary to implement the amended Directive, changes to the Planning Regulations were made which extended EIA to a limited number of other types of project and changed a number of procedures relating, for example, to consultation and participation where further documentation was required by the LPA. The 1999 Regulations, necessitated by the requirements of the amended Directive (see Chapter 4), also reflected the findings of the various commissioned studies, of other research, and of experience gained. Proposed changes to EIA procedures are circulated to consultees by DTLR, made available to the public, posted on DTLR's web site, and frequently modified as a result of comments received, in much the same way as a proposed project subject to EIA (see, for example, DETR, 1997a,b).

In brief, while the only formal requirement for EIA system monitoring involves the provision of copies of ESs for central government, partial lists of ESs exist, monitoring and review of the EIA system has taken place, and significant amendments have been made.

The Netherlands

The Netherlands EIA system is subject to several monitoring provisions. Perhaps the most important is the requirement of the Environmental Management Act 1994 for the EIA system to be reviewed every five years (section 7.2(5)).[5] To undertake these reviews, an evaluation committee (ECW) was set up to advise the Ministry of Housing, Spatial Planning and the Environment (VROM) and the Ministry of Agriculture, Nature Management and Fisheries, on how the EIA system (and the Environmental Protection/Management Act more generally) was working (see Chapter 5). It commissions background studies and conducts interviews with government officials, developers, consultees and interest groups. Many of the recommendations of its first report (ECW, 1990) were duly implemented. Its second report, which relied on commissioned studies on the effect of EIA on decisions, monitoring in EIA, and on the relationship between EIA and permitting, was published in 1996 (ECW, 1996).

The second most important system monitoring provision in the Act is the requirement for the EIA Commission (EIAC) to make an annual report on its work (section 2.18). These annual reports provide a summary of EIA activity during the year and a complete list of all the proposals which have been referred to EIAC. The reports also contain discussions about aspects of EIA practice. A copy of nearly every: set of guidelines; EIS; EIAC review; decision; and auditing report is available in the EIAC library in Utrecht, principally (but not exclusively) for internal use. More than 1,000 EIAs have been initiated and more than 500 decisions have been taken since 1986.

In addition to these requirements, other major studies of the Dutch EIA system have been undertaken (see, for example, Mostert, 1995; Arts, 1998). No formal records of financial costs or time (see Chapter 18) are kept, though estimates have been made, for example, by the Ministry of Transport. The Act provides for a minimum of 21 weeks to elapse between the submission of the notification of intent and the decision, though EIA cases (especially those concerned with complex or controversial proposals) have sometimes only been decided several years after initiation.

Despite the amount of information available about the EIA system in the Netherlands, the results of individual EIAs have not always been sufficiently fed back into the system. Thus, while EIAC is able to utilise experience in drawing up its recommended scoping guidelines and, to a lesser extent, in reviewing EISs, the Evaluation Committee on the Environmental Management Act was anxious that EIAC 'should be used more than has hitherto been the case as a "knowledge bank" ' (ECW, 1996, p. 21). Many developers and consultants were failing to utilise the available information fully, and some competent authorities, and EIAC, had not been utilising sufficiently the experience of similar projects gained elsewhere (ECW, 1996). There is, therefore, some scope to improve practice in the well-monitored Dutch EIA system.

Canada

The Canadian Environmental Assessment Act 1992 contains three main provisions for system monitoring: first, the maintenance of a public registry; second, annual performance reports; and third, a comprehensive review after five years. In addition to these requirements, a commitment was made when the Act came under force that the Auditor-General would review the Act's efficiency and effectiveness after three years.

These provisions follow the tradition of Canadian EIA system monitoring. Numerous reviews of the Canadian Environmental Assessment and Review Process and the quality of EA were undertaken (including several damning internal evaluations) (see, for example, Beanlands and Duinker, 1983; MacLaren and Whitney, 1985; Fenge and Smith, 1986; CEARC, 1988; Federal Environmental Assessment Review Office, 1988; Weston, 1991; Smith, 1993; Sadler, 1995; Sadar and Stolte, 1996; Hazell, 1999).

In principle, the first requirement, to maintain the public registry (see Chapter 16) should provide the basis for statistical summaries of operations, since the various individual project entries could be aggregated to provide an up-to-date national picture. However, record-keeping utilising the electronic index system, although making it easier to access available information, has proved to be partial (see Chapter 16) and therefore less helpful in system monitoring than anticipated.

A register of all comprehensive study and panel review reports is kept and published by the Canadian Environmental Assessment Agency (CEAA). The reports and all the public comments are available for inspection at one of the Agency's seven offices, allowing the diffusion of best practice.

The second requirement is that an annual report must be made to parliament 'on the activities of the Agency and the administration and implementation of this Act and Regulations during that year' (section 71(1)). This annual performance report describes the results and achievements of the Agency against previously stated commitments such as delivering 'environmental assessments that are effective, timely efficient, involve public participation and support the principles of sustainable development' (CEAA, 2000,

p. 12). In addition, it contains a statistical summary of environmental assessments, financial performance figures and much other valuable information. Annual performance reports, which have been strengthened as a result of criticisms by the Commissioner of the Environment and Sustainable Development (CESD, 1998) provide valuable information for system monitoring.

To meet the commitment that the Auditor-General review the Act, the Commissioner undertook an audit based on 187 screenings associated with 11 responsible authorities. These were chosen to reflect a cross-section of government activities but, where possible, he 'selected screenings for larger and potentially more environmentally significant projects' (CESD, 1998, pp. 6–12). The audit concluded that there were several opportunities for improved EA, including increased attention to the EA of policies and programmes, that good practice needed to be shared and that the Agency could play a more pivotal role (CESD, 1998, pp. 6–28).

To meet the third requirement, that a comprehensive quinquennial review be submitted to parliament (section 72 (1)), the Minister of the Environment (MoE) commissioned 12 independent studies, including one on multi-jurisdictional EAs, in order to prepare an initial discussion paper (MoE, 2001). About 800 people attended consultation sessions in 19 locations; another 350 invitees participated in one-day workshops in seven cities, and more than 100 formal submissions were made to the Agency. The multi-stakeholder Regulatory Advisory Committee set up to advise the Minister on implementation of the Act, made numerous recommendations for improvement (Regulatory Advisory Committee, 2000). A review web site was set up, which was visited by 14,000 people, consultations with the provinces and with federal departments and agencies took place, and discussions were held with aboriginal groups. In short, the review was truly 'comprehensive'. It concluded that there were many strengths in the current EA process, not least promoting sustainable development, involving the public, and forcing federal departments to protect the environment. These provided the basis of three goals for enhancing its operation:

1. Making EA a certain, predictable and timely process.
2. Improving the quality of EAs.
3. Strengthening opportunities for public participation.

Proposals for the achievement of these modest but politically realistic goals were advanced (MoE, 2001) and a bill to amend the Act accordingly was drafted (Government of Canada, 2001).

While there are no formal records of the financial costs of EA, nor of its time requirements, studies for the five-year review provide some evidence about these (see Chapter 18). There can be no doubt that lessons from specific EAs have been fed back into the EA system as a result of administrative experience with the Act, court cases and the development of technical expertise (see, for example, MoE, 2001). The history of Canadian EA indicates that further revisions to improve the Act and its Regulations will take place as experience with the amended legislation is gained.

Commonwealth of Australia

Neither the Environment Protection (Impact of Proposals) Act 1974 nor the Administrative Procedures (Commonwealth of Australia, 1995) required any review or monitoring of the EIA system. In practice, a limited amount of monitoring occurred. A record of all the EISs and public environment reports (PERs) produced was kept by Environment Australia and is publicly available. However, no record was kept of other EIA documents, apart from assessment reports. The annual reports of the Department of the Environment present summary statistics about the numbers and types of assessment undertaken. Environment Australia holds a central collection of EISs and PERs and a publicly accessible collection is maintained by the Environment Department's library in Canberra. No record is kept of the costs or time requirements of the EIA system, though anecdotal evidence of, and opinions about, costs and delays abound (see, for example, Anderson, 1994; Harvey, 1998).

Australia's Commonwealth EIA system has been in place since 1974. Since that time there have been only a handful of major challenges in the courts (see Chapter 6). There have been innumerable detailed reviews and critiques either of the EIA system or dealing with the EIA system. Reviews (many of which have involved consultation and participation) have been sponsored by environmental organisations, by business and by government (see, for example, Jambrich et al., 1992; Commonwealth Environment Protection Agency – CEPA, 1994; Harvey, 1998).

The most exhaustive review, conducted between 1993 and 1995, involved the production of seven consultancy reports on public participation, indigenous people's involvement, public inquiries, social impact assessment, cumulative and strategic impact assessment, state EIA processes and overseas EIA processes. Following the production of a discussion paper (CEPA, 1994), workshops were held in the various states and territories to discuss the proposals. Unfortunately, despite agreement by both business and conservation interests, these system-strengthening proposals foundered on the rock of interparty, intergovernmental and interdepartmental rivalries, like many previous recommendations to improve the Commonwealth EIA system (Carbon, 1998). Some (but by no means all) of the proposals were evident in the consultation paper issued by the Liberal government (Hill, 1998) and subsequently enacted in the Environment Protection and Biodiversity Conservation Act 1999 (EPBC Act).

The EPBC Act contains a number of reporting and review requirements. In particular:

- All Commonwealth bodies must report annually on their implementation of ecologically sustainable development.
- The Commonwealth must prepare a state of the environment report every five years.
- The list of matters of national environmental significance must be reviewed every five years (Environment Australia, 1999a).
- An annual report on the operation of the EPBC Act must be prepared by Environment Australia and tabled by the Minister.

These reports and reviews are subject to auditing by the Auditor-General as well as to parliamentary and public scrutiny. Despite the advances demonstrated by these requirements, there is no provision relating specifically to the monitoring of the EIA system.

This is a missed opportunity since various methods of strengthening the monitoring of the Commonwealth EIA system have been proposed, especially the provision of better environmental data and a national inventory and depository of Commonwealth and state EIA documents (Anderson, 1994). The reform agenda (CEPA, 1994, p. 52) proposed the annual publication of an 'EIA performance statement' detailing, *inter alia*, EIA activities at the Commonwealth level, project monitoring results and developments in state EIA systems. Some of these recommendations are met by Environment Australia's EPBC Act web site. In particular, better environmental data are available, there are notices about all EIA documents (including preliminary documentation) and the texts of some documents are published. However, this does not provide a complete record of Commonwealth EPBC Act documents and, despite linkages to state EIA web sites, it does not paint a comprehensive national EIA picture.

New Zealand

The Ministry for the Environment (MfE) is required to monitor the effect and implementation of the Resource Management Act 1991 (section 24(f)), and every local authority is supposed to monitor its functions under the Act (see Chapter 14). Although several clarifying amendments to the Act have been made, there is no duty to review its operation in a specified number of years. Nor does the Act contain any specific monitoring requirements, such as a duty to collect EIA documentation at a central point of reference, or to record assessment of environmental effects (AEE) reports. Since 1996, there has been an annual survey of local authority performance under the Act (see, for example, MfE, 2001c[6]). This provides valuable statistical evidence about numbers of consents notified and granted, time-frames, public participation, cost recovery, etc. A public awareness survey was undertaken in 2000 which revealed that less than 20 per cent of the population were readily aware of the Act. Because of the way in which EIA is integrated into the whole New Zealand environmental management system (Dixon and Fookes, 1995; Bartlett, 1997), it is very difficult to disentangle information which relates specifically to it.

The Parliamentary Commissioner for the Environment (PCE) helps to ensure that lessons learned from specific AEEs are fed back into the system since his reports are published and disseminated (see PCE, 1995, 1996a,b, 1998). The Ministry for the Environment plays an informal but important role in the dissemination of information about AEE. This type of information exchange is common in New Zealand where professional networks are very small. The Ministry also occasionally makes submissions in individual consent cases where it feels that important issues are involved (see Chapter 13). However, the major element in EIA system development is probably the Environment Court, whose decisions are widely read by environmental professionals.

Some monitoring of the EIA system has taken place independently. In particular, Morgan's (1995, 2001) and the PCE's (1995, 1996a,b) studies of the operation of Resource Management Act EIA procedures threw light on current practice. Unsurprisingly, they reported that procedures for major developments were operating as they had under the previous Environmental Protection and Enhancement Procedures, with the retention of consultants, scoping, the preparation of professional AEE reports, and widespread consultation. However, smaller proposals were creating difficulties since large numbers of short AEE reports were being prepared by inexperienced applicants and reviewed by newly appointed and overwhelmed council officials (with inadequate supervisory support, Dixon *et al.*, 1997). It was hoped that many of these AEE reports would not be required once screening criteria were set down in operative regional and (especially) district plans, but there was little evidence that this screening was taking place in 2000 (Morgan, 2000b).

Frieder's (1997) report on the Resource Management Act (RMA) did not dwell on EIA. She believed that the Ministry was inadequately resourced, that there 'is a growing gap between the RMA implementers in the regions and the national policy hub in Wellington' (p. 43), and that a preoccupation with cost-minimisation was jeopardising good practice. She argued for increased monitoring of the Act.

The monitoring of local authority performance (and other MfE activities) was commenced as a result of complaints about delays, costs and uncertainties by a small but highly influential part of the business community. This pressure also led to the commissioning of the idiosyncratic and hostile report on the Act by one of its leading critics and of critiques of his polemic by three practitioners (McShane, 1998). The Minister for the Environment came forward with a number of proposals to amend the Act in 1998 (MfE, 1998). These included consent processing by consultants rather than councils, direct appeal to the Environment Court, tightening of the definition of 'environment' and deletion of the Fourth Schedule. The Ministry newsletter *Update*, (October 1999) reported that nearly 750 submissions were received, many of which approved the proposal changes while supporting technical amendments to strengthen the Act.

One of the most damning critiques was advanced by the Parliamentary Commissioner for the Environment (1998) who felt that undue weight was being given to the Act's implementation costs with insufficient accorded to achieving sustainable management. The process of review of the Act was described as inadequate. It is clear that many others agreed, since the proposals emerging from the Local Government and Environment Select Committee in 2001 were much more modest, being largely confined to technical amendments which were likely to be enacted in 2002. The criticisms of inadequate Ministry for the Environment implementation of the Act, especially in its early years, resulted in initiatives such as environmental legal aid (see Chapter 16) and the preparation of various guidance documents (Chapter 15).

South Africa

There is no provision in either the Environment Conservation Act 1989 or the 1997 EIA Regulations for any review of the EIA system or for the keeping of documents relating to EIA. (Relevant authorities must, however, keep a register of all applications received (Republic of South Africa, 1997, Regulation 4(5).) Despite this lacuna, there is great interest in South Africa in the implementation of the Regulations, and in their relationship to the wider canvas of integrated environmental management.

The national Department of Environmental Affairs and Tourism (DEAT) appears not to keep any record of EIA documents or copies of EIA reports. The Department sees this as the responsibility of the provincial governments or, where EIA responsibilities are delegated, of the appropriate local government, but this view is not shared by these bodies, despite the requirement that they maintain a register of applications. At present, practice in the keeping both of records and of EIA reports is variable, but is often rudimentary. The fragmentation of EIA responsibilities, the understaffing of relevant authorities and the unaccountable bureaucratic culture in South Africa all militate against adequate EIA system monitoring. Documents are, however, generally publicly available to the persistent enquirer and it is intended to make use of the internet to afford greater accessibility in the future.

No annual reports on EIA activities are required or are likely to be prepared. Equally, no records of the financial costs or time requirements of EIA are kept, though some anecdotal evidence exists. Unfortunately, the cutback in central government financing of environmental protection has meant that a previously effective intergovernmental EIA liaison committee's meetings have been suspended, so that provincial governments are not learning sufficiently from each other. This function does not appear to have been taken over by the Committee for Environmental Coordination (see Chapter 6).

It is probable that the active group of over 250 EIA practitioners and academics in South Africa (many of whom are members of the South African chapter of the International Association for Impact Assessment) provide the best means of EIA system monitoring, through their meetings, conferences, training and other activities. Their work tends to be somewhat ad hoc, however, and there is clearly a need for more formal EIA system monitoring if experience is to be used effectively in improving EIA in the future.

The 1997 EIA Regulations were promulgated hurriedly and the White Paper on environmental management policy was frank about the need for early revision:

> The EIA regulations legislated only the scoping and EIA portions of the integrated environmental management (IEM) procedures. This is a major limitation of the current regulations. (Republic of South Africa, 1998, p. 73)

The discussion document on IEM spelt out how the IEM procedure could be applied to land-use zoning plans, to new activities, to existing activities and to activities within an IEM-approved land use zoning plan (but not to economic policies) (DEAT, 1998a). This document was rapidly followed by draft legislation (DEAT, 1998b) and by the National Environmental Management Act 1998 (NEMA) which was intended to provide the legal framework for an enhanced EIA system.

It is expected that NEMA will be modified in order to, *inter alia*, strengthen its EIA provisions (DEAT, 2000a – see also Chapter 5) and that NEMA EIA regulations to replace the 1997 EIA Regulations will be promulgated in 2002. The enhancement of the current EIA system, based upon the 1997 Regulations and the accompanying guidelines, is likely to be more appreciable if it is informed by sound information about existing performance.

Summary

The EIA systems in the Netherlands, Canada and Australia all meet the EIA system monitoring criterion (Table 17.1). The EIA system in the United States partially meets the criterion and those in the UK, New Zealand and South Africa are adjudged not to meet it. There is, however, substantial documentation and monitoring information available relating to these EIA systems

It is probably no coincidence that the EIA systems which are monitored all possess a single body with overall responsibility for EIA as well as a legal duty either to review or to oversee the EIA system. Of the three, only the Australian EIA system does not possess a legally imposed quinquennial review requirement relating specifically to EIA. Both the Canadian Environmental Assessment Agency and the Dutch EIA Commission (which are solely concerned with EIA) possess adequate staff resources to undertake EIA system monitoring. This is also true of the Environment Australia which must report on the operation of the Environment Protection and Biodiversity Conservation Act 1999 annually and on other matters every five years. The Canadian and Australian initiatives with internet-accessible documentation could provide models for EIA system monitoring worthy of consideration elsewhere.

It is the major task of the Council on Environmental Quality to oversee the US EIA system, but it has recently had to make do with far fewer staff resources than it needs to undertake its oversight role effectively. In addition, the Council is no longer required to publish an annual report on the operation of the system. Fortunately, the Environmental Protection Agency has the resources necessary to undertake an EIA system monitoring function in the United States. The UK, New Zealand and South African requirements do not include a formal duty to review or monitor the EIA systems and the jurisdictions do not allocate sufficient resources to the central bodies responsible for EIA to permit effective system monitoring to take place. The first steps in improving EIA system monitoring are clearly to make it a legal requirement and to provide the resources available to undertake it.

Table 17.1   The treatment of system monitoring in the EIA systems

Criterion 12: Must the EIA system be monitored and, if necessary, be amended to incorporate feedback from experience?

Jurisdiction	Criterion met?	Comment
United States	Partially	Council on Environmental Quality charged with general oversight of EIA implementation but no longer reports annually. Numerous reviews undertaken and many amendments to agency regulations made
UK	No	No formal general requirement to monitor but some records published. EIA system reviews undertaken, and changes made to improve operation
The Netherlands	Yes	EIA Commission prepares annual report. Independent comprehensive quinquennial EIA system review undertaken
Canada	Yes	Act contains annual report and five-year review requirements and public registry should facilitate monitoring of EA system
Commonwealth of Australia	Yes	Annual report on 1999 Act required but no specific requirement for monitoring or periodic review of EIA system. Numerous reviews undertaken prior to Act
New Zealand	No	Duty to monitor operation of Act as a whole but not to collect data, review or amend EIA system. Parliamentary Commissioner audits aspects of EIA system
South Africa	No	No formal EIA system monitoring or review requirements. Few records kept, so proposed EIA system changes unlikely to be based on experience to date

Notes

1. *NEPA News*, a quarterly newsletter published by the Natural Resources Council of America, Washington, DC, provides valuable regular information about some of the activities of CEQ.
2. Much relevant information about EISs can be obtained from the EIA Office of Federal Activities web site at www.es.epa.gov/oeca/ofa/
3. The NEPAnet web site is at www.ceq.eh.doe.gov/nepa/nepanet.htm
4. The Department of Energy also publishes *Lessons Learned*, a quarterly newsletter on NEPA implementation, and hosts two relevant web sites (see Jessee, 1998).
5. Notwithstanding this requirement, no review was proposed or under way in mid-2001.
6. Available on the Ministry for the Environment web site at www.mfe.govt.nz

Chapter 18

Benefits and costs of EIA systems

As mentioned in Chapter 1, there has been, as yet, no reliable quantification of the effectiveness of EIA, and it may be that it can only be measured subjectively and qualitatively by examining the attitudes and opinions of those involved. Certainly, it is impossible to establish precisely either the benefits or the costs of EIA. Nevertheless, while the existence of firm justification remains scarce, the continued diffusion of EIA requirements around the world (Sadler, 1996) demonstrates the prevailing belief that EIA is an effective and efficient environmental management tool, i.e. that the benefits conferred by the EIA process outweigh the costs.

This chapter discusses the various ways of attempting to evaluate the benefits and costs of EIA systems. It relies upon the results of any EIA system reviews, upon any published studies of EIA system effectiveness and upon the opinions of the participants in the EIA process. A set of evaluation criteria for the benefits and costs of EIA systems is put forward. These criteria are then employed in the analysis and comparison of the EIA systems in the United States, UK, the Netherlands, Canada, Commonwealth of Australia, New Zealand and South Africa.

Benefits and costs of EIA systems

As Sadler (1996, p. 41) has pointed out, the purpose of undertaking EIA system monitoring (see Chapter 17) is to establish the effectiveness of the EIA system. It is important to note that judgements about EIA effectiveness are subjective, with different stakeholders holding different perspectives about what constitutes an effective EIA process. For example, from a proponent's perspective, it may be that the costs associated with potential delays to project authorisation as a result of conducting the EIA are more significant than benefits associated with the avoidance of long-term deterioration in environmental quality. As Sadler (1996) has noted, comparing the benefits and costs of EIA is a matter of judgement which depends on how the various benefits and costs are weighted. Both are extraordinarily difficult to measure.

Glasson *et al.* (1999) noted that the benefits associated with EIA are mostly unquantifiable. Benefits vary between projects in the same way as the cost of EIA but are more difficult to estimate (Canter, 1996). EIA is intended to improve the quality of decisions having environmental implications by amending the behaviour of proponents, consultants, consultees, the public and the decision-making authorities. Examples of such changes in behaviour include an increase in public participation in decision making, increased coordination between the authorities responsible for environmental protection, and rising environmental awareness among proponents. Sadler (1996) has referred to such changes as the indirect benefits of EIA practice. It is generally accepted that such changes take time but that they have taken place in the more mature EIA systems (see, for example, in relation to the United States: Wandesforde-Smith and Kerbavaz, 1988; Caldwell, 1989; Council on Environmental Quality – CEQ, 1990).

The crucial question in relation to the efficacy of EIA is whether the quality of decisions has actually increased and whether they have become more acceptable as a consequence of its use. Sadler (1996) has described this direct benefit of EIA, influence on decision making, as the litmus test of EIA effectiveness. Examples of direct benefits include the increased use of modification or mitigation (see Chapter 16), the use of more stringent conditions upon permissions and the refusal of potentially environmentally damaging proposals which might previously have been approved. Once again, it is very difficult to obtain concrete evidence of such changes. A study of the costs of the EIA of wastewater treatment facilities carried out by the US Environmental Protection Agency demonstrated that EIA 'was effective in (1) causing major changes in projects, (2) providing more protection for the environment' and in improving opportunities for public participation in the decision-making process (CEQ, 1990, p. 31). There is other evidence of influence on decisions from, for example, the UK and the Netherlands (see below). It is, nevertheless, generally necessary to rely mainly on stakeholders' opinions about the effectiveness of EIA systems in improving the environmental quality of decision making. Such opinions will, no doubt, rely on anecdotal evidence from particular examples of the use of EIA, some of which have found their way into the literature (see, for example, CEQ, 1990; Sadler, 1996).

For a variety of reasons, the costs of EIA systems are difficult to distinguish from other costs incurred in obtaining approvals. As Sadler (1996 p. 58) stated, estimating the cost of compliance with EIA process requirements 'has typically been a creative accounting exercise.' The chief reason for this difficulty is the integration of EIA into decision-making processes:

> The costs associated with [EIA] activities become harder to identify as environmental considerations are better integrated into planning and decision making. Thus, a 'successful' EIS [environmental impact statement] program is defined, in part, by its inability to be evaluated accurately in terms of economic efficiency. (Hart, 1984, p. 340)

Hart (1984, pp. 348–9) distinguished four principal elements of the cost of the EIA process:

- Costs of document preparation, review, circulation, and administration of the law.
- Costs of delay (inflation and forgone opportunity).
- Costs of uncertainty (due to risk of failure).
- Costs of mitigation (to moderate impacts).

While the costs of EIA programme administration and document preparation, circulation and review are not easy to calculate, they are less difficult than those associated with delay, uncertainty and mitigation.

Apart from the costs involved in preparing the EIA documentation, the proponent may have to pay a fee to the decision-making authority for EIA report review (as in the United States) or may have to pay the authority's review consultants (as in New Zealand). Decision-making authorities may have to maintain a specialist EIA staff unit, or to shift personnel from other activities. Similarly, consultee organisations and the public will have to expend resources if they are to participate effectively. While the additional costs attributable to EIA are not known, it is widely held that the 'costs of environmental review are generally insignificant when compared to other accepted planning, design and regulatory costs' (Hart, 1984, p. 340).

The costs associated with EIA vary considerably between projects (Glasson et al., 1999) but appear to range from about 0.1 per cent to 1 per cent, with 0.5 per cent as a commonly quoted figure (Hollick, 1986). Canter (1996, p. 30) noted that although the cost of an environmental impact statement was usually 1 per cent or less of total project costs, this proportion could vary, usually falling with increasing project costs. In some instances, as in the US study of the costs of the EIA of wastewater treatment facilities mentioned above, it has been claimed that the benefits of EIA included 'cost-savings that were the result of project changes prompted by the EIS process' (CEQ, 1990, p. 31).

Closely allied to the question of cost is that of delay. Glasson et al. (1999) observed that much of the early resistance to EIA was due to perceived delays occurring as a result of the process. Most EIA systems specify the times within which the various stages of the EIA process should be completed. There may be lengthy periods for public participation at several stages in the EIA process (as in the United States) as well as a specified period during which the decision, based upon the EIA report, should be made (as in the UK). The time taken by the decision-making authorities has frequently exceeded that specified and this has been a major source of complaint by proponents. However, such complaints have proved somewhat difficult to sustain because the delays have frequently been partly attributable to inadequacies in the information provided by proponents.

Delays have been a constant source of complaint since the US National Environmental Policy Act 1969 came into effect. However, many proponents, especially experienced proponents, have built sufficient lead-times into their project planning procedures to accommodate the EIA process. Others submit 'draft' EIA reports for informal scrutiny and amendment before formal submission to try to avoid the problems of delay caused by complying with subsequent requests for further information. While proponents have complained of delay in EIA procedures, they have frequently not implemented approvals promptly once they have been granted. Thus, Sewell and Korrick (1984) found that 15 per cent of projects examined were abandoned and 32 per cent were still ongoing six years later. They confirmed the findings of an earlier US study of delays by suggesting that EIA imposed little or no significant time penalty on proponents. All the evidence is that these findings still apply.

Nevertheless, despite the absence of reliable information about their extent, delays remain a recurring source of anxiety in many jurisdictions (for example, in the Commonwealth of Australia). These concerns are frequently coupled with complaints about 'moving goal-posts': unreasonable requests for further information or for changes to the design of the action. One method of overcoming these problems is the agreement of an action-specific timetable between the proponent and the decision-making authority specifying the circumstances in which requests for further information are to be made.

Glasson *et al.* (1999) observed that there are differing perceptions about whether or not delays associated with EIA are excessive. On the one hand, there are undoubtedly cases where EIA delays project authorisation and these delays can be costed through calculations of profits lost or increases in wage bills borne by the proponent. On the other hand, when the interests of a wider range of stakeholders are taken into account, the issue of delays becomes less pertinent. The cost of the delay may be more than compensated by the avoidance or reduction of the costs of adverse environmental impacts as a result of the EIA.

Since it should be the aim of any EIA system to maximise environmental benefits, to minimise environmental costs, and to minimise the costs to the proponent, a final question about the effectiveness of EIA must be posed: is there any evidence that EIA has led to any improvement in the quality of the environment generally? Because EIA is only one of an array of environmental management measures, it is extremely difficult to distinguish its effect from those of other anticipatory controls, pollution controls, environmental standards, environmental designations, etc. It is, therefore, doubtful whether evidence of general environmental improvement attributable to EIA (as opposed to anecdotal evidence of the effectiveness of EIA in improving, or preventing the deterioration of, a particular local environment) can ever be adduced. This does not mean that this ultimate criterion should not be advanced as a yardstick, simply that it is likely to remain theoretical. It has therefore not been included with the other criteria summarised in Box 18.1. These criteria are now employed to analyse the benefits and costs of each of the seven EIA systems.

Box 18.1   Evaluation criteria for the costs and benefits of EIA systems

Are the discernible environmental benefits of the EIA system believed to outweigh its financial costs and time requirements?

- Does empirical evidence exist that the EIA process has significantly altered the outcome of decisions?
- Do the participants in the EIA process believe that the environmental quality and acceptability of decisions are improved by it?
- Do the participants in the EIA process believe that it has altered the behaviour of proponents, consultants, consultees, the public and the decision-making authorities?
- Do the financial costs of the EIA process to proponents, consultees, the public and the decision-making authorities exceed those which would have been incurred in any event?
- Do the times required to complete the various stages of the EIA process exceed those specified?

United States

The costs of EIA in the United States are substantial, and must exceed those that would have been incurred had the National Environmental Policy Act 1969 (NEPA) never been passed. Despite the fact that environmental impact statements occasionally cost millions of dollars, the expense of EIA is generally seen as 'part of the cost of doing business'. This, perhaps, is why authoritative EIA costs are so elusive: they are inextricably tied to other related costs. The National Academy of Public Administration (NAPA) found that the cost of preparing both environmental assessments and EISs had dropped between 1994 and 1998; both amounted to about 0.2 per cent of the cost of the actions assessed. The average cost of a project specific EIS was $1.5 million, and of a programmatic EIS a prohibitive $12.5 million (NAPA, 1998).

While the Council on Environmental Quality has only a small staff, total employment in EIA must run to several thousands in the United States. The Environmental Protection Agency alone employs several hundred personnel to meet its EIA commitments.

Generally, EIAs take longer to complete than the times specified in the Regulations. However, many EIAs take no more than 12–18 months, though occasional cases may take 30 months from initiation to the record of decision. Programmatic EISs take about five months longer than site-specific EISs (NAPA, 1998). The participants in the EIA process have adapted to these times, but proponents still resent unexpected delays (CEQ, 1997d). Many of the participants in the EIA process firmly believe that their own behaviour and that of others has been affected by the EIA process. There have been numerous independent confirmations of these changes of behaviour, most notably in the study by Taylor (1984). (Unfortunately, CEQ's (1997d) effectiveness study did not fully address this issue.) The participants in the EIA process also believe that both the environmental quality and the acceptability of decisions have been improved by NEPA, though few are entirely satisfied with the process or the product. In particular, it is widely felt that projects are now much better designed and impacts better mitigated than was previously the case (Dickerson and Montgomery, 1993), that certain projects have been abandoned as a result of EIA and that projects have been modified to minimise or avoid environmental impacts before they occur (Clark and Canter, 1997).

The various participants in the EIA process have almost all supported it when its future was threatened (for example, during the Reagan regime and early in the Clinton era – see Chapter 2). In this they have reflected the continuing US public's profession of environmental concern (CEQ, 1998). Despite this consensus view, it is not possible to produce unambiguous evidence that the EIA process has significantly altered the outcome of decisions. EIA has been in being for three decades and parallel changes in environmental management have made it impossible to unravel the effect of EIA from all the other factors determining the outcome of decisions. In the last analysis, as explained in Chapter 1, the opinions of the participants in the EIA process are probably the only measure of success or failure.

Bear (1989) was able to find very little empirical evidence of the effect of EIA on decision making. She cited reduced EIA litigation as one indicator of success, but this is, at best, ambiguous. This is, perhaps, not surprising as Bartlett and Kurian (1999) have described EIA as a particularly complex, subtle policy tool that focuses on the reform of processes for making individual decisions and policy choices. Clark (1993, p. 4) was remarkably frank about CEQ's view of the success and failures of EIA:

> Certainly, many environmental impact statements (EISs) are too long, take too long to prepare, cost too much, and many times do little to protect the environment. Some EISs are prepared to justify decisions already made, many agencies fail to monitor during and after the project, some agencies do not provide adequate public involvement, and few agencies assess the cumulative effects of an action.

CEQ (1997d) also expressed an opinion on these matters. Others have criticised weaknesses in the treatment of cumulative impacts, biological diversity and global climate change in EIA (see, for example, Hildebrand and Cannon, 1993), though improvements in these areas have recently been made.

Yost (1990) believed that the EIA provisions in NEPA had been successful because they were in essence procedural, but that the Supreme Court had undone the promise of NEPA (see Chapter 13). Caldwell (1998) took a similar view, stating that NEPA was a Congressional declaration of national policy, not a regulatory statute. He argued that:

> The purpose of the Act is not to write impact statements. To regard the action-forcing provisions of Section 102 ... as the essence of the Act is to misinterpret its purpose. (pp. xvi–xvii)

Nevertheless, he believed that EIA had both a discovery and disclosure function which had caused proposals to be reconceived, revised or withdrawn. This view is widely held: the benefits of EIA in the United States are generally perceived to be less than many had hoped but still to outweigh its costs.

UK

Under the Planning Regulations, most of the cost of EIA is borne by the developer and by local planning authorities (LPAs). The Department of Transport, Local Government and the Regions (DETR) devotes six staff to EIA policy work but no estimates of personnel involved in EIA more generally (in consultancies and LPAs etc.) exist. Of a sample of 40 cases, only 15 per cent of the developers prepared the environmental statement (ES) themselves. Consultants were used in the remainder of cases, appearing frequently to charge fees in the range £10,000–£100,000 for their services (Department of the Environment – DoE, 1991a: see also DoE, 1996; Jones *et al.*, 1998; Glasson *et al.*, 1999). The Department of Environment, Transport and the Regions (DETR) suggested that an appropriate median figure for the cost of undertaking EIAs under the new Regulations might be £35,000 (DETR, 1997a). These sums equate to 0.1–1.0 per cent of project costs in most cases (European Commission, 1996a; DETR, 1997b).

While consultants would often have been employed in the absence of EIA, there has undoubtedly been an increase in consultancy activity as a result of its introduction. Many developers feel that EIA has caused a slight increase in the cost of obtaining planning permission (Glasson *et al.*, 1999).

The costs of consultants used by LPAs to evaluate ESs ranged from less than £1,000 to over £20,000 in addition to LPA staff time (DoE, 1991a; Glasson *et al.*, 1999) with over half expending less than £5,000 (Leu *et al.*, 1995). Consultees also incur considerable staff costs in dealing with EIA (DoE, 1996).

The mean ES preparation time for a sample of 40 ESs was about 30 weeks, within a range of 3–100 weeks (Jones *et al.*, 1998). As mentioned in Chapter 6, the Regulations extend the time allowed to the LPA to reach a decision to 16 weeks for applications involving EIA. Various studies have indicated that the mean time to determine applications involving EIA is about 40 weeks, considerably more than for those unaccompanied by an ES (DoE, 1991a, 1996). Many developers and consultants believe that EIA slows the decision-making process, but others take a more positive view. About 50 per cent of planning officers are of the opinion that EIA makes little difference to the time taken to decide planning applications (DoE, 1991a; Glasson *et al.*, 1999). There is some evidence that the time taken by LPAs to reach a decision is inversely proportional to the quality of the ES (Lee and Brown, 1992; European Commission, 1996a).

There is little evidence, to date, that EIA has led to a reversal of the outcome of decisions (see Chapter 13). However, as mentioned in Chapter 15, a study of 40 EIAs showed that modifications were made to two-thirds of the projects as a result of EIA. Furthermore, only one-fifth of the developers or consultants concerned felt that there had been no benefits associated with EIA (Jones *et al.*, 1998). In general, it is believed that the benefits of EIA in the UK outweigh its costs:

> the developers, competent authorities and statutory consultees all agreed that the benefits of the individual EIA had outweighed the costs.... [The] key environmental benefits include the avoidance of environmentally sensitive areas, improvements in project design to reduce potential environmental damage at source, higher standards of mitigation, and the provision of a better framework for environmental monitoring. (European Commission, 1996a, p. 96)

Generally, the principal stakeholders express the opinion that EIA is a worthwhile and helpful procedure which makes a positive contribution to informed decision making in the planning system. If EIA has not yet greatly increased public participation it does appear to have led to an increase in coordination between the relevant agencies and in effectiveness of consultation (European Commission, 1996a; Jones *et al.*, 1998; Glasson 1999b; Glasson *et al.*, 1999).

The Netherlands

There is no doubt that the substantial sums of money spent by the government, by proponents and by third parties in the EIA process considerably exceed those which would have been expended on proposals in any event. It is believed that the total costs of EIA are generally limited to 0.01–0.5 per cent of the cost of large projects but may constitute 0.1–3.0 per cent of the cost of smaller projects (consultancy costs for EIAs are usually €50,000–€250,000) (Commission of the European Communities, 1993). Extra costs therefore tend to impinge most on small projects: the additional costs of EIA for large projects are small and there are no proponent fees for EIA. There have been no significant complaints about the cost of EIA in the Netherlands. In some instances it is felt that EIA may have saved money by demonstrating that certain lower-cost alternatives or mitigation measures may be environmentally preferable (Scholten and van Eck, 1994; Mostert, 1995).

In addition to the costs of the EIA Commission (about €3 million annually) and the costs of three people working on EIA in the Ministry of the Environment, the Ministry of Agriculture employs one person, the Ministry of Transport has an EIA staff of seven, and most of the twelve provinces employ one or more persons on EIA. The total government annual budget on EIA (including the EIA Commission) is less than €4 million, including the research it funds. This is a reduction on previous spending levels.

In general, the time limits imposed on the EIA Commission are met, especially where informal liaison occurs. There have been cases where guideline recommendations have been late but significant slippage is rare. The most time-consuming part of the EIA process results from the activities of the developer in preparing the EIS (and in supplying any supplementary information that may be required). The competent authority may sometimes take a long time to make a complex or politically difficult decision. EIA has probably resulted in a reduction in the number of legal challenges to projects, thus saving time. Indeed, some EISs have been submitted voluntarily as a means of overcoming objections to projects. This is seen as an indication of the success of the EIA process. The Evaluation Committee on the Environmental Protection Act (ECW, 1990) felt that EIA often resulted in a streamlining of existing procedures.

The Ministry of Housing, Spatial Planning and the Environment (VROM, 1994b, pp. 26–7) reported that EIA:

> effectiveness strongly depends on the manner in which the instrument is used and on the circumstances in which it is applied. . . . An EIA performs better with a good report and in situations where the competent authority is also the initiator.

Mostert (1995) described the consensus culture of the Netherlands. There is certainly a consensus that the EIA system has changed the behaviour of the participants. Thus, Scholten and van Eck (1994) and van Eck and Scholten (1996) used examples to demonstrate that EIA had both direct effects (for example, the introduction of additional mitigation measures) and indirect effects (for example, the internalisation of environmental awareness in developer organisations). Mostert (1995, pp. 166, 167) concluded that:

> EIA is a valuable addition to the range of instruments for environmental management, but there is much room for improvement. . . . [The] environmental movement appreciates EIA. Although they sometimes suspect bias and an unwillingness to change decisions, still they would not want to live without EIA. Initiators and competent authorities sometimes have a more negative impression of EIA: EIA

costs money, leads to delays and may necessitate modifications of proposals that are already optimal from an economic and/or administrative-political point of view.

However, over the years, as their experience has grown, more proponents (and non-governmental organisations) have come to accept the need for EIA in the Netherlands.

In some cases projects have been cancelled and in others a less damaging alternative has been chosen (van Eck and Scholten, 1996). In almost every case, more consideration has been given to the impacts on the environment than would have been given without EIA. The outcome of the EIA process is therefore often different from the initial proposal. The Evaluation Committee on the Environmental Management Act found that 'incorporating an EIA in the licensing procedure not only generates more knowledge and information, but also makes for a more informed decision' (ECW, 1996, p. 8).

In their study of 100 EIAs for ECW, ten Heuvelhof and Nauta (1997) reported that, when weighed against the input in time, money and effort, EIA was considered to convey a large net beneficial impact in 14 per cent of cases, a reasonable net benefit in 26 per cent of cases, a small net benefit in 30 per cent of cases and no net benefit in 30 per cent of cases (see also van Eck and Scholten, 1996; de Jong, 1997; Arts, 1998). Ten Heuvelhof and Nauta (1997) attributed the fact that 70 per cent of EIAs had a net (if often minor) beneficial impact to the interdisciplinary consensus evolved in the lengthy process of undertaking EIA, to the negotiation which took place during the EIA process, and to the role of the EIA Commission in delivering quality in the EIA process.

It is small wonder that the Evaluation Committee on the Environmental Management Act (ECW, 1996, p. 19) concluded, on the basis of this and other commissioned research, that:

> The EIA scheme is a reasonably effective instrument. The satisfactory results mean that there is no real need for major changes in the law.

One minor change which is anticipated relates to the introduction of a simplified form of EIA for projects proposed specifically to improve the environment. As elsewhere, the most important factor is the attitude of the participants. Where developers and competent authorities take a positive approach to EIA, real benefits often ensue.

Canada

There is no doubt that the financial costs of the environmental assessment (EA) process in Canada are very considerable. A study undertaken as part of the five-year review of the Canadian Environmental Assessment Act indicated that the cost of a screening was, on average, 3.9 per cent of capital costs (i.e. about C$200,000) whereas comprehensive studies and panel reviews averaged 2.4 per cent (about C$750,000) and 1.1 per cent (about C$8.5 million) respectively (CEAA, 1999c, p. 23). It is, perhaps, surprising that 'none of proponents interviewed considered assessment costs to be a significant business impact' (Hagler Bailly, 1999, p. E-iv).

It is clear that some EA costs would have been incurred in any event during project planning. However, costs for panel review projects have sometimes run into tens of millions of Canadian dollars. A proportion of this cost has been attributable to the public hearings and to participant funding but the major component has been the cost of preparing EA reports, which has frequently exceeded C\$1 million.

Large public and private teams are dedicated to the EIA industry in Canada: government officials, consultants, developers, researchers and academics are involved. The level of resource devoted to EIA per head in Canada is probably the highest in the world, notwithstanding swingeing budget cuts. The Canadian Environmental Assessment Agency (CEAA) employs approximately 100 staff (see Chapter 5), Environment Canada over 20 EA staff, provinces such as Alberta over 30, and some Crown corporations over 20.

No time limits are specified under the Canadian Environmental Assessment Act but the federal EA coordination regulations specify that decisions about whether or not EA is required or not should be made within 30 days. Other time limits are also specified in these regulations. A study by David Redmond and Associates (1999) revealed that screenings for small projects typically took two months, and involved 6 person-days of consultancy input, and that those for medium-sized projects took an average of five months (involving 27 days' input). Screenings for large projects usually took a year from initiation to decision.

Unsurprisingly, panel reviews take longer. Average times for scoping hearings for panel review projects are about 2 weeks and for the hearings on the project about 2–6 weeks. The typical panel review takes 12–18 months (with EIS preparation taking most of the time). The briefest panel review has taken less than a year, but the longest has taken several years as a result of the demands of the proponent's project planning process. Canadian EA is therefore sometimes very time consuming and delays are widely seen as the most serious problem associated with the Act. Thus, Hagler Bailly (1999, p. E–iv) stated that:

> project start-up costs occurred when some aspect of the Act was mismanaged during the assessment process. These were identified . . . as being the one potential significant business impact resulting from the assessment process.

There is virtual unanimity that EA has altered the behaviour of the participants in the EA process. Although the performance of proponents has not always changed, Hagler Bailly (1999, p. E–iv) reported that only in 15 per cent of cases did proponents incur significant costs without commensurate benefits: 'almost without exception, all stakeholders felt strongly that the Act provided a range of benefits commensurate with its costs'.

The expertise of consultees, especially of federal departments like Environment Canada, has increased substantially. The same is true of environmental consulting firms. The behaviour of the public has certainly changed. The public has become increasingly professional and adept in using participant funding to mount a strong case to argue for the adoption of a preferred alternative or a particular mitigation measure. As a result, the acceptability to the public of decisions which implement panel recommendations is thought to have increased: a considerable benefit for both proponents and governments.

Evidence of the benefits of federal EA abounds in Canada both in relation to the formal panel review process and to self-assessment (Hagler Bailly, 1999; CEAA, 2000). The quality of the constructed project is generally believed to be better as a result of changes occasioned by the EA process. While a few projects have been abandoned or found to be unnecessary as a result of EA, its main benefit has been in mitigating the environmental effects of proposals.

One of the main problems of EIA in Canada has been overlap between the requirements of the federal and provincial EA systems and the confusion, conflict, duplication and delay this engenders. The Act attempts to reduce this inefficiency by providing for interjurisdictional agreements leading to, for example, joint panel reviews.

In 1998, the federal and provincial governments (except Quebec) ratified the Sub-agreement on Environmental Assessment to provide greater consistency and predictability. The Agency has concluded bilateral arrangements with four provinces on EA co-operation. It has also set up regional offices to act as 'single windows' for EA processes. These measures, together with the federal coordination regulations, designed to minimise delays where more than one federal authority is involved, have led to reductions in the problems caused by overlaps. The Commissioner of the Environment and Sustainable Development (1998, pp. 6–23) found that 25 of the 187 federal screenings studied required provincial EAs. Perhaps surprisingly, he reported that 'in most of the 25 cases, there was more evidence of federal–provincial cooperation than of duplication of effort'. David Redmond and Associates (1999) confirmed these findings: just over 10 per cent of the federal screenings they studied were subject to another EA regime and three-quarters of these were harmonised.

Nevertheless, a study of the effect of EA on competitiveness found that Canada performed weakly against a group of other countries, broadly on a par with the United States, but behind the UK (RIAS and Gartner Lee, 2000). The Minister of the Environment (MoE, 2001, p. 13) reported that industry representatives felt that, in joint EAs,

> project proponents cannot always be certain of the information requirements they would need to meet in assessments or even when a final decision would be made. The result can be delays in project planning and increases in costs.

He proposed that a federal coordinator for each EA, together with a strengthening of federal–provincial arrangements, be provided for in the Act (Government of Canada, 2001; MoE, 2001). In an interesting separate initiative, the multi-stakeholder Canadian Standards Association attempted (unsuccessfully) to set up a strong set of EA standards to which federal and provincial agencies could adhere (Hazell, 1999, p. 166).

Commonwealth of Australia

There can be no doubt that the financial costs of Commonwealth EIA exceed those which would have been incurred had no EIA been required. Significantly, however, there has been virtually no debate about the financial costs of the EIA process in a system where an EIA for a highly significant project can cost more than A$1 million (see Chapter 13) and where 35 governmental staff are employed on EIA at Commonwealth level alone. As one authoritative review put it:

> The direct compliance cost of the assessment process is not a significant problem for large companies, especially if environmental assessment is integrated with feasibility studies. These costs represent only a small proportion of total project costs. (Bureau of Industry Economics, 1990, p. v)

The duration of the EIA process under the repealed Impact of Proposals Act was the main cause of complaint (Jambrich *et al.*, 1992). There were several stages where delays could arise. These delays were exacerbated by uncertainty about whether or not an EIS or public environment report (PER) would be required. Delays could be expensive if proponents failed to plan adequately for EIA (Harvey, 1994) or if belated demands were made by government (Bureau of Industry Economics, 1990). As mentioned in preceding chapters, as well as complaints about lack of certainty, there were many other criticisms of the previous EIA system.

Despite the weaknesses in the previous EIA system, there has been little debate about the overall value of the Commonwealth EIA system, which is perceived to have improved the prior evaluation of project environmental impacts, principally because it has raised the level of awareness and knowledge about the environment among agencies and the public alike (Australia and New Zealand Environment and Conservation Council – ANZECC, 1991). This increased awareness is widely believed to have altered the behaviour of the main proponents in the EIA process (though not, perhaps, of the political decision-makers) and to have led to environmentally better and more acceptable actions. However, prior to the enactment of the current legislation, many stakeholders felt that the benefits of EIA did not outweigh its costs as much as they had done a decade previously.

There have also been substantial criticisms of the overlap between Commonwealth and state EIA systems but these have diminished as a result of the activities of the ANZECC (1991, 1996). These resulted in a National Agreement on EIA which was intended to eliminate duplication between jurisdictions (ANZECC, 1997).

As elsewhere, empirical evidence of decision modification is almost impossible to obtain. It is necessary, as usual, to rely upon opinions. The Ecologically Sustainable Development Working Group Chairs (1992, p. 213) have summarised the Commonwealth and state EIA situation succinctly:

> Recommendations for improving specific aspects of the EIA regime ... included the need for greater public involvement, extension to some areas not already

covered (for example, mineral exploration activities), the adoption of more effective conflict resolution procedures and the need for a more structured follow up of conditions of approval through formalised environmental monitoring and auditing arrangements where these do not already exist. Nevertheless, . . . given a more effective better resourced approach to evaluating EIAs, the present system does not require fundamental structural change.

This view was orally endorsed in 1998 by a senior business appointee of the new Liberal government after being asked to conduct yet another review (see Chapter 17) of the efficiency of the Commonwealth EIA system. The Commonwealth reform paper (Commonwealth Environment Protection Agency, 1994), the Ministerial consultation paper (Hill, 1998) and the Council of Australian Governments agreement (COAG, 1997) proposed time limits to increase certainty. COAG (1997) and Hill (1998) also proposed Commonwealth accreditation of state EIA procedures.

These recommendations were implemented in the Environment Protection and Biodiversity Conservation Act 1999 (EPBC Act). The ANZECC principles of Commonwealth accreditation of state processes through bilateral assessment agreements and case-by-case accreditation intended to ensure efficient, timely and effective EIA processes and approval of actions were embodied in the Act. The *Australian Financial Review* (editorial, 25 June 1999) stated approvingly that the EPBC Act:

> represents a considerable step forward because it rationalises government responsibility for environmental assessment and establishes a much more transparent assessment process with specific time-frames and upfront certainty.

Conservation groups, on the other hand, 'are concerned that accreditation of State and Territory processes may reduce the overall level of protection for the Environment' (Environment Australia, 1999b, Regulation Impact Statement). Provided that this does not happen, stakeholders should once again agree that the benefits of EIA clearly outweigh its costs.

New Zealand

It was perhaps inevitable that the euphoria following the enactment of the revolutionary and ground-breaking Resource Management Act 1991 should have evaporated, to be replaced by a more sober realisation of the problems inevitably associated with the implementation of such ambitious legislation. Morgan (1995, p. 340) reported that there was virtual unanimity about the merits of EIA despite problems with the numerous requirements of the Resource Management Act 1991 (RMA):

> Interviewees were almost unanimous in their support for EA and the principles in the RMA on which EA is built, at least with respect to the resource consents process. It is seen as an essential part of an integrated approach to planning natural resources and managing the environment in New Zealand.

This consensus was shattered by the vociferous complaints about costs, delays, uncertainties and lack of local authority expertise by a small group of business critics (for example, McShane, 1998) and their sympathetic reception by a neo-liberal government for which environmental protection was a low priority (see Chapters 5 and 17). The withdrawal of Ministry for the Environment leadership and support for the Act's implementation shortly after it came into force because of severe budget cuts has been widely lamented. For example, the Parliamentary Commissioner for the Environment (PCE, 1998, p. 4) stated that the lack of investment by government on the management of changes in approach required 'probably cost local authorities, businesses and communities millions of dollars'. The Ministry for the Environment's (MfE's) newsletter (*Update*, February 1998) was remarkably candid:

> With hindsight, we acknowledge that more should have been done at an earlier stage, by both central and local government, to help people to use the legislation effectively. ... We have very clearly got the message from business that we need to provide more assistance with training for people involved in Resource Management Act decision-making.

Apart from the anticipated changes to the Act, one outcome of the complaints was that the Ministry was given an enhanced budget to increase the effectiveness of the Act. This resurgence of Ministry activity (apparent, for example, in the publication of guidance (see Chapter 5), in system monitoring (Chapter 17) and in environmental legal aid to community groups (Chapter 16)) has been welcomed by many EIA stakeholders.

Notwithstanding the problems, it is apparent that the majority of participants in the EIA process (environmentalists, business, government, local councils, consultancies and academics) believe that the benefits of EIA outweigh the costs. As the Ministry newsletter (*Update*, October 1999) put it, in response to the amendments to the Act proposed in 1998: '[s]ubmissions from all sectors showed a commitment to the Resource Management Act, with many noting that the Act is fundamentally sound'.

While there is little empirical evidence available about the financial costs of EIA, the Parliamentary Commissioner for the Environment (1998, p. 4) felt that there had been an undue focus on: 'the time taken to process resource consents and the costs of so doing' in bringing forward the 1998 amendments:

> The merits of advancing sustainable development and improving environmental management appear to be largely forgotten ... the costs to the environment, and to future generations, of any changes cannot yet be evaluated. (PCE, 1998, pp. 4, 5)

Regional and territorial councils have had difficulties in meeting the detailed time constraints imposed by the Act since nearly 20 per cent of all resource consents and nearly 40 per cent of notified consents were not processed on time in 1999/2000 (MfE, 2001c). However, while some delays have arisen because local authorities have protected their positions by requesting further information rather than applying appropriate professional judgement, delays are diminishing (MfE, 2001c).

It appears that behavioural changes are being brought about by the all-pervasive EIA system but that the attitudes of many local council officials and elected members (PCE, 1998) and the public (Morgan, 2000b, 2001) to EIA show considerable scope for further development. As the Parliamentary Commissioner put it, there has been a:

> lack of accountability of some councils in fulfilling their responsibilities under the RMA. The intent of the RMA can be thwarted by councillors and staff who ignore community preferences for resource management. (PCE, 1998, p. 2)

It is widely believed in New Zealand that the Mark I EIA system (MfE, 1987) did alter the outcome of many decisions, and the general belief is that the much more widely applicable Mark II EIA system is doing the same but that much progress remains to be made.

South Africa

In the past, integrated environmental management (IEM) was essentially voluntary: most EIAs were undertaken for large projects proposed by public or semi-public bodies and by large private sector companies conscious that an environmentally responsible image was necessary to maintain international trading links.

It is known that some EIAs have been protracted, taking perhaps 6–24 months (Weaver *et al.*, 2000), but few data on time-scales exist. However, the White Paper on environmental management policy presented the results of a business survey which included some information about the cost of EIA undertaken under the voluntary IEM procedures:

> Twenty five per cent spent less than 1% of establishment costs on environmental impact assessments for new activities and 13% between 2 to 4%. Sixty per cent were unsure what percentage of costs went to environmental impact assessments. (Republic of South Africa, 1998, p. 75)

Most practitioners felt this to be an overestimate of the financial costs of EIA, which they believe to have been 0.2–0.4 per cent of project costs (see, for example, Weaver *et al.*, 2000).

Overall, almost all stakeholders believe that EIA undertaken under the voluntary IEM procedure and subsequent EIA Regulations altered their behaviour and that of others. They are almost unanimous in expressing the view that the environmental quality and acceptability of decisions is improved by EIA, since mitigation of environmental impacts is so important in South Africa. There is a large body of anecdotal evidence about the effect which EIA has had on the nature of the approved development as a result of such mitigation of impacts and about how practice has improved since the publication of the IEM guidelines (Department of Environmental Affairs, 1992).

The compulsory nature of the requirements of the 1997 EIA Regulations has attracted some hostility from developers, many of whom perceive EIA to be a burden. In particular, the EIA Regulations have resulted in entangling the projects of many small developers who had no previous experience of EIA. They have often resented the requirement to retain independent consultants and have accepted the cheapest tender, sometimes resulting in inadequate and unsuccessful scoping reports (Duthie, 2001). As elsewhere, it is the possibility of the delay engendered by EIA (and the consequent uncertainty about dead-lines) rather than the financial cost of EIA which disturbs developers most (Ridl, 1994). This problem has been exacerbated by the absence of time limits from the EIA Regulations and by the tendency for public participation to be an open-ended and lengthy process.

The public and the environmental groups tend to see EIA as a means of delaying and improving projects. The empirical evidence of EIA under the IEM procedure suggests that, while delays were common, stoppage or with-drawal was rare, but not unknown (see Chapter 13).

Consultants, of whom there is a substantial and growing number in South Africa, not unnaturally see EIA as a worthwhile process. They also tend to approve of the requirement to employ independent consultants (see Chapter 11): this has ensured that there has been a surfeit of work, a shortage of experi-enced EIA consultants and a growth in the retention of inexperienced prac-titioners. Unfortunately, this has had a deleterious effect on the quality of some EIA reports (Duthie, 2001).

The provincial relevant authorities perceive EIA to be a valuable environ-mental management tool, though many believe that too many projects are being assessed.[1] They, unlike the consultancy sector, have suffered from low staffing levels and a high staff turnover (many public sector EIA professionals have transferred to consultancy) (Weaver et al., 2000; Duthie, 2001). They are, therefore, encouraging the delegation of powers to local government, the use of discretion in the notification of minor projects and the use of exemp-tions. They are also seeking powers to cover the costs of, for example, the inde-pendent review of EIA reports, by charging developers.

The forthcoming amendment of the National Environmental Management Act 1998 and the promulgation of accompanying EIA regulations to replace those currently in force (see Chapter 5) provide a unique opportunity to address the legitimate concerns of the various stakeholders in the South African EIA process.

Summary

It is, perhaps, a testament to the inherent effectiveness and efficiency of EIA that, despite the marked differences in each of the seven EIA systems, there should be an unanimity of view that the benefits of all the EIA systems outweigh their costs (Table 18.1). It is nevertheless quite clear, from the criticisms of the seven EIA systems that have been listed in this chapter and earlier in the book, that the effectiveness of EIA can be substantially improved (see Chapter 20).

It is noticeable that complaints about delays in project approvals as a result of EIA have been most vociferous in the United States, Canada and Australia. These are the jurisdictions with the most formalised EIA systems and those which require proponents to undertake the greatest number of EIA steps. However, while complaints about delay are probably least in the UK and South African EIA systems, they do occur in these jurisdictions, as well as in the Netherlands and New Zealand.

Table 18.1   The benefits and costs of the EIA systems

Criterion 13: Are the discernible environmental benefits of the EIA system believed to outweigh its financial costs and time requirements?

Jurisdiction	Criterion met?	Comment
United States	Yes	Probable consensus view by proponents, consultees and the public that benefits of EIA exceed its substantial time and other costs
UK	Yes	Consensus (but not unanimity) as to increasing utility of EIA in improving project design and mitigation measures
The Netherlands	Yes	Majority belief that benefits of EIA outweigh its financial and time costs
Canada	Yes	Probable consensus that benefits of EA system exceed often high costs. Significant mitigation has occurred but uncertainties and delays sometimes unacceptable to proponents
Commonwealth of Australia	Yes	Belief that EIA delivers real environmental benefits by mitigating impacts generally outweigh previous complaints about uncertainty and delays which may not apply to 1999 Act
New Zealand	Yes	Consensus that complaints about uncertainty, costs and delays outweighed by environmental benefits of EIA
South Africa	Yes	Most stakeholders believe that EIA is cost-effective, though some developers resent financial costs and (especially) time requirements of EIA

It is clear that delay is the major criticism in most EIA systems. Delay is the main reason why some members of the development industry in, particularly, the Commonwealth of Australia and New Zealand do not share the general view about the net benefits of EIA. The most significant advance towards the unanimous acceptability of EIA could be made by reducing the delays engendered by it and by explaining fully that these delays are generally attributable much more to proponents than they are to EIA agencies. Needless to say, such efficiency gains must not be made at the expense of EIA effectiveness.

Note

1. Personal communication by Andries van der Walt of unpublished research by the Environmental Assessment Research Group at Potchefstroom University.

Chapter 19

Strategic environmental assessment

It is apparent from Chapter 18 that the benefits of project EIA are generally believed to outweigh its costs. The widespread acceptance of the utility of EIA in improving the quality of decisions about proposed projects has led to active consideration of, and growing practice in, strategic environmental assessment (SEA) or the environmental assessment of policies, plans and programmes (PPPs). This is a consequence of the emerging awareness that project EIA may occur too late in the planning process to ensure that all the alternatives and impacts relevant to sustainable development goals are adequately considered. The application of SEA to the development of PPPs is widely perceived to have the potential to streamline and strengthen project EIA and to contribute towards the aims of sustainable development.

This chapter explains the need for (and nature of) SEA and discusses the implementation of SEA. It then reviews the current provisions for, and practice in, SEA in the United States, UK, the Netherlands, Canada, Commonwealth of Australia, New Zealand and South Africa.

Environmental assessment of PPPs

The aim of SEA is to provide decision-makers and affected stakeholders with timely and relevant information on the potential environmental impacts of a PPP in order to modify the PPP to make it environmentally sounder. Like EIA, SEA is a process. (Figure 19.1 shows the SEA process for transport infrastructure plans.) As is the case with project EIA, SEA is inextricably linked to decision making. Sadler and Verheem (1996) described the SEA process as a decision-aiding rather than decision-taking tool. Partly because the advantages of project EIA are so widely recognised, the desirability of taking the environment into account earlier in the planning process has been generally accepted throughout the world. Buckley (1998b) described SEA as the single most important direction in the field of environmental assessment.

There are several perceived benefits concerning the application of SEA to the formulation of PPPs. The theoretical benefits of SEA centre on the ability of the process to help PPPs reflect sustainable development concerns. Sadler and Verheem (1996, p. 13) believed that:

> SEA represents a promising approach to incorporating environmental and sustainability considerations into the mainstream of development policy making.

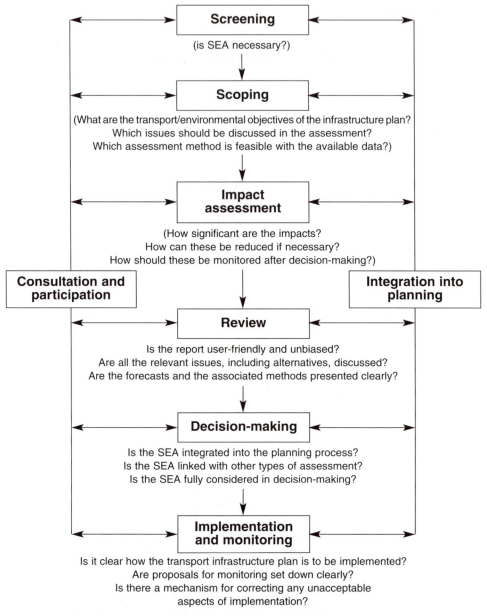

Figure 19.1   Steps in the SEA process for transport infrastructure plans
Source: European Commission, 1999b, opp. p. 12.

It has been suggested that, through the use of SEA, the concept of sustainable development could be incorporated as an integral part of the development of all policies and then 'trickled down' through plans to programmes, and finally to the project level (Therivel *et al.*, 1992).

The other widely acknowledged benefit of the SEA process is that it can streamline and strengthen project EIA practice. Whereas project EIA essentially reacts to proposed developments and their environmental impacts, the SEA process has the potential to be more proactive. SEA facilitates the earlier consideration of environmental impacts, the examination of a wider range of potential alternatives, the generation of standard mitigation measures and the opportunity to address a wider range of impacts, including those that are cumulative, synergistic, indirect, long range, delayed and global. In addition, the SEA of PPPs enables the EIA process to be streamlined by focusing it on the most significant issues. SEA also has the potential to appraise the impacts of non-project activities, for example, farming activities and taxation policies. The various arguments advanced in favour of the environmental assessment of PPPs are summarised in Box 19.1.

When certain alternatives and significant environmental impacts cannot be adequately assessed at the project level, it may well be possible to assess them at the programme, plan or policy level, utilising a form of SEA basically similar in nature to that employed for projects. Thus, SEA would involve screening, scoping, prediction, consultation, public participation, mitigation of impacts and monitoring (UN Economic Commission for Europe – UNECE 1992) and

---

Box 19.1   Potential benefits of strategic environmental assessment

- Encourages the consideration of environmental objectives during policy, plan and programme-making activities within non-environmental organisations.
- Facilitates consultations between authorities on, and enhances public involvement in, evaluation of environmental aspects of PPP formulation.
- May render some project EIAs redundant if impacts have been assessed adequately.
- May leave examination of certain impacts to project EIA.
- Allows formulation of standard or generic mitigation measures for later projects.
- Encourages consideration of alternatives often ignored or not feasible in project EIA.
- Can help determine appropriate sites for projects subsequently subject to EIA.
- Allows more effective analysis of cumulative effects of both large and small projects.
- Encourages and facilitates the consideration of synergistic effects.
- Allows more effective consideration of ancillary or secondary effects and activities.
- Facilitates consideration of long-range and delayed impacts.
- Allows analysis of the impacts of policies which may not be implemented through projects.

Source: Wood and Djeddour, 1992, p. 7.

meet certain SEA principles (Box 19.2; see also Marsden, 1998a). SEA could deal with the alternatives and significant impacts not covered adequately in project level EIA in more detail than impacts that can be dealt with later.

The assessment of alternatives and significant impacts need not be confined to one level of action. Different types of alternatives or different types of impact can be assessed at different tiers in the planning process, provided that

---

Box 19.2   Principles for SEA

An SEA process ensures that:

Screening — an appropriate environmental assessment is carried out for all strategic decisions with potentially significant (positive or negative) environmental consequences by the agencies initiating these decisions

Publication — it is clear to all parties affected by the decision how the assessment results were taken into account when coming to a decision

Monitoring — sufficient information on the actual impacts of implementing the decision is gained to judge whether the decision should be amended

Timing — the results of the assessment are available sufficiently early to be used effectively in the preparation of the strategic decision

Environmental scoping — all relevant environmental information is provided, and all relevant information is excluded, to judge whether an initiative should go ahead or whether the objectives of the initiative could be achieved in a more environmentally friendly way

Socio-economic scoping — sufficient information on other factors, including socio-economic considerations, is available, either parallel to, or integrated in, the assessment

Views of the public — sufficient information is available on the views of the public affected by the strategic decision early enough to be used effectively in the preparation of the strategic decision

Documentation — the results of the assessment are identifiable, understandable and available to all parties affected by the decision

Quality review — the quality of process and information is safeguarded by an effective review mechanism

Source: Verheem and Tonk, 2000, p. 179.

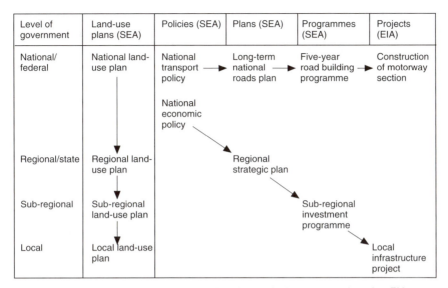

Level of government	Land-use plans (SEA)	Policies (SEA)	Plans (SEA)	Programmes (SEA)	Projects (EIA)
National/ federal	National land-use plan	National transport policy → National economic policy	Long-term national → roads plan	Five-year road building → programme	Construction of motorway section
Regional/state	Regional land-use plan		Regional strategic plan		
Sub-regional	Sub-regional land-use plan			Sub-regional investment programme	
Local	Local land-use plan				Local infrastructure project

Figure 19.2   Chronological sequence of actions within a comprehensive EIA system
Sources: adapted from Lee and Wood, 1978a, and Lee and Walsh, 1992.

such assessment takes place as early as feasible in this process. In many countries where there is already a project-level EIA system, the most sensible course of action might be to supplement these EIAs with higher-tier SEAs largely confined to issues, such as cumulative impacts, which cannot be adequately assessed at the project level.

Such an approach is feasible because, in principle, there exists a tiered forward-planning process which starts with the formulation of a policy at the upper level, is followed by a plan at the second stage, and by a programme at the end. A policy may thus be considered as the inspiration and guidance for action, a plan as a set of coordinated and timed objectives for implementing the policy, and a programme as a set of projects in a particular area. The tiered system can apply at the national level and also may apply at regional and local levels (Figure 19.2). It can apply to sectoral actions and to physical planning actions (Wood, 1988a; Lee and Walsh, 1992; Therivel *et al.*, 1992; Wood and Djeddour, 1992; Sadler and Verheem, 1996; Therivel and Partidario, 1996; Sheate, 1996; Glasson *et al.*, 1999; Partidario, 1999, 2000; Partidario and Clark, 2000). Although, in practice, such a tiered planning may not always exist (Noble, 2000a), tiering is nevertheless an important element of SEA. Sadler and Verheem (1996) felt that SEA and EIA should be consistent with, and reinforce, each other and that SEA should provide the frame of reference for EIA.

The SEA process is applicable to various sectoral activities, especially (but not only): agriculture, forestry, fishing, extractive industry, energy industry, manufacturing industry, transport, non-transport infrastructure, housing, the environment, and recreation and tourism. It can also be applied to policies, plans and programmes concerning research and development generally.

Therivel (1998) found that, in the UK, in addition to the environmental appraisal of land-use plans, a small number of SEAs were being undertaken for other public sector PPPs.

Many alternative arrangements of juxtaposed land uses, and some significant synergistic and cumulative impacts, cannot be satisfactorily considered in sectoral or project environmental assessments, because of the effects of new activities in other sectors, or because of the cumulative effects of many activities not subject to project EIA. They can, however, be considered in the SEA of land use and other spatial plans. Indeed, it could be argued that the environmental assessment of land-use plans is the most easily implementable of all types of SEA. Certainly, in California most SEAs have been related to land-use plans (Bass and Herson, 1999). This has also been the case in the UK.

Evolution and diffusion of SEA

SEA practice has received considerable impetus from a number of international organisations. The World Conservation Strategy pinpointed the need to integrate environmental considerations with development in 1980 (International Union for Conservation of Nature and Natural Resources, 1980). This theme has become an accepted part of World Bank (1987) policy, which stated that environmental issues must be addressed as part of overall economic policy rather than project by project. (Subsequently, the World Bank has required regional and sectoral environmental assessments in some developing countries.) The same philosophy was echoed in the Brundtland report (World Commission on Environment and Development, 1987) and at the United Nations summit on the environment held in Rio de Janeiro in 1992. The United Nations Economic Commission for Europe (UNECE, 1992) has recommended the extension of EIA principles to policies, plans and programmes. The Commission of the European Communities' fifth environmental action programme (CEC, 1992, p. 66) stated that:

> Given the goal of achieving sustainable development it seems only logical, if not essential, to apply an assessment of the environmental implications of all relevant policies, plans and programmes.

The European Commission has long espoused the desirability of extending EIA from projects to higher tiers of action and promulgated a directive on the assessment of certain plans and programmes in 2001 (European Commission, 2001a; see also Chapter 3).

In the United States, where the National Environmental Policy Act 1969 provided for SEA from the outset, there has been an upsurge in interest in programmatic environmental impact statements, as SEA reports are termed (Sigal and Webb, 1989; Bass and Herson, 1999). California probably has the most developed SEA system in the world and several hundred SEAs have been undertaken to date. An advantage of assessing the impacts of higher-tier actions in California has been that certain aspects of subsequent projects have not then needed to be assessed in detail (Bass and Herson, 1999). In Canada, it is a requirement that submissions to cabinet should contain an assessment of any likely significant environmental effects. In the UK local planning authorities are effectively required to undertake environmental appraisal of development plans, and in New Zealand local authorities are also obliged to undertake a form of SEA of their plans and policies.

In recent years, the most appropriate methodological approach to the SEA process has been debated at some length. There has been considerable support for the need to develop flexible approaches to SEA designed to suit the character of the PPP being assessed (Sadler, 1996; Department of the Environment, Transport and the Regions – DETR, 1998d). Verheem and Tonk (2000, p. 181) believed that '[t]he most effective form of SEA should be chosen according to the context in which it should operate'. Brown and Therivel (2000) referred to the need to 'graft' the SEA process on to established PPP development procedures. Partidario (2000) suggested that there was no one universal approach to SEA. Rather, SEA was to be seen as a family of tools, not as a single assessment approach. The existence of a number of types of SEA, differing according to the action being assessed or the types of impact considered (Buckley, 1998b; Noble, 2000a), supports this idea. As with project EIA, the skill of the assessor lies in selecting an appropriate mix from all the different approaches, tools and techniques available.

The main elements of the EIA process and its most tangible output – the EIA report – are, in principle, as applicable to policies, plans and programmes as they are to projects. However, in practice it is likely that the scope and purpose of the environmental assessment of PPPs will be different from that of projects in five main ways:

1. The precision with which spatial implications can be defined is less.
2. The amount of detail relating to the nature of physical development is less.
3. The lead-time is greater.
4. The decision-making procedures and the organisations involved may differ, requiring a greater degree of coordination.
5. The degree of confidentiality may well be greater.

These variations indicate that the nature of strategic environmental assessment will differ in detail from the nature of project EIAs. The potential for the application of EIA techniques, methods and approaches within the SEA process in certain situations is considerable but complicated (Kleinschmidt and Wagner, 1998). Sadler and Verheem (1996, p. 105) pointed out that, while EIA processes and methods can be useful for some PPPs, for less specific PPPs: 'modifications are necessary to take account of the greater degree of generality and uncertainty encountered in policy and plan making'. Brown and Therivel (2000, p. 186) believed that project EIA approaches, of themselves, were not enough: 'new methodologies and procedural requirements, specifically for SEA, will be required.'

A number of methodological and procedural constraints to the development of SEA have been identified (Therivel *et al.*, 1992; Sadler and Verheem, 1996; Therivel and Partidario, 1996; Curran *et al.*, 1998; DETR, 1998d; Kleinschmidt and Wagner, 1998). Methodological difficulties in undertaking

SEA appear to be secondary to political difficulties, namely the reluctance of politicians and senior bureaucrats in powerful departments voluntarily to cede any role in the making of decisions to external environmental authorities by permitting the SEA of their activities. As Buckley (2000, p. 209) noted:

> Even the most avowedly democratic of governments are notoriously shy of subjecting their actions, decisions and expenditure to public scrutiny by the citizens whose interests they are supposed to serve.

For SEA to take place, it appears that there needs to be both a specific and unambiguous requirement to undertake it and either a legal system to enforce it (as in the United States or California) or an environmental authority with sufficient strength of purpose, and professional competence, to ensure that the requirement is discharged. While many jurisdictions possess permissive powers for SEA to be undertaken, few outside the United States possess the necessary clear legal triggering mechanism (Kornov and Thissen, 2000).

Very few SEAs have related to policies, as opposed to plans or programmes (Therivel *et al.*, 1992, p. 71; Renton and Bailey, 2000). The European SEA Directive omits policies, and most other jurisdictions possess powers that are discretionary in application (if they have any SEA powers at all). There have been few attempts to enforce the SEA provisions in the United States in relation to policies. It is therefore perhaps understandable that Buckley (2000) found that policy SEA generally only took place where the policy was directly related to the environment. He noted, nevertheless, that if governments were serious about commitments to sustainable development, policy SEA had to become a reality.

As well as the development of policy SEA, and the clarification of the role of tiering in environmental assessment (see above), a number of other steps is also necessary to overcome some of the constraints to improving SEA practice.

- Increasing the general understanding of SEA: for example, the types of action to which SEA could usefully be applied and the types of impact that SEA should cover, and the relationship of SEA to existing EIA and sustainable development policies.
- Clarifying procedural issues: for example, the decision points in a planning process at which SEA should be applied (generally as early as possible), and the means by which SEA findings should be integrated with other policy and planning considerations in decision making.
- Clarifying methodological issues by adapting existing methods (including EIA methods) and developing new methods for SEA use.
- Strengthening the capacity for the practical application of appropriate SEA methods: for example, undertaking 'trial runs', developing review mechanisms, diffusing examples of good SEA practice, preparing SEA guidance and providing training in its use.
- Reviewing existing environmental data sources to assess their potential use in SEA and prioritising measures for correcting any deficiencies.
- Making transparency a key procedural consideration and providing for consultation and public participation at appropriate times during the SEA process.

In many ways, the current situation regarding SEA resembles that before and during the early years of EIA implementation when similar fears about delays, duplication and difficulty were raised. Experience has shown that project-level EIA is feasible, that EIA has altered decision making to give more weight to the environment, and that EIA costs very little in relation to the costs of implementing the actions assessed and that it has delivered net benefits. SEA, if implemented judiciously and at the appropriate level in the various planning processes, will undoubtedly establish itself, in the same way as EIA, as a cost-effective tool of environmental management.

For the sake of symmetry, the evaluation criterion, 'does the EIA system apply to significant programmes, plans and policies, as well as to projects?', that was advanced in Box 1.2 is used in this chapter. This criterion implies some legal provision for the SEA of PPPs and evidence of SEA practice. However, no sub-criteria have been advanced, as they were in Chapters 6–18. Indeed, those criteria could, with appropriate adaptation, be applied to SEA in the same way as to EIA but a detailed analysis of the various aspects of SEA could easily involve the writing of another book. Rather, the current provisions for, and practice in, SEA in the seven jurisdictions are now summarised.

United States

The National Environmental Policy Act 1969 (NEPA) requires EIA for 'major Federal actions significantly affecting the quality of the human environment' (Box 2.1). The phrase 'major Federal action' has subsequently been defined in the Council on Environmental Quality (CEQ) Regulations as including projects and programmes, rules, regulations, plans, policies or procedures, and legislative proposals advanced by federal agencies (40 CFR 1508.18). SEA procedures are not distinguished from project EIA procedures in the United States. The environmental impact statements (EISs) for policies, plans and programmes are called, variously, programmatic, regional, cumulative or generic EISs (or sometimes simply EISs) (Sigal and Webb, 1989).

CEQ indicated as early as 1972 that programmatic EISs should be prepared for federal programmes that might involve numerous actions to ensure that cumulative impacts were addressed. Such EISs would be broad in nature and cover 'basic policy issues that would not have to be repeated in subsequent impact statements for individual actions within a program' (Sigal and Webb, 1989). This was subsequently termed 'tiering'. Tiering involves the preparation of an EIS to cover general issues, and particularly alternatives, in a broader policy or programme-activated analysis. EISs at subsequent stages then incorporate by reference the general discussions from the broader EIS, while concentrating on the issues specific to the action being evaluated (Bass *et al.*, 2001, p. 123).

The CEQ Regulations and published advice on EIA (CEQ, 1981a) deal specifically with tiering. Programmatic EISs may be required for broad federal actions which can be grouped geographically (for example, covering a metropolitan area), generically (for example, actions having similar methods of implementation) or by stage of technological development (for example, federally assisted research on new energy technologies) (Regulations, section 1502.4). Programmatic EISs are considered to be particularly relevant for actions that are complex or give rise to environmental effects of unknown extent where the comparative analysis of alternatives can highlight potential problems (Mandelker, 2000, chapter 9.02). An example of geographical tiering would be the preparation of a forest plan EIS, followed by a watershed programme EIS for part of the forest and then an individual timber harvest or road EIS or environmental assessment within that watershed (Bass and Herson, 1999). The programmatic EIS should be prepared at a point in the agency planning process when it can highlight potential environmental problems and allow a wide range of alternatives to be evaluated.

Sigal and Webb (1989) estimated that around 35 programmatic EISs per annum were issued between 1979 and 1987, of which the majority related to policies and programmes but a significant minority were regulatory. The largest categories were resource management, pest control and flood control. Other important types of programmatic EIS related to wilderness, permits, water development, technology development and mineral/timber decisions. Bass and Herson (1999), using a broader definition of programmatic, suggested that the number of such EISs submitted was over 100 per annum during 1994–96. However, Clark (2000) estimated that only about 10 programmatic EISs were submitted annually, partly because of the time and cost involved in their preparation (see Chapter 18). The number of legislative EISs was running at 1–8 per annum in the late 1990s.[1] Although SEA practice is clearly developing and becoming more effective in the United States, it still has not been formally applied at the broadest level of policy. While inevitably somewhat general, and containing less detail and quantification than project EISs, the quality of programmatic EISs is undoubtedly increasing. Sigal and Webb (1989) stressed the role of public participation and, especially, of public meetings in the SEA process.

Sigal and Webb (1989) pointed out that agencies have frequently resisted preparation of programmatic EISs because of perceptions of cost, lack of understanding of timing and scope, and fear of litigation. (The United States assessed the environmental effects of the North America Free Trade Agreement not under NEPA but under an executive order to avoid litigation.) Programmatic EISs do tend to take longer and cost more than project EISs (National Academy of Public Administration, 1998). However, they believed that there was a growing realisation that a well-prepared, timely programmatic EIS can highlight and anticipate potential environmental problems, prevent future delays, assist in long-range planning and prevent or simplify litigation.

There is increasing CEQ emphasis on SEA as it is widely believed that EIA 'is often triggered too late to be fully effective' (CEQ, 1997d). At least at the plan level, there is considerable SEA experience and, according to Bass and Herson (1999, p. 298), federal agencies have developed innovative approaches to produce EISs which are generally 'comprehensive, interdisciplinary, quantitative and analytical'. There is also considerable expertise in the SEA of land-use plans in several states (Pendall, 1998), especially California (Olshansky, 1996; Bass and Herson, 1999).

UK

As mentioned in Chapter 7, there is no formal requirement for the strategic environmental assessment of PPPs in the UK. However, it has been accepted since 1990 that a systematic approach is needed to take account of the environmental impacts of government policies during their formulation. An official guide to incorporating environmental factors in policy appraisal suggested an iterative process involving consideration of the environment from the outset, identification of policy options, seeking advice, identification of impacts, analysis of environmental effects, and monitoring and evaluation (Department of the Environment – DoE, 1991b). This guide, intended principally for central government and (to a lesser extent) local government, was found often to have been ignored (DETR, 1997c). In response to this finding and in accordance with the government's avowed intention to put 'sustainable development at the heart of every Government Department's work' (Her Majesty's Government, 1999) an updated, simpler summary version was published in 1998 (DETR, 1998b). The main steps in the process outlined in this guidance are shown in Box 19.3.

Since the Town and Country Planning (Development Plan) (England) Regulations 1999 came into force, the long-established practice that local planning authorities (LPAs) take environmental (together with economic and

---

Box 19.3   UK policy environmental appraisal checklist

1. **What does the policy or programme aim to achieve?** (Include possible trade-offs, conflicts and constraints.)
2. What are the **options** for achieving your objectives?
3. What **impacts** will these have on the environment at home and abroad? (Consider both direct and indirect costs and benefits; possible mitigation measures; and need for risk assessment.)
4. How **significant** are the impacts? How large are they in relation to the other costs and benefits of the policy concerned?
5. How far can the costs and benefits be **quantified** without disproportionate effort?
6. What **method will be used to value the costs and benefits?** (Options include using monetary values, calculating physical quantities and systematic ranking or listing of impacts.)
7. **What is the preferred option and why?**
8. What arrangements are in place for effective **monitoring** and **evaluation?** (What data will be needed and when?)
9. How will the appraisal be **publicised?**

Source: DETR, 1998b.

social) considerations into account in preparing their land-use plans has been a statutory requirement. It has been expected (but not legally required) that LPAs should appraise the environmental implications of policies and proposals in land-use plans since 1992.

DETR updated and strengthened its expectations regarding the appraisal of the environmental impacts of land-use plans in 1999. *Inter alia*, it stated:

> Development plans should be drawn up in such a way as to take environmental considerations comprehensively and consistently into account . . .
> The appraisal should:

- apply to all types of plan;
- apply to all policies and proposals;
- be part of the plan preparation process; and
- be a process of identifying, quantifying (where appropriate), weighing up and reporting on environmental effects of those policies and proposals. (DETR, 1999a, paras 4.4, 4.17)

LPAs were required to include in their plan documentation an explanation of how the outcomes of the iterative environmental appraisal process, which was to be applied at every stage of development plan preparation, had informed the policies and proposals in the plan. LPAs were also asked to develop appraisal methodologies to encompass environmental, economic and social (i.e. sustainability) issues (DETR, 1999a).

In 1993, a good practice guide to the environmental appraisal of development plans was published (DoE, 1993). This suggested that, following the characterisation of the environment and scoping, a policy impact analysis should form the core of the environmental appraisal. The use of a policy impact matrix was recommended so that an explicit identification of the impact of each policy option on each aspect of the environmental stock could be made. This matrix could then be used to record whether there was a positive (enhancing), negative (harmful), or neutral impact. This analysis was intended to work in three ways:

- . . . in a iterative way with the initial evaluation of impacts providing an indication of ways in which the policy needs to be refined;
- . . . [by] provid[ing] a basis for comparing relative performance when choices need to be made;
- . . . as a tool of policy-checking or review, in the consideration of proposed modifications after plan examination or inquiry. (DoE, 1993, p. 22)

Figure 19.3 shows a worked example of the use of the policy impact matrix, with the elements of the environmental stock expressed as criteria on one axis and individual policies on the other axis. A typical plan will have scores of policies to be tested in this way.

Further valuable unofficial guidance for local authorities, resulting from direct experience of carrying out environmental appraisal, was published in 1996 (Bedfordshire County Council and Royal Society for the Protection of Birds, 1996). A package for reviewing LPA environmental appraisals has been developed by Lee *et al.* (1999; see also Simpson, 2001).

The government has commissioned research into methods of environmental appraisal (DETR, 1998c) and has organised seminars on SEA (DETR, 1998d, 2000b). Therivel (1998) reported that about three-quarters of LPAs had begun to carry out an environmental appraisal – the proportion was much higher by mid-2001. These and other UK SEAs (for example of European funding applications and of water resource programmes) were characterised 'by their qualitative, subjective approach, by the resulting report's brevity ... and by their relatively low cost' (Therivel, 1998, p. 40). Most LPAs followed the 1993 guide to environmental appraisal, with little adaptation of methodologies or approach to local circumstances (Curran *et al.*, 1998). Russell (1999, p. 529) suggested that LPAs needed 'additional support in the form of guidance, information and resources if they were to use the system to best effect'. Many LPAs have moved recently towards sustainability appraisal involving the integration of economic and social impacts. It is clear that practice has developed very rapidly over the past decade:

> The SEAs have generally raised awareness of environmental issues and are increasingly resulting in changes to the relevant policy, plan, or program. What was 'best practice' a few years ago is now normal practice, ... SEA is becoming a mainstream activity with considerable benefits in Britain. (Therivel, 1998, p. 56)

The Netherlands

The Netherlands has experience of SEA dating back to the early 1980s (Wood, 1988a). As a result, the EIA system was designed to encompass SEA. Article 3 of the amended Environmental Impact Assessment Decree makes provision for certain types of plan and programme to be subject to EIA. These relate to: structure plans for electricity supply, industrial and drinking water supply, landscaping, nature conservation and outdoor recreation; provincial waste management proposals; mineral extraction plans; and certain types of land-use plans. The Decree does not apply to national policy plans. In addition, since 1994, the Dutch government has required a different form of SEA (the E-test) to be applied to new legislation.

Experience of the SEA of over 40 plans and programmes containing proposals in specific locations has been gained to date (Verheem and Tonk, 2000). Decisions have been reached on many of these plans and programmes and several have been implemented. The SEA process prescribed in the Act is the same as for project EIA, involving screening (in some cases), scoping, documentation, reviewing, decision making and monitoring. The EIA Commission (EIAC) plays the same pivotal role in these SEAs as in EIA (Verheem, 1999).

A SEA was carried out, on a voluntary basis, of the National Programme on Waste Management 1992–2002. The EIS compared the environmental impact of the various options for waste disposal, including incineration and various land disposal methods, by using a tiering approach leading to specific projects, together with a number of environmental indicators such as energy use, land use and pollutant emissions (Verheem, 1996). Other SEAs have been undertaken on the national structure plan for electricity supply utilising

	Criteria	Global sustainability					
		1	2	3	4	5	6
Policies		Transport energy: Efficiency: trips	Transport energy: Efficiency: models	Built environment Energy: efficiency	Renewable energy potential	Rate of CO_2 'fixing'	Wildlife habitats
1	To provide a network for open space corridors	●	✔	●	●	✔	✔
2	To concentrate residential development on an existing public transport corridor of the city	●	✔	✔	●	●	✔?
or							
3	To concentrate residential development on a new rural 'green' settlement (c. 8,000 pop)	X	X	✔	✔?	✔?	✔?

Context: District-wide plan for a city of 150,000 and its hinterland
Illustrative policies: 1 For open space; 2 and 3 Represent options
for the location of housing

Suggested impact symbols:

●   No relationship or insignificant impact

X   Significant adverse impact

Figure 19.3   Policy impact matrix for environmental appraisal of UK land use plans
Source: DoE, 1993, p. 24.

	Natural resources			Local environmental quality				
7	8	9	10	11	12	13	14	15
Air quality	Water conservation and quality	Land and soil quality	Minerals conservation	Landscape and open land	Urban environmental 'liveability'	Cultural heritage	Public access open space	Building quality
✔	●	✔	●	✔?	✔	✔	✔	●
X	●	●	●	✔?	✔	✔?	X	✔
●	✔?	X	✔?	X	✔	✔?	✔	X

✔?	Likely, but unpredictable impact
✔	Significant beneficial impact
?	Uncertainty of prediction or knowledge

different fuels and different sites for power stations, on the national plan on drinking water and on Schiphol Airport (European Commission, 1997b; Verheem, 1999; EIAC, 2001b).

SEAs have also been carried out for several regional physical plans, especially where they involved site selection for major housing areas (European Commission, 1997b). Several SEAs for local land-use plans, which frequently also involve site selection for development, have also been undertaken (Verheem, 1992, 1999; ten Holder and Verheem, 1996, 1997; European Commission, 1997b). Some difficulties were encountered in undertaking these SEAs, especially in relation to the poor quality of information and uncertainty inherent in SEA (van Eck, 1994). However, the EIA approach for projects generally worked satisfactorily for plans and programmes, largely as a result of the concrete, quasi-project nature of the proposals, the preparation of guidelines, the formulation of alternatives and the participation of non-governmental organisations and the public (Verheem and Tonk, 2000).

Despite its merits, however, SEA and EIA based on the provisions of the Act were not felt to be resulting in comprehensive assessment of, for example, infrastructure proposals (Niekerk and Arts, 1996; Niekerk and Voogd, 1999). In particular, this form of SEA was not considered to be appropriate for the assessment of draft regulations which are less subject to public scrutiny. For this reason, the E-test for new regulations was developed (Verheem, 1992, 1999; de Vries and Tonk, 1997; Coenen, 1999; Verheem and Tonk, 2000).

By 1998, more than 80 regulations (about 5 per cent of those proposed) had been subjected to the E-test (de Vries and Tonk, 1997). This SEA procedure is not formalised in law, does not involve public participation and does not rely on EIAC to ensure quality. It is based on answering four questions which are deliberately kept simple:

> What are the consequences of the draft legislation:
> 1. for energy consumption and mobility?
> 2. for the consumption and stocks of raw materials?
> 3. for waste streams and atmospheric, soil and surface water emissions?
> 4. for use of available physical space? (Verheem, 1999, p. 189)

A small interministerial joint support centre was set up to screen legislative proposals, to determine which of the four E-test questions should be answered and to review the draft results (a paragraph or two) prepared by the responsible ministry before the proposal was forwarded for decision to ministers.

Of 83 proposals selected for E-testing, 69 were found to have a potential environmental effect and, of these, three were modified and the explanatory memorandum was amended in all the remaining cases (de Vries, 1998). Like the EIA-based SEA procedure, therefore, the E-test is judged to be working satisfactorily, largely as a result of the sharing of experience, the Dutch culture of consensus, and the support centre's review of quality with the attendant threat of delay (de Vries, 1996; de Vries and Tonk, 1997; Verheem, 1999; Verheem and Tonk, 2000).

Canada

There has been considerable interest in SEA in Canada for some time (Federal Environmental Assessment Review Office, 1988; Bregha *et al.*, 1990). The arguments for and against the codification of SEA requirements in law culminated in the government's commitment, in the 1990 cabinet directive, that a non-legislated environmental assessment (EA) process was to apply to proposals for policy or programme initiatives submitted to the cabinet for consideration. This EA process was seen as a demonstration of Canada's commitment to sustainable development, but it was felt by its many opponents that the theory and practice of SEA (for example, in determining compliance) was insufficiently developed to enact it. However, environmental commentators (see, for example, Gibson, 1993) regretted the lost opportunity to legislate for SEA by extending the Canadian Environmental Assessment Act 1992 and thereby 'integrate environmental factors into government discussion making at all levels' (Hazell, 1999, p. 53).

The government decided that:

> a public statement outlining the anticipated environmental effects of a policy or programme initiative, which would be determined through an environmental assessment, would, as appropriate, accompany that announcement of the initiative. (Federal Environmental Assessment Review Office, 1993)

This statement was to be a means of demonstrating that the assessment had been undertaken. In practice, not only were few public statements published, but few SEAs were undertaken. Hazell (1999, p. 134) was certain that 'Cabinet secrecy has virtually guaranteed a failure to comply'. Those public statements which were made generally consisted of a sentence or paragraph stating that the initiative would have no significant adverse effects. However, some were highly significant. Such an example was the environmental review of the North American Free Trade Agreement (Government of Canada, 1992). Although this SEA was criticised for its lack of detail and failure to influence the outcome of the agreement, it provided a useful indication of what could be achieved in such assessments (Hazell, 1999). There have been other examples of successful application of the cabinet directive, for example, in relation to international trade, agriculture and park management plans (Commissioner of the Environment and Sustainable Development – CESD, 1998; Partidario and Clark, 2000).

Notwithstanding these successes, Marsden (1998b) pointed out significant weaknesses in the implementation of the cabinet directive and CESD (1998, pp. 6–26) reported that departments had been slow to comply. He quoted a Canadian Environmental Assessment Agency (CEAA) study that indicated that 60 per cent of departments had conducted EAs of policy and programme proposals but found during his investigation that some 'senior officials . . . were not aware of the existence of the Cabinet directive or did not know how it was being implemented' (CESD, 1998, pp. 6–27). Despite the development of an SEA training package, CESD (1998) recommended that additional measures be taken to improve compliance. The Commissioner returned to this theme in his 1999 report:

Compliance with the directive has been slow and uneven across departments. . . . In their first sustainable development strategies, only 12 of 28 departments mentioned strategic environmental assessment. (CESD, 1999, p. 6)

As a result of these pressures, the cabinet directive was updated in 1999 to strengthen SEA:

by clarifying obligations of departments and agencies and linking environmental assessment to the implementation of Sustainable Development Strategies. (CEAA, 1999e, p. 3)

In contrast to guidance drafted in the early 1990s but never released because of its perceived complexity, the guidelines on implementing the 1999 directive are based on 'current, proven, good practices within federal departments' (CEAA, 1999e, p. 10). The guiding principles of the new approach to SEA are:

early integration;
examination of alternatives;
flexibility;
self-assessment;
appropriateness of level of analysis;
accountability; and
use of existing mechanisms. (CEAA, 1999e)

It is suggested that a preliminary scan should be undertaken of a proposal to determine if further EA will be necessary (Box 19.4).

It is perhaps surprising that the Minister of the Environment (MoE, 2001) made no mention of SEA in his summary of the five-year review of the Act or in his proposals for strengthening it (Government of Canada, 2001). It appears that the revised non-legislated SEA cabinet directive may be proving no more influential than the original version.

Commonwealth of Australia

Although its wording was rather opaque, it was clearly intended that the Environment Protection (Impact of Proposals) Act 1974 would apply to policies and other Commonwealth actions as well as to projects (Fowler, 1982, p. 17). As the responsible Minister stated in parliament in debating the bill: 'the Bill will affect many facets of the Australian Government's activities' (Cass, 1974, p. 4082). In practice, the degree of discretion provided by the Act and the reluctance of politicians and administrators to extend EIA to non-project actions restricted its coverage almost entirely to projects (Harvey, 1998). Of the 215 Commonwealth directives that EIA reports be prepared between 1974 and the repeal of the Act (see Chapter 5), only seven woodchip EISs related obviously to PPPs. There were, however, several examples of recommendations being made under the provisions of the Act in relation to programmes without requiring the preparation of formal EIA reports. These included the assessment of Commonwealth-funded highway construction programmes, of systematic applications to the Australian overseas aid programme, of a national park

Box 19.4   Canadian advice on screening in SEA

Undertaking a preliminary scan – Useful tools and criteria

To conduct a scan of the proposal, the analyst may use a variety of tools, including available matrices, checklists and experts both within the department and from other departments. The following considerations also may be of assistance in conducting the preliminary scan:

1. The proposal has outcomes that affect natural resources.
2. The proposal has a known direct or a likely indirect outcome that is expected to cause considerable adverse impacts on the environment.
3. The outcomes of the proposal are likely to affect the achievement of an environmental quality goal (e.g., reduction of greenhouse gas emission or the protection of an endangered species).
4. The proposal is likely to affect the number, location, type and characteristics of sponsored initiatives which would be subject to project-level environmental assessments, as required by the *Canadian Environmental Assessment Act* or an equivalent process.
5. The proposal involves a new process, technology or delivery arrangement with important environmental implications.
6. The scale or timing of the proposal could result in significant interactions with the environment.

Source: CEAA, 1999e, p. 11.

management plan and of an international trade agreement. Harvey (2000) outlined how an SEA was used to overcome weaknesses in project-level EIA for coastal developments in South Australia, and Renton and Bailey (2000) found that some SEA of policies was being undertaken in Australia and that several, quite different, approaches were being employed.

This gradual growth in practice was matched by an increase in commitment to employ SEA. The Australian and New Zealand Environment and Conservation Council (ANZECC, 1991, p. 9) agreed that the principles and objectives of its national approach to EIA 'would be as effectively, and often more effectively, applied to [the assessment of] policies and major programmes' as to projects, with major benefits. The 1992 Australian Intergovernmental Agreement on the Environment, part of which drew upon the ANZECC national approach agreed between its Commonwealth and state governments, specifically mentioned impact assessment in relation to programmes and policies (Commonwealth Environment Protection Agency – CEPA, 1992).

Despite commissioning research on SEA (Court *et al.*, 1996) the reform discussion paper (CEPA, 1994) noted that SEA was unfinished business to be returned to later. The ministerial consultation paper proposed a limited form of tiered SEA in relation to a policy, plan or strategy which provided for the carrying out of a series of individual actions which might require EIA (Hill,

1998, p. 16). These proposals were duly given legislative force in the Environment Protection and Biodiversity Conservation Act 1999 (EPBC Act). Part 10 of the Act provides an important permissive power for SEA:

> The Minister may agree in writing with a person responsible for the adoption or implementation of a policy, plan or program that an assessment be made of the relevant impacts of actions under the policy, plan or program that are controlled actions. (section 146(1))

The EPBC Act includes specific provisions for public consultation on the draft terms of reference for, and the draft report on, any assessment. In a procedure reminiscent of the project EIA process under the Impact of Proposals Act, the Environment Minister makes recommendations, including recommendations for modifications of the PPP, to the proponent. The EPBC Act also includes requirements for the mandatory SEA of Commonwealth fishery management plans. The inclusion of these provisions in the EPBC Act is bound to lead to increased demands for SEA of policies, plans and programmes and it would therefore be expected that increased SEA activity would ensue.

New Zealand

The 1974 Environmental Protection and Enhancement Procedures applied in principle, but not in practice, to the 'management policies of all departments which may affect the environment' and to certain other government higher-tier actions (Ministry for the Environment – MfE, 1987, p. 1). In the absence of any other forum, EIA often provided a focus for debate on the environmental impacts of policies in New Zealand (Morgan, 1988), principally as a result of the comments by the then recently appointed Parliamentary Commissioner for the Environment (PCE) (see Chapter 5) in published 'audits' and of debate at hearings before the Planning Tribunal (now the Environment Court). The Ministry for the Environment reports to cabinet on the environmental implications of national policies proposed by other ministries.

Building upon this experience, the Resource Management Act 1991 has ensured that environmental assessment applies to policy statements and plans prepared under its provisions (but not to other PPPs). SEA is not a separate and distinguishable process but applies, in a tiered fashion, to national policy statements, to regional policy statements and to regional and district plans.

The Act obliged central government to prepare a coastal national policy statement (and it can produce others), regional councils to prepare regional policy statements and regional coastal plans (and they may prepare other types of regional plan) and territorial (city and district) authorities to prepare district plans. In each case the focus is upon effects rather than on uses, and an environmental evaluation of the policy statement or plan must be carried out (sections 62(1)(g) and 75(1)(g)).

These policy statements and plans also provide a framework for the implementation of project EIA. Section 75(1)(f) of the Act requires district plans to specify the information that should be submitted with an application for a resource consent, including the circumstances in which the powers to ask for further environmental information may be used. Plans, which are themselves public documents, should also set out the consultation requirements for any EIA required for a consent. In this way the plan can be used to ensure the EIA requirements are better tailored to selected types of activities according to local circumstances and are focused on the relevant environmental issues.

Section 32 of the Resource Management Act requires decision-makers to have regard to the extent to which objectives, policies, plans or rules are needed, alternative means of achieving aims and the reasons for and against the method proposed and the principal alternatives to it (including the no-action alternative). In addition, councils are required to subject their plans and policies to an 'evaluation which ... is appropriate to the circumstances, of the likely benefits and costs of the principal alternative means' (section 32(1)(b)).

Benefits and costs are divided into environmental on the one hand and social and economic on the other (MfE, 2000b). This onerous requirement was intended to ensure that proposed plans and policy statements used robust measures to achieve better environmental outcomes while keeping costs to the community to a minimum. Although section 32 demands that a formal integrated assessment be carried out, it is in effect more a regulatory impact assessment requirement than a mandatory SEA provision.

As mentioned in Chapter 5, several guides relating to the environmental assessment of policy statements and plans have been issued (e.g. MfE, 1992a, 2000b). A guide to good practice under section 32 has also been issued which, *inter alia*, advocated describing the options, defining the evaluation framework by identifying relevant costs and benefits (including social and environmental impacts), analysis involving comparison of options, and decision making (MfE, 1993). A further guide was intended to offer advice on 'what to do, how to do it, when to do it, who should do it' (MfE, 2000b, p. 3). However, these guides say little about the environmental assessment of the effects of proposed plans and policy statements.

About 50 per cent of the proposed plans and policy statements were operative in mid-2001 (MfE, 2001c) and the final plans will not be in place for some years (see Chapter 5). Dixon *et al.* (1997) reported that, unsurprisingly, territorial councils were lagging behind regional councils in the production of plans. Although SEA is supposedly being incorporated into the plan preparation and policy analysis process by land-use planners, separate SEA reports are not being prepared and the quality of the assessments has generally been weak (see also, Dixon and Fookes, 1995; Salmon, in McShane, 1998). Dixon *et al.* (1997) suggested there has been an overemphasis on procedure at the expense of substance (consultation rather than analysis) and that managerialism and an emphasis on efficiency has jeopardised the unique integration (Bartlett, 1997) for which the Act strives. In practice, there appears to have been little connection between the provisions contained within plans and policy statements and the granting of resource consents (PCE, 1995; Dixon *et al.*, 1997). Horton and Memon (1997) felt that coastal SEA in New Zealand gave priority to economic development rather than to the environment. Fookes (2000, p. 91) concluded that:

> There is a prevailing view amongst practitioners that there is little in the current planning documents or in Section 32 reports that suggests any systematic analysis.

Fookes also questioned whether local politicians 'adequately appreciate the questions they should ask'.

Generally, it appears that expertise is weak in the SEA of plans and policy statements, and that there is a need for the development of, and research into, appropriate SEA methods. Further, the monitoring demanded by section 35 of the Act (see Chapter 14) has been neglected, public scrutiny has been lacking and there remains a need for a wider range of professional skills to be employed in the environmental assessment of plans. There has been a lack of central government guidance (only one national policy statement has been issued), insufficient local government financial and staff resources, have been deployed, and duplication of effort has taken place (Dixon *et al.*, 1997).

The anticipated amendments to the Resource Management Act (see Chapter 5) are not expected to affect its SEA procedures. However, the preparation of a national strategy for sustainable development and the increasing informal use of SEA with regional growth strategies, structure plans, design guidelines and stormwater strategies are challenging the integrated system envisaged in the Act. As a result, discussion is taking place about how SEA might be applied more systematically (i.e. not just under the provisions of the Act) across all levels of government in both formal and informal contexts.

Despite signs of progress, it is apparent that the generally small regional and territorial authority professional staffs are struggling to come to terms not only with the EIA provisions of the Act but also with its SEA provisions and that there is a need for further SEA guidance, for case studies and for training. The decisions of the Environment Court are providing some clarification but there is a need for the Ministry for the Environment to increase its informal work with practitioners and the rigour of its formal plan submissions and to generate more national policy statements. The Parliamentary Commissioner for the Environment may also need to play a more active role if SEA practice is to improve. These actions would help to realise Fookes's (2000, p. 92) aspiration:

> The hope is that second generation policy statements and plans will better reflect the statutory requirements of, and experience with, Strategic Environmental Assessment.

South Africa

The integrated environmental management (IEM) procedures extended from projects to policy and planning proposals, including various land-use plans, the establishment of townships, land acquisition for protected areas and government policies on the use of natural resources (Department of Environmental Affairs, 1992, vol. 1). While SEAs of such plans and policies were never formally carried out under the voluntary IEM procedures, there has been considerable non-mandatory SEA practice in South Africa (Council for Scientific and Industrial

Research, 1997; Republic of South Africa, 1998; Department of Environmental Affairs and Tourism – DEAT, 2000b; Rossouw *et al.*, 2000; Wiseman, 2000).

Notwithstanding this SEA practice, the 1997 EIA Regulations in South Africa apply only to projects. The deletion of non-project activities from the list of actions to be assessed was seen as another example of the hurried implementation of the EIA powers in the Environment Conservation Act 1989 (see Chapter 5). The need for SEA is widely acknowledged in South Africa since previous EIA practice, and the existing EIA Regulations, have failed to take account of cumulative effects and sustainability issues more generally (Weaver, 1996; Rossouw *et al.*, 2000; Wiseman, 2000).

There has never been any doubt about the government's intention to integrate IEM principles and methodologies into the approval of certain plans and economic policies. Thus, the White Paper on environment management policy stated clearly that:

> Integrated Environmental Management (IEM) will be a prerequisite for government approval of all activities with potentially adverse environment impacts.... Local governments will be required to incorporate IEM into Integrated Development Plans and Land Development Objectives.... Economic policies and strategies and spatial development plans impact on the environment and must be dealt with in the context of IEM. (Republic of South Africa, 1998, p. 56)

The discussion document on IEM (DEAT, 1998a) was less broad-ranging, but made it clear that it was intended that SEA should be employed to evaluate the potential environmental impact of land-use zoning plans and schemes.

In fact, the Development Facilitation Act 1995 contains powers for the environmental assessment of development proposals which could easily be extended to SEA (Glazewski, 2000). In any event, the proposal was that new activities complying with the provisions of an approved land-use plan would be exempted from EIA requirements (DEAT, 1998a, p. 36). This proposed application of SEA to land-use plans led to a severe difference of opinion between environmental assessors and the South African planning profession which saw the imposition of SEA as unnecessarily duplicating the current planning consideration of environmental impacts, with the added disadvantage of consideration being limited to, for example, nature conservation impacts (see, for example, Gasson and Todeschini, 1997). Wiseman (2000) recognised the need to avoid duplication between governments and legislative requirements but believed that an agreed SEA process could resolve this problem.

The National Environmental Management Act 1998 (NEMA) draws no distinction between the environmental assessments of policies, plans, programmes and projects (section 1(1)(i)). It is clear that:

> not only are actual activities included in the ambit of the legislation, but also any planning or policy development activity which may precede the actual physical commencement of the activity. (Glazewski, 2000, p. 288)

A set of guidelines on SEA, setting out principles, issues, case studies and the key elements (Figure 19.4) of the SEA process for plans and programmes has been issued. The guidelines make it clear that, at present, 'no

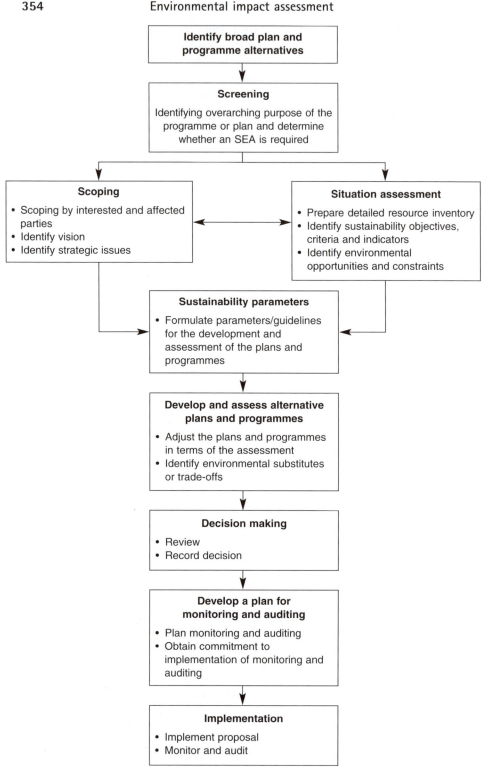

Figure 19.4 The South African SEA process
Source: DEAT, 2000b, p. 18.

particular body has legal responsibility for undertaking or appraising an SEA' (DEAT, 2000b, p. 5). Nevertheless, it is likely that formal SEA will be introduced in regulations made under NEMA in 2002. Whether or not SEA is confined to land-use plans (DEAT, 1998a), there will need to be a major improvement in the capacity of consultancies, provincial authorities and local authorities to deal with it since, at present, SEA skills reside largely in a small segment of the consultancy sector in South Africa.

Summary

Table 19.1 summarises the SEA situation in the seven jurisdictions. The United States, the Netherlands, Australia and New Zealand all make some

Table 19.1   The treatment of SEA in the EIA systems

Criterion 14: Does the EIA system apply to significant programmes, plans and policies, as well as to projects?

Jurisdiction	Criterion met?	Comment
United States	Yes	1969 Act provides clear legal provisions for SEA. SEA practice developing steadily: over 100 programmatic EISs now prepared annually
UK	Partially	No formal requirement for SEA but 'expected' for land-use plans and must be implemented by 2004. Guidance exists. Substantial local planning authority SEA practice
The Netherlands	Yes	EIA Decree defines 'proposal' to include certain plans and programmes. E-test on legislative proposals. SEA practice substantial
Canada	Partially	Revised non-legislative SEA process applies to cabinet and other proposals. SEA research and training undertaken, guidance prepared. Growing SEA practice
Commonwealth of Australia	Partially	Act provides for discretionary SEA and requires SEA of Commonwealth fisheries plans. Few SEA reports published but SEA practice slowly growing
New Zealand	Partially	1991 Act requires SEA of regional policies and plans and of district plans. Practice often weak
South Africa	No	Firm proposals to extend environmental assessment to land-use plans and possibly to policies and programmes. Some limited informal SEA practice

legal provision for the SEA of policies, plans or programmes but practice is insufficiently developed in the antipodean countries to meet the evaluation criterion fully. However, the United Kingdom, Canada and South Africa currently make no legal provision though there has been considerable SEA practice in the first two countries and some experience of SEA in South Africa. The need to implement the European directive on SEA by 2004 will ensure that the UK meets the criterion sooner rather than later.

Many of the points made in earlier chapters and in Chapter 20 in relation to the different stages of the EIA process could be applied to SEA. There has been more progress in SEA than in any other aspect of environmental assessment since the first edition of this book was published. It is clear that SEA practice is developing rapidly around the world and that SEA is beginning to catch up with EIA, though it still has a considerable distance to go. The increasing emphasis on cumulative and indirect impacts and on the sustainability of development make further acceleration of progress on SEA, and the sharing of the burgeoning experience to date, imperative.

Note

1. Office of Federal Activities, Environmental Protection Agency, Washington DC, personal communication (July 2001).

Chapter 20

Conclusions

As mentioned in Chapter 1, several developments in EIA have taken place as experience with its use has grown. These may be summarised as:

- increasing emphasis on the relationship of EIA to its broader decision-making and environmental management context and recognition of its subjective and political nature;
- increasing codification of EIA requirements;
- increasing refinement of EIA requirements;
- increasing concern with the quality of EIA;
- increasing emphasis on maximising the benefits and minimising the costs of EIA;
- increasing recognition of the need to link EIA and ongoing environmental management;
- increasing recognition that some form of strategic environmental assessment is necessary.

These developments are evident in each of the seven EIA systems reviewed.

This final chapter draws together the main threads of the earlier chapters by summarising the performance of the EIA systems in the United States, UK, the Netherlands, Canada, Commonwealth of Australia, New Zealand and South Africa against the evaluation criteria listed in Box 1.2. Table 20.1 summarises the overall performance of the seven EIA systems against these criteria. Several suggestions for improving the various EIA systems by overcoming the shortcomings identified, and for improving EIA generally, are advanced. These suggestions derive directly from the comparative review of the EIA systems and provide one of the principal justifications for such a study.

United States

The US EIA system meets 10 of the 14 evaluation criteria and partially meets three others for environmental impact statements (EISs). Its performance in relation to the numerous environmental assessments (EAs) is much weaker. The main shortcomings of the system relate to its coverage (which is confined to federal actions), to lack of centrality to decision making (notwithstanding the requirement to publish a record of decision), to the monitoring and mitigation of impacts and to system monitoring. Other

Table 20.1 The overall performance of the EIA systems

Evaluation criterion	Criterion met within jurisdiction						
	United States	UK	Netherlands	Canada	Australia	New Zealand	South Africa
1. Legal basis	●	●	●	●	●	●	●
2. Coverage	◐	●	●	○	◐	●	●
3. Alternatives	●	◐	●	●	●	◐	●
4. Screening	●	●	●	●	●	●	●
5. Scoping	●	◐	●	●	●	◐	●
6. EIA report preparation	●	◐	●	◐	●	○	◐
7. EIA report review	●	◐	●	◐	◐	◐	○
8. Decision making	◐	◐	◐	◐	●		○
9. Impact monitoring	○	○	◐	◐	●	○	○
10. Mitigation	●	●	●	●	●	◐	●
11. Consultation and participation	●	◐	●	●	●		◐
12. System monitoring	◐	○	●	●	●	○	●
13. Benefits and costs	●	●	●	●	●	●	●
14. Strategic EA	●	◐	●	◐	◐	◐	○

● Yes   ◐ Partially   ○ No

weaknesses relate to the lack of oversight of EAs, to lengthy descriptive and derivative EISs and to the court-driven procedural nature of the system. Because the system is operated by federal agencies, the general level of expertise is high but 'agencies often do not assign their most effective and efficient personnel to NEPA tasks' (Offringa, 1997, p. 295). There is still a perceived need for training and guidance (Council on Environmental Quality – CEQ, 1997d). The roles of CEQ, of the Environmental Protection Agency (EPA) and of public interest groups in maintaining and refining the system and in ensuring that federal agencies perform are pivotal.

The National Environmental Policy Act 1969 (NEPA) has become part of the US environmental furniture – a comfortable support, which is now taken for granted. NEPA has prevented inappropriate projects from being implemented and improved others but broadening the coverage of NEPA to cover other actions would provoke a constitutional outcry. Increasing the centrality of EIA to the decision-making process could be achieved by amending NEPA to prohibit an action from being taken unless all feasible mitigation measures were included in the proposal (Blumm, 1990). Such a solution could be applied following both an EIS and a finding of no significant impact (FONSI). This would go some way to achieving the initial intention of NEPA's authors (Yost, 1990; Caldwell, 1998). However, any current attempt to modify NEPA would expose it to the risk of being weakened, not strengthened, by Congress.

A further improvement would be to ensure that EISs and environmental assessments are made shorter and more readable and thus more accessible to decision-makers (Clark and Canter, 1997; Blumm, 1990). Other improvements include mandatory public review of FONSIs, the provision of guidance on EAs generally and on the role of public participation in their preparation in particular (Blumm, 1990; Blaug, 1993). This would supplement EPA's sourcebook providing guidance on the whole EIA process (EPA, 1993) and the guidance on cumulative impact assessment (CEQ, 1997a; EPA, 1999a). More broadly, an early, open and collaborative approach to the EIA process needs to be adopted to achieve greater effectiveness and efficiency (Clark and Richards, 1999). Rendering post-decision monitoring compulsory is a necessary reform (Blumm, 1990). Additional auditing studies are also needed to ensure that innovative mitigation measures are effective (Carpenter, 1997).

There is also a need to improve the quality of predictions by developing methodologies (for example, for cumulative impacts and effects on ecosystems), by preparing further technical CEQ and other guidance, by the training of appropriate personnel and by peer review (Dickerson and Montgomery, 1993; Canter and Clark, 1997). The referral of more cases to CEQ and an extension of CEQ's role in EIA oversight might also be helpful in increasing the effectiveness of EIA (Blumm, 1990). However, as the Environmental Law Institute (1995b) and Caldwell (1998) have pointed out, both strengthening the number and status of CEQ staff and an increasing presidential commitment to CEQ's activities and recommendations are needed to achieve the full benefits of EIA in the United States.

UK

The UK EIA system fully meets five and partially meets another seven of the 14 evaluation criteria employed in this comparative review (Table 20.1). The transition to a modified Mark I system with the implementation of the new Regulations has led to marked improvement, but the UK EIA system still meets fewer criteria than many of the other EIA systems evaluated in this book.

In comparison with many other systems, the UK's is a fairly weak, modified first-generation EIA system, with alternatives, screening, partial scoping, environmental statement (ES) publication and public participation provisions integrated into existing town and country planning decision-making processes, but without early participation, true centrality of EIA to the decision, third-party appeal or monitoring provisions. Obviously, experience of EIA has been gained by local planning authorities (LPAs), developers and consultants as the diffusion of practice has taken place, and the quality of EIA practice has improved (Lee and Brown, 1992; Jones *et al.*, 1998; Glasson *et al.*, 1999). However, while the range of experience within consultancies is growing, local authority experience of EIA is still surprisingly limited in many cases.

The shortcomings of the UK EIA system relate to mandatory scoping, to the use made of EIA in decision making, to project monitoring, to consultation and participation, to formal system monitoring and to strategic environmental assessment (see also Sheate, 1996; Glasson, 1999b; Glasson *et al.*, 1999). They are a reflection of the UK's implementation almost to the letter of the compromise requirements of the amended European Directive (Chapter 4). In these and other aspects of the EIA process, practice varies considerably (Leu *et al.*, 1996), from the exemplary to the unprofessional.

These shortcomings mean that the aims of EIA – better quality project planning and better quality decision making – are still not being universally achieved. If weaknesses continue to be evident as practice under the new Regulations evolves, then more radical changes than those made to implement the amended Directive will be necessary. As a first step (Jones *et al.*, 1998) measures relating to the following could be taken:

- Better diffusion of EIA information and, in particular, of ESs.
- Provision of project-specific guidance.
- Provision of guidance on methods of public participation.
- Further on-the-job training provision.
- Improvement of LPA procedures for coping with EIA.
- Improved incorporation of ES mitigation measures into planning conditions.
- Better provision of information to the public.
- Briefing of planning inspectors on the acceptability of ESs.
- Research into the operation of the various stages of the EIA process.

If practice subsequently failed to improve sufficiently then the EIA system itself would need to be strengthened further, as has happened over the years in many mature EIA systems (for example, the United States, the Netherlands,

Canada, Commonwealth of Australia and New Zealand) which now satisfy far more criteria than does the UK's. As Leu *et al.* (1996, p. 11) have stated: 'the proactive imposition by central government of strong legislative and procedural control is required if the highest standards of EIA practice are to be universally applied.' Changes to the EIA system (and in particular to mandatory scoping, to formal ES review, to the centrality of EIA to decision making, to project monitoring and to formal provision for SEA) should be designed to ensure that the evaluation criteria employed in this review are met more fully than at present.

The Netherlands

The Dutch EIA system meets almost every one of the evaluation criteria utilised in this review (Table 20.1). The only criteria which are not met relate to decision making and to monitoring (even here the legal provisions meet the criteria: it is practice which falls short). While the mitigation of environmental impacts is not separately specified in the law there appears to be no inherent weakness in the treatment of mitigation in the EIA system. Mitigation is subsumed under the very extensive coverage of alternatives in the Dutch EIA system and, in particular, in the environmentally preferable alternative. However, since Article 5(2) of the European EIA Directive, as amended, requires the EIS to include 'a description of the measures envisaged in order to avoid, reduce and, if possible, remedy significant adverse effects' (see Chapter 3) it is somewhat surprising that the omission of mention of mitigation measures in the Dutch EIA legislation persists.

Major omissions in the implementation of the monitoring and auditing provisions of the Environmental Management Act 1994 exist, though practice has sometimes been good and is slowly improving. It would assist the operation of the EIA system, and the knowledge base for EIA, if these provisions were strengthened, perhaps by requiring the submission of satisfactory monitoring information when permits are renewed, or by increasing linkages with other legislative controls. This would improve the feedback of knowledge into the EIA system, where some weaknesses in utilising previous experience are apparent (Arts, 1998). This is, perhaps, symptomatic of the failure, in some instances, fully to integrate the results of the EIA into the proponent's own planning and project development at a sufficiently early stage to influence project design. It may also reflect the willingness of many competent authorities to leave too much of the operation of the EIA process to highly competent consultancies and to the influential EIA Commission and not to make EIA truly central to their decisions.

While many excellent examples of the use of EIA to guide development exist (see, for example, Verheem *et al.*, 1998), Arts (1998, p. 137) provided an accurate summary of indigenous views about the Dutch EIA system:

> The potential of the Dutch EIA system is not fully used; EISs are often too thick, containing redundant, irrelevant information but sometimes lacking essential information – e.g., about alternatives, uncertainties; sometimes EIAs are started too late;

the practical applicability is limited; EIA is often viewed as overly cumbersome, overdone, inflexible, expensive, time consuming, causing delays etcetera.

In 1990, the Evaluation Committee on the Environmental Protection Act (ECW, 1990, p. 13) could 'express its satisfaction about the way the regulation has operated in the first years since its introduction. Experience to date gives ground for optimism'. This optimism was tempered in ECW's second report (1996) when it stated that 'EIA is an instrument which functions reasonably well' (p. 8) but that, paradoxically, 'despite the impact which EIA is found to be having, the instrument does not always have a favourable image' (p. 10).

Further improvements in the effectiveness of the Dutch EIA system can therefore be anticipated, including the introduction of more flexible assessments (for example, simplified EIA utilising the notice of intention), selective monitoring and greater integration with other legislative requirements (for example, land-use planning). The Dutch EIA system has sometimes been criticised as overly cumbersome, expensive, time-consuming and limited in application: a Rolls Royce where a Ford would suffice. The Dutch would generally argue that the wealth of their concern about the environment of their small and vulnerable country justifies maintaining the quality of the best EIA system their money could purchase. They would also argue that the revised EIA system is likely to be flexible enough to allow them only to use a car when absolutely necessary: that detailed EIA is needed to address only potentially acute environmental problems.

Canada

The federal Canadian EA system delineated by the Canadian Environmental Assessment Act 1992 meets eight, and partially meets another five, of the evaluation criteria used in this review. Table 20.1 summarises the strengths and weaknesses of this second generation federal EA system. The main weaknesses are:

- limitations on coverage of actions and (for interjurisdictional reasons) on the coverage of environmental effects;
- limitations on treatment of screening reports;
- lack of means for ensuring that EA centrally influences decisions;
- lack of effective mechanism for ensuring compliance with mitigation and monitoring requirements;
- lack of formal strategic environmental assessment (SEA) requirement.

Deficiencies relating to, *inter alia*, alternatives, coverage of socio-economic and cultural impacts and to the discretionary nature of the Act have also been identified (Gibson, 1992; Hazell, 1999).

Quite apart from the strengths identified in Table 20.1, Hazell (1999, pp. 143, 144) has identified three more subtle successes of the Act:

- First, the very fact of [the Act's] statutory requirements has changed the structure of federal bureaucracy in significant ways ...

- Second ... it has set a standard by which environmental assessment processes of provincial and aboriginal governments will be and are being judged....
- The third and greatest success of the *Canadian Environmental Assessment Act* is that it is a tool for democracy and open government.

Because of their public participation provisions, panel reviews have generally been seen as providing an excellent means of cooperatively identifying and mitigating the environmental effects of a handful of projects. Screening, on the other hand, has been seen as a largely clandestine and ineffective means of assessing the effects of the vast majority of projects. The panel review has been an elaborate ice castle constructed on a screening iceberg more than 99 per cent of which has been submerged in cold, impenetrable waters.

There is a strong case for excluding many of the nearly 6,000 screenings per annum from the EA process, to concentrate resources on an increased number of comprehensive studies and panel reviews of the larger projects with more serious impacts. There is an equally strong case for strengthening the leadership of the Canadian Environmental Assessment Agency since, as Nikiforuk (1997, p. 25) has stated: 'when all government departments are responsible authorities, it seems all are equally irresponsible'. The Commissioner of the Environment and Sustainable Development (1998, pp. 6–28) recommended that 'the Agency could be more forceful in expressing its concerns'.

There is optimism that the revisions to the Act will lead to improvements in EA practice in Canada. While the problems of lack of environmental political clout at federal cabinet level and of constitutional debate with the provinces about the extent of federal jurisdiction in environmental matters remain, the revisions strengthen provisions for bilateral agreements and coordination (and hence facilitate harmonisation) and for exposing more of the EA system to public scrutiny. These provisions provide a firm foundation for the reinforcement of an EA system already widely perceived as effective.

How well the revised Act works in practice will, like the non-legislated environmental assessment of policies, plans and programmes, depend on the efforts and resourcefulness of the Agency and on the growing influence and expertise of environmental professionals within the responsible authorities and proponent organisations. Above all, however, the strength of the Canadian system depends upon the bedrock of public environmental concern, vigilance and participation.

Commonwealth of Australia

The new Commonwealth EIA system delineated by the Environment Protection and Biodiversity Conservation Act 1999 (EPBC Act) fully meets ten of the evaluation criteria and partially meets another four (Table 20.1). The main weaknesses of the legislation relate to coverage of impacts, to decision making, to monitoring and to system monitoring. There can be no doubt that this EIA system is stronger than that set out by the repealed Impact of Proposals Act which fully met seven of the criteria and partially met two others. Numerous criticisms were made of the previous EIA system, especially of uncertainty (principally caused by the use of discretion, duplication of rules between the Commonwealth and the states, changes of rules, etc.) and of delay (Bureau of Industry Economics, 1990; Australia and New Zealand Environment and Conservation Council, 1991).

In response to these criticisms the 1992 Intergovernmental Agreement on the Environment (Commonwealth Environment Protection Agency – CEPA, 1992) stated that time schedules for all stages of Commonwealth and state EIA processes would be set for proposals and that the Commonwealth would accredit state EIA systems, thus removing much of the remaining need for joint working on EIA (Fowler, 1996). The comprehensive public review of the Commonwealth EIA system proposed measures to resolve almost all the short-comings identified (CEPA, 1994), many of which were adopted by the Environment Minister (Hill, 1998) and subsequently became law.

While the EPBC Act provides a strong basis for the Commonwealth EIA system, much discretion remains and the effectiveness of the Act is critically dependent on how it is implemented by Environment Australia and by the Environment Minister. For example, the coverage of impacts and actions is, appropriately, confined to matters of national environmental significance, but other impacts of Commonwealth-assessed actions, and other actions, will need to be assessed by the states. Screening involves both the coordination and the accreditation of states and other Commonwealth agencies, and it is this which has evoked most concern (see, for example, Fowler, 1996; Ogle, 2000; Padgett and Kriwoken, 2001; Scanlon and Dyson, 2001).

The need to broaden the range of screening triggers to incorporate greenhouse, National Heritage Places and perhaps certain forestry and other issues has also been stressed (Ogle, 2000). The provision of reasons for level of assessment and approval decisions has been demanded by Hughes (1999). Equally, the effectiveness of the discretionary SEA powers in the EPBC Act will depend heavily on the willingness of Environment Australia and of the Environment Minister to press for them to be used.

The Ecologically Sustainable Development Working Group Chairs (1992, p. xxii) recommended that:

> a high priority be accorded to work on the evaluation and development of techniques for environmental impact assessment, particularly on aspects of the technique associated with the framing of the terms of reference, social impact assessment, economic analysis, cumulative effects and post-approval assessment.

James (1995) echoed these comments, suggesting that more rigorous predictive modelling was necessary. The need for increased training and communication (CEPA, 1994) continues to apply.

Practice will reveal the extent to which the Act's discretionary impact monitoring and inadequate EIA system monitoring provisions (and those governing its cooperation arrangements (Scanlon and Dyson, 2001)) need to be revisited. However, the crucial measure, increasing the centrality of EIA to decision making, requires both public and political will to utilise fully the powers afforded by the EPBC Act.

New Zealand

The complex and ambitious Resource Management Act 1991 (RMA) is one of the first attempts in the world to promote sustainable management. EIA and SEA are not separate procedures but are integrated with project authorisation and plan-making activities which are both based on effects (rather than land use). It is, perhaps, all the more surprising that New Zealand should have relied quite so completely on local discretion in its Mark II EIA system despite its strong local government tradition. While the temptation to devolve responsibilities proved irresistible to a reformist, monetarist, central government (Bührs and Bartlett, 1993; Memon, 1993) the implementation gap has proved to be enormous, and was exacerbated by the budget cuts that hampered the Ministry for the Environment in providing the intended support.

New Zealand is famous for 'do it yourself' activities, but to have left local authorities with tiny professional staffs and little or no experience of EIA to evolve screening, scoping, review and decision-making procedures individually seems courageous, especially in a country with such a small cadre of EIA experts. The huge task faced by planners in coping with EIA (and other aspects of the Resource Management Act requiring significant professional reorientation (Montz and Dixon, 1993; Morgan, 1995; Dixon *et al.*, 1997)) was exacerbated by several factors. As well as the early curtailment of formal Ministry for the Environment support, these included simultaneous local government reorganisation, the separation of much local authority policy making from consent processing, and the transfer into management positions of many professional staff possessing EIA expertise.

It can be seen from Table 20.1 that the New Zealand EIA system satisfies five, and partially meets another six, of the evaluation criteria. This may, to some extent, be an inevitable consequence of making the EIA system applicable to such an extraordinary range of activities, from the trivial to the highly significant. The EIA system necessarily provides considerable flexibility to ensure that the scale of the assessment of environmental effects (AEE) is appropriate to the likely severity of the impacts. In some instances (for example, impact monitoring) a failure is recorded because the mandatory requirement is inadequately or insufficiently carried out in practice.

Overall, however, there are weaknesses in scoping, in EIA report preparation, in the quality and review of EIA reports, in the centrality of EIA to the decision, in monitoring, in public participation and in EIA system monitoring. Numerous suggestions have been made to overcome these shortcomings. In particular, Morgan (1995) and the Parliamentary Commissioner for the Environment (PCE, 1995, 1996b) suggested greater consultation in scoping and in AEE report preparation. Dixon (1993) has stressed the need to develop EIA prediction methods and to enhance skills in, for example, the assessment of cumulative impacts. Improvements in local authority monitoring and enforcement procedures and methods have also been suggested (PCE, 1996a; Morgan, 2001). Equally, there would be many benefits in diffusing information and practice through better EIA system monitoring and by, for example, publicising and making available examples of good EIA practice.

Because of the lack of EIA expertise of most applicants, many of their consultants, and many local authority officers and elected members there is a pressing need for training and for more specific guidance on the different stages in the EIA process for the various participants. (Helpful advice has been published, but much of this tends to be procedural rather than substantive.) Training and encouragement to adapt are particularly necessary in overcoming the continuing resistance of traditional land-use planners to the EIA procedures which require reorientation of practice from control of use to control of effects (Dixon and Fookes, 1995, Fookes, 2000; Morgan, 2001). Increased priority within local authorities, greater professionalism, growing experience, and Environment Court decisions will also help to improve practice. In addition, Parliamentary Commissioner interventions, the implementation of regional policy statements and plans and district plans and minor (not radical) amendments to the Resource Management Act will all help to realise the undoubted potential of this innovative EIA system. However, as elsewhere, public awareness and pressure on politicians to give appropriate weight to EIA are the most important factors in increasing the effectiveness of EIA in New Zealand. As the Parliamentary Commissioner (PCE, 1998, p. 7) stated, enhancement of environmental management requires:

> A real commitment and investment by the Government to make the RMA a key component of contributing to sustainable development in measurable ways.

South Africa

The South African EIA system, while formally introduced only in 1997, is based on part of the integrated environmental management (IEM) principles enunciated almost ten years previously, refined in 1992 and enacted (but not implemented) in 1998 (see Chapter 5). The South African EIA system currently meets seven of the 14 evaluation criteria, partially meets two and fails to meet five. South African EIA is therefore no pastiche (Table 20.1). It is not surprising that South Africa, simultaneously a developed and a developing country, should fail to meet more criteria than any of the other systems. The principal weaknesses relate to EIA report review, to the centrality of the full range of impacts to decision making, to impact monitoring, to EIA system monitoring and to the environmental assessment of programmes, plans and policies.

The White Paper summed up the EIA situation in South Africa under the voluntary IEM procedure bleakly in its overview of environmental management policy generally:

> There is a widespread view that environmental issues in South Africa have had low priority, being narrowly defined as relating mainly to nature conservation. This is reflected by a failure to integrate environmental concerns into economic planning and decision making at all levels in society. Sustainable development and effective integrated environmental planning and management are seriously impeded by:

- fragmented policy and ineffective legislation
- uncoordinated planning
- ineffective enforcement of regulations
- institutionalised conflicts of interest in regulating environmental impacts and promoting resource exploitation
- confusion about the assignment of functions at different levels of government
- limited capacity and resources in government and civil society, and
- limited public participation. (Republic of South Africa, 1998, p. 73)

Weaver (1996) identified a number of additional weaknesses in South African EIA practised under the IEM procedure, including poor treatment of alternatives (see Chapter 8), poor framing of conditions (Chapter 13) and inadequate monitoring (Chapter 14).

Weaver felt that various steps were needed to overcome these weaknesses in the new EIA system. These included: the empowerment of decision-makers, the development of seamless linkages between EIA, auditing and environmental management systems, the development of effective community participation, seeking 'the middle road between "science and values" to reduce adversarial situations', and training in EIA (Weaver, 1996, p. 115; see also Khan, 1998). The introduction of time limits, increased public access to the EIA process and the extension of environmental assessment to policy, plans and programmes are also necessary.

Granger (1998) and Duthie (2001) have expressed justifiable concern about the lack of financial and staff capacity within provincial and local government to deal effectively with applications involving EIA (see Chapter 5). It is apparent that staffing levels at central government level are also inadequate to oversee the national EIA system. The International Development Research Centre recommended that if EIA was to be strengthened:

> A strong and comprehensive office for environmental assessment and review, with legislative 'teeth' and competent professional staff, will be needed, both in the interests of environmental protection and in the interests of a stable regulatory environment for the private sector. (Whyte, 1995, p. 36)

Although concurrent responsibility for the environment has resulted in the delegation of many EIA functions, the need for a strong national focus for EIA remains. Generally, the capacity of governments at national, provincial and local levels needs to be increased to reduce the consultants' (and hence, indirectly, the developers') stranglehold on South African EIA. As Duthie (2001, p. 222) has stated:

> Political support at provincial level for the EIA function is absolutely vital to its success. Without this support the EIA function will not be able to gather enough management support and resources to do its work.

The National Environmental Management Act 1998 (NEMA) provides for its EIA requirements to be implemented by regulation. The opportunity exists in the forthcoming amendments to NEMA and in the promulgation of EIA regulations to replace those currently in place to overcome most of the problems identified. Inevitably, initial uncertainties and difficulties associated with the implementation of NEMA's requirements will be resolved as practice develops and experience is gained. Nevertheless, it is already apparent that political will and institution strengthening, as well as guidance, training and research, will be needed if EIA is to take its rightful place as a powerful weapon in South Africa's environmental management armoury and to confirm that it is no postiche.

Improving EIA

It can be seen from Table 20.1 that the EIA systems do not perform equally well, and that certain shortcomings become more evident when they are seen overall, rather than partially as in the earlier chapters. Overall, EIA is falling short of its potential as a tool to help achieve sustainable development (Sadler, 1996, 1998).

Certain general shortcomings in the current state of EIA practice can be observed. These may be summarised as:

- Weaknesses in coverage.
- Weaknesses in integrating EIA into decision making.
- Weaknesses in EIA report preparation.
- Weaknesses in EIA report review.
- Weaknesses in impact monitoring and enforcement.
- Weaknesses in system monitoring.
- Weaknesses in SEA.

A number of specific measures can be used to strengthen the different EIA systems by introducing or bolstering appropriate procedural requirements (see above). There continues to be a need, in each EIA system, for four other elements to strengthen EIA practice: guidance, training, research, and emphasis on the achievement of sustainable development.

The existence of published guidance on the EIA systems as a whole is clearly useful to those responsible for preparing EIA reports, to those reviewing them and making decisions, to those consulted and to the public. Guidance materials can include manuals, leaflets, computer programs, web sites and videos. Such guidance provides a valuable aid in undertaking any stage of the EIA process. The provision of guidance of this type tends to vary from EIA system to EIA system, just as the provision of more detailed guidance on the different stages of the EIA process varies. While the practitioners most closely involved do not tend to need general guidance unless changes to EIA procedures are introduced, there are always those new to EIA or some aspect of the EIA process who do require relevant guidance.

The provision of EIA training for EIA project managers, for technical specialists and for others involved in the EIA process is an effective method of increasing the standard of practice even in mature EIA systems. While EIA training need not be provided only by the agency responsible for EIA, encouragement of, and participation in, such training by the agency is clearly desirable. The involvement of responsible authorities in EIA training tends to vary from jurisdiction to jurisdiction. A variety of different training methods is appropriate in most EIA systems: courses, manuals, guides, case studies, web sites, videos, computer programs, etc.

There is a continuing need for research on various aspects of EIA, both general and specific. Generally, research is sorely needed on the treatment of alternatives, on scoping, on forecasting, on review methods, on the integration of EIA into decision making, on monitoring, on public participation, on system monitoring and on strategic environmental assessment. Such research needs to be concerned with both substantive (methodological) and procedural issues. Although research on EIA methods is likely to be of most application across EIA systems, the results of procedural research tend also to be widely disseminated since there is a desire to share knowledge and insights. Clearly, each EIA system is likely to have its own specific research needs in addition to those identified here.

Several of the EIA systems reviewed in this book have progressed from Mark I to Mark II versions (the Australian, Canadian and New Zealand systems) or are about to do so (South Africa). The EIA systems in the United States, UK and the Netherlands have been refined. The steps outlined above should enable progress to be made towards Mark III versions. However, it is necessary not only to improve the efficiency of EIA procedures but also to ensure that they receive the public and political endorsement necessary for them more fully to deliver their aims (Figure 20.1).

There remains widespread concern that, despite the establishment and refinement of EIA systems, the achievement of sustainable development goals remains elusive. While the linkage between EIA and sustainable development is widely accepted, principles, criteria, thresholds and limits concerning aspects

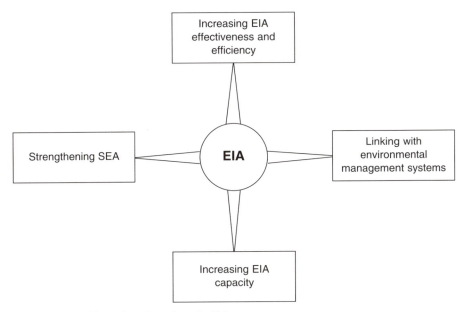

Figure 20.1   Emerging directions in EIA

of sustainable development have not been incorporated into EIA practice sufficiently, resulting in continuing incremental environmental deterioration. Much could be achieved by increasing the weight given to environmental resources and capacities in existing EIA systems to enable the individual effect of a project on cumulative impacts (Rees, 1995), climate change and biodiversity losses to be estimated and tracked (Figure 20.1). The effectiveness of these systems would be bolstered if their aim was raised to achieve more ambitious goals such as 'no net environmental deterioration' or 'net environmental gain' and to provide the basis for more informed application of the precautionary principle in decision making.

The development of Mark III systems placing EIA more firmly in the context of other policy and environmental instruments (Sadler, 1996; Ugoretz, 2001) to provide more adaptive environmental management (Holling, 1978; Noble, 2000b) is necessary. The strengthening of SEA to make it a more effective instrument is gathering pace but requires further effort. Much is being achieved in strengthening EIA capacity, especially in developing countries, but much remains to be done (Lee and George, 2000).

Notwithstanding these emerging directions in EIA, the crucial factor in environmental protection remains the political decision that EIA informs (Bartlett and Kurian, 1999). The influence of EIA on decision making is by far the most important evaluation criterion employed in this review and is the only one not to be met by any of the EIA systems. Only if the public and politicians truly will the ends, can EIA deliver a truly effective means of achieving more sustainable development.

References

Anderson, E. (compiler) (1994) *Australian Country Paper.* 7th Tripartite Workshop on Environmental Impact Assessment. Canberra: Commonwealth Environment Protection Agency.

Arnstein, S.R. (1969) A ladder of citizen participation in the USA. *Journal of the American Institute of Planners* **35**: 216–24.

Arts, E.J.M.M. (1998) *EIA Follow-up: On the Role of Ex-post Evaluation in Environmental Impact Assessment.* Groningen: Geo Press.

Arts, E.J.M.M. and Nooteboom, S. (1999) Environmental impact assessment monitoring and auditing. In Petts, J. (ed.) *Handbook of Environmental Impact Assessment,* vol. 1. Oxford: Blackwell.

Australia and New Zealand Environment and Conservation Council:

ANZECC (1991) *A National Approach to Environmental Impact Assessment in Australia.* Canberra: ANZECC.

ANZECC (1996) *Guidelines and Criteria for Determining the Need for and Level of Environmental Impact Assessment in Australia.* Canberra: ANZECC.

ANZECC (1997) *Basis for a National Agreement on Environmental Impact Assessment.* Canberra: ANZECC.

Avis, A.M. (1994) Integrated environmental management and environmental awareness in South Africa: where are we going? *Journal of Environmental Planning and Management* **37**: 227–42.

Bailey, J.M. and Hobbs, V. (1990) A proposed framework and database for EIA auditing. *Journal of Environmental Management* **31**: 163–72.

Bailey, J.M., Hobbs, V. and Saunders, A. (1992) Environmental auditing: artificial waterway developments in Western Australia. *Journal of Environmental Management* **34**: 1–13.

Barker, A. and Wood, C.M. (1999) An evaluation of EIA system performance in eight EU countries. *Environmental Impact Assessment Review* **19**: 387–404.

Barker, A. and Wood, C.M. (2001) Environmental assessment in the European Union: perspectives, past, present and strategic. *European Planning Studies* **9**: 243–54.

Barker, J. (1996) Implementing environmental impact assessment recommendations: the practical implications. Paper to International Association for Impact Assessment (South Africa Chapter) Conference, Cape Town. Mill Street: IAIA-SA.

Bartlett, R.V. (1989) Impact assessment as a policy strategy. In Bartlett, R.V. (ed.) *Policy through Impact Assessment.* New York: Greenwood Press.

Bartlett, R.V. (1997) Integrated impact assessment: the New Zealand experiment. In Caldwell, L.K. and Bartlett, R.V. (eds) *Environmental Policy: Transnational Issues and National Trends.* Westport, CT: Quorum.

Bartlett, R.V. and Baber, W.F. (1989) Bureaucracy or analysis: implications of impact assessment for public administration. In Bartlett, R.V. (ed.) *Policy through Impact Assessment.* New York: Greenwood Press.

Bartlett, R.V. and Kurian, P.A. (1999) The theory of environmental impact assessment: implicit models of policy making. *Policy and Politics* **27**: 415–34.

Bass, R.E. (1998) Evaluating environmental justice under the NEPA. *Environmental Impact Assessment Review* **18**: 83–92.

Bass, R.E. and Herson, A.I. (1999) Environmental impact assessment of land-use plans: experience under the National Environmental Policy Act and the California Environmental Quality Act. In Petts, J. (ed.) *Handbook of Environmental Impact Assessment,* vol. 2. Oxford: Blackwell Science.

Bass, R.E., Herson, A.I. and Bogdan, K.M. (1999) *CEQA Deskbook: A Step-by-step Guide on how to Comply with the California Environmental Quality Act,* 2nd edn., Point Arena, CA: Solano Press.

Bass, R.E., Herson, A.I. and Bogdan, K.M. (2001) *The NEPA Book: A Step-by-step Guide on how to Comply with the National Environmental Policy Act,* 2nd edn. Point Arena, CA: Solano Press.

Bates, G.M. (1995) *Environmental Law in Australia,* 4th edn. Sydney: Butterworths.

Baxter, W., Ross, W. and Spaling, H. (1999) To what standard? A critical evaluation of cumulative effects assessments in Canada. *Environmental Assessment* 7(2): 30–2.

Beanlands, G. (1988) Scoping methods and baseline studies in EIA. In Wathern, P. (ed.) *Environmental Impact Assessment: Theory and Practice.* London: Unwin Hyman.

Beanlands, G.E. and Duinker, P.N. (1983) *An Ecological Framework of Environmental Impact Assessment in Canada.* Hull, Quebec: Federal Environmental Assessment Review Office.

Bear, D. (1989) NEPA at 19: a primer on an 'old' law with solutions to new problems. *Environmental Law Reporter* **19**: 10060–9.

Bear, D. (1994) Using the National Environmental Policy Act to protect biological diversity. *Tulane Environmental Law Journal* **8**: 77–96.

Bear, D. and Blaug, E. (1991) Recent EIA developments in the United States of America. *EIA Newsletter* **6**: 18–19. (EIA Centre, Department of Planning and Landscape, University of Manchester.)

Bedfordshire County Council and Royal Society for the Protection of Birds (1996) *Step-by-step Guide to Environmental Appraisal.* Bedford: Bedfordshire CC.

Bell, S. and McGillivray, D. (2000) *Environmental Law,* 5th edn. London: Blackstone.

Bingham, G. (1986) *Resolving Environmental Disputes: A Decade of Experience.* Washington, DC: Conservation Foundation.

Bisset, R. (1981) Problems and issues in the implementation of EIA audits. *Environmental Impact Assessment Review* **1**: 379–96.

Bisset, R. (1988) Developments in EIA methods. In Wathern, P. (ed.) *Environmental Impact Assessment: Theory and Practice.* London: Unwin Hyman.

Bisset, R. and Tomlinson, P. (1988) Monitoring and auditing of impacts. In Wathern, P. (ed.) *Environmental Impact Assessment: Theory and Practice.* London: Unwin Hyman.

Biswas, A.K. and Agarwala, S.B.C. (eds) (1992) *Environmental Impact Assessment for Developing Countries.* Oxford: Butterworth-Heinemann.

Bjarnadóttir, H. (ed.) (2001) *A Comparative Study of Nordic EIA Systems: Similarities and Differences in National Implementation.* Stockholm: Nordregio.

Blackmore, R., Wood, C.M. and Jones, C.E. (1997) The effect of environmental assessment on UK infrastructure project planning decisions. *Planning Practice and Research* **12**: 223–38.

Blaug, E.A. (1993) Use of the environmental assessment by federal agencies in NEPA implementation. *The Environmental Professional* **15**: 57–65.

Blumm, M.C. (1990) The National Environmental Policy Act at twenty: a preface. *Environmental Law* **20**: 447–83.

Bond, A. (1997) Environmental assessment and planning: a chronology of development in England and Wales. *Journal of Environmental Planning and Management* **40**: 261–71.

Bond, A. and Wathern, P. (1999) Environmental impact assessment in the European Union. In Petts, J. (ed.) *Handbook of Environmental Impact Assessment*, vol. 2. Oxford: Blackwell.

Bregha, F., Bendickson, J., Gamble, D., Shillington, T. and Weick, E. (1990) *The Integration of Environmental Considerations into Government Policy.* Hull, Quebec: Canadian Environmental Assessment Research Council.

Broili, R.T. (1993) Identification, tracking and closure of NEPA commitments: the Tennessee Valley Authority Process. In Hildebrand, S.G. and Cannon, J.B. (eds) *Environmental Analysis: The NEPA Experience.* Boca Raton, FL: Lewis.

Brown, A.L. and McDonald, G.T. (1995) From environmental impact assessment to environmental design and planning. *Australian Journal of Environmental Management* **2**: 65–77.

Brown, A.L. and Therivel, R. (2000) Principles to guide the development of strategic environmental assessment methodology. *Impact Assessment and Project Appraisal* **18**: 183–9.

Buckley, R. (1991a) Auditing the precision and accuracy of environmental impact predictions in Australia. *Environmental Monitoring and Assessment* **18**: 1–23.

Buckley, R. (1991b) Environmental planning legislation: court back-up better than Ministerial discretion. *Environmental and Planning Law Journal* **8**: 250–7.

Buckley, R. (1998a) Improving the quality of environmental impact statements (EISs). In Porter, A.L. and Fittipaldi, J.J. (eds) *Environmental Methods Review: Retooling Impact Assessment for the New Century.* Atlanta, GA: Army Environmental Policy Institute.

Buckley, R. (1998b) Strategic environmental assessment. In Porter, A.L. and Fittipaldi, J.J. (eds) *Environmental Methods Review: Retooling Impact Assessment for the New Century.* Atlanta, GA: Army Environmental Policy Institute.

Buckley, R. (2000) Strategic environmental assessment of policies and plans: legislation and implementation. *Impact Assessment and Project Appraisal* **18**: 209–15.

Bührs, T. and Bartlett, R.V. (1993) *Environmental Policy in New Zealand: The Politics of Clean and Green.* Auckland: Oxford University Press.

Bureau of Industry Economics (1990) *Environmental Assessment – Impact on Major Projects.* Research Report 35. Canberra: AGPS.

Burger, A. and McCallum, A. (1997) The negative impacts of public involvement: observations and lessons from practice. Paper to International Association for Impact Assessment (South Africa Chapter) Conference, KwaMaritane. Mill Street: IAIA–SA.

Caldwell, L.K. (1989) Understanding impact analysis: technical process, administrative reform, policy principle. In Bartlett, R.V. (ed.) *Policy through Impact Assessment.* New York: Greenwood Press.

Caldwell, L.K. (1998) *The National Environmental Policy Act: An Agenda for the Future.* Bloomington, IN: Indiana University Press.

Caldwell, L.K., Bartlett, R.V., Parker, D.E. and Keys, D.L. (1983) *A Study of Ways to Improve the Scientific Content and Methodology of Environmental Impact Analysis.*

Document No. PB 83–222851, Springfield, VA, National Technical Information Service.

Canadian Environmental Assessment Agency:

CEAA (1994) *The Canadian Environmental Assessment Act: Responsible Authority's Guide.* Hull, Quebec: CEAA.

CEAA (1995) *The Citizen's Guide: Canadian Environmental Assessment Process.* Hull, Quebec: CEAA.

CEAA (1996) *A Guide on Biodiversity and Environmental Assessment.* Hull, Quebec: CEAA.

CEAA (1997a) *Environmental Impact Statement (EIS) Guidelines for the Review of the Voisey's Bay Mine and Mill Undertaking.* Hull, Quebec: CEAA.

CEAA (1997b) *Guide to the Preparation of a Comprehensive Study.* Hull, Quebec: CEAA.

CEAA (1997c) *Procedures for an Assessment by a Review Panel.* Hull, Quebec: CEAA.

CEAA (1998a) *Addressing 'Need for', 'Purpose of', 'Alternatives to' and 'Alternative Means' under the Canadian Environmental Assessment Act.* Operational Policy Statement. Hull, Quebec: CEAA.

CEAA (1998b) *Guide to Information Requirements for Federal Environmental Assessment of Mining Projects in Canada.* Discussion draft. Hull, Quebec: CEAA.

CEAA (1999a) *Cumulative Effects Assessment: Practitioners Guide.* Hull, Quebec: CEAA.

CEAA (1999b) *Participant Funding Program.* Hull, Quebec: CEAA.

CEAA (1999c) *Performance Report for the Period ending March 31, 1999.* Hull, Quebec: CEAA.

CEAA (1999d) *Report on the Proposed Voisey's Bay Mine and Mill Project.* Hull, Quebec: CEAA.

CEAA (1999e) *Strategic Environmental Assessment: The 1999 Cabinet Directive on the Environmental Assessment of Policy, Plan and Program Proposals; Guidelines for Implementing the Cabinet Directive.* Hull, Quebec: CEAA.

CEAA (2000) *Performance Report for the Period ending March 31, 2000.* Hull, Quebec: CEAA.

Canadian Environmental Assessment Research Association (1988) *Evaluating Environmental Impact Assessment: An Action Prospectus.* Hull, Quebec: CEARC.

Canter, L. (1993) The role of environmental management in responsible project management. *The Environmental Professional* **15**: 76–87.

Canter, L. (1996) *Environmental Impact Assessment*, 2nd edn. New York: McGraw-Hill.

Canter, L. (1997) Cumulative effects and other analytical challenges of NEPA. In Clark, E.R. and Canter, L. (eds) *Environmental Policy and NEPA: Past, Present and Future.* Boca Raton, FL: St Lucie Press.

Canter, L. and Clark, E.R. (1997) NEPA effectiveness – a survey of academics. *Environmental Impact Assessment Review* **17**: 313–27.

Carbon, B. (1998) One thousand days at the Federal Environment Protection Agency. *Australian Journal of Environmental Management* **5**: 7–8.

Carpenter, R.A. (1997) The case for continuous monitoring and adoptive management under NEPA. In Clark, E.R. and Canter, L. (eds) *Environmental Policy and NEPA: Past, Present and Future.* Boca Raton, FL: St Lucie Press.

Cass, W.H. (1974) Speech by the Minister for the Environment and Conservation, *Parliamentary Debates: House of Representatives* **92**: 4081–3, 26 November 1974, AGPS, Canberra.

Catlow, J. and Thirwall, C.G. (1976) *Environmental Impact Analysis*. Research Report 11. London: Department of the Environment.

Chadwick, A. and Glasson, J. (1999) Auditing the socio-economic impacts of a major construction project: the case of Sizewell B nuclear power station. *Journal of Environmental Planning and Management* 42: 811–36.

Clark, B.D., Chapman, K., Bisset, R. and Wathern, P. (1976) *Assessment of Major Industrial Applications: A Manual*. Research Report 13. London: Department of the Environment.

Clark, B.D., Chapman, K., Bisset, R., Wathern, P. and Barrett, M. (1981) *A Manual for the Assessment of Major Development Proposals*. London: Department of the Environment/HMSO.

Clark, E.R. (1993) The National Environmental Policy Act and the role of the President's Council on Environmental Quality. *The Environmental Professional* 15: 4–6.

Clark, E.R. (1997) NEPA: the rational approach to change. In Clark, E.R. and Canter, L. (eds) *Environmental Policy and NEPA: Past, Present and Future*. Boca Raton, FL: St Lucie Press.

Clark, E.R. (2000) Making EIA count in decisionmaking. In Partidario, M.R. and Clark, E.R. (eds) *Perspectives on Strategic Environmental Assessment*. Boca Raton, FL: Lewis.

Clark, E.R. and Canter, L. (eds) (1997) *Environmental Policy and NEPA: Past, Present and Future*. Boca Raton, FL: St Lucie Press.

Clark, E.R. and Richards, D. (1999) Environmental impact assessment in North America. In Petts, J. (ed.) *Handbook of Environmental Impact Assessment*, vol. 2. Oxford: Blackwell.

COAG (1997) *Heads of Agreement on Commonwealth/State Roles and Responsibilities for the Environment*. Canberra: Environment Australia/Department of the Environment and Heritage.

Cocklin, C., Parker, S. and Hay, J. (1992) Notes on cumulative environmental change: concepts and issues. *Journal of Environmental Management* 35: 31–49.

Coenen, F. (1999) New developments in decision making and the place of environmental assessment in the Netherlands. *Evaluation Environnementale des Plans et Programme. Amenagement et Nature* 134: 27–35.

Coenen, R. (1993) NEPA's impact on environmental impact assessment in European Community member countries. In Hildebrand, S.G. and Cannon, J.B. (eds) *Environmental Analysis: The NEPA Experience*. Boca Raton, FL: Lewis.

Commissioner of the Environment and Sustainable Development:

    CESD (1998) Environmental assessment – a critical tool for sustainable development. Chapter 6 of *Report to the House of Commons*. Ottawa: Office of the Auditor General of Canada.

    CESD (1999) Greening policies and programmes: supporting sustainable development decisions. Chapter 9 of *Report to the House of Commons*. Ottawa: Office of the Auditor General of Canada.

Commission of the European Communities:

    CEC (1977) European Community Policy and Action Programme on the Environment for 1977–1981. *Official Journal of the European Communities* C139: 1–46, 13 June 1977.

    CEC (1979) *State of the Environment: Second Report*. Brussels: CEC.

    CEC (1980) Proposal for a Council Directive concerning the assessment of the environmental effects of certain public and private projects. *Official Journal of the European Communities* C169: 14–22, 9 July 1980.

CEC (1982) Proposal to amend the proposal for a Council Directive concerning the assessment of the environmental effects of certain public and private projects. *Official Journal of the European Communities* **C110**: 5–11, 1 May 1982.

CEC (1985) Council Directive of 27 June 1985 on the assessment of the effects of certain public and private projects on the environment. *Official Journal of the European Communities* **L175**: 40–8, 5 July 1985.

CEC (1992) *Fifth Environmental Action Programme: Towards Sustainability.* Brussels: CEC.

CEC (with the assistance of Lee, N. and Jones, C.E.) (1993) *Report from the Commission of the Implementation of Directive 85/337/EEC and Annexes for all Member States.* COM(93) 28, vol. 13. Brussels: CEC.

CEC (1994) Proposal for a Council Directive amending Directive 85/337/EEC on the assessment of the effects of certain public and private projects on the environment. *Official Journal of the European Communities* **C130**: 8–13, 12 May 1994.

Commonwealth of Australia (1995) *Environment Protection (Impact of Proposals) Act 1974 Administrative Procedures.* Order under section 6 of the Act, 5 May 1995, Canberra: AGPS.

Commonwealth Environment Protection Agency:

CEPA (1992) *Inter-governmental Agreement on the Environment.* Canberra: CEPA.

CEPA (1994) *Public Review of the Commonwealth Environment Impact Assessment Process.* Discussion paper. Canberra: CEPA.

Council on Environmental Quality:

CEQ (1974) *Environmental Quality 1974: Fifth Annual Report.* Washington, DC: USGPO.

CEQ (1978) Regulations for implementing the procedural provisions of the National Environmental Quality Act. *40 Code of Federal Regulations* 1500–1508. (Reproduced in Mandelker, 2000.)

CEQ (1981a) Memorandum: forty most asked questions concerning CEQ's National Environmental Policy Act Regulations (40 questions). *46 Federal Register* 18026 (23 March 1981), as amended by *51 Federal Register* 15618 (25 April 1986). (Reproduced in Bass *et al.*, 2001, and in Mandelker, 2000.)

CEQ (1981b) *Memorandum: Scoping Guidance.* Washington, DC: CEQ. (Reproduced in Bass *et al.*, 2001.)

CEQ (1990) *Environmental Quality 1989: Twentieth Annual Report.* Washington, DC: USGPO.

CEQ (1997a) *Considering Cumulative Impacts Under the National Environmental Policy Act.* Washington, DC: CEQ.

CEQ (1997b) *Environmental Justice: Guidance under the National Environmental Policy Act.* Washington, DC: CEQ.

CEQ (1997c) *Environmental Quality: Twenty-fifth Anniversary Report.* Washington, DC: USGPO.

CEQ (1997d) *The National Environmental Policy Act: a Study of its Effectiveness after Twenty-five Years.* Washington, DC: CEQ.

CEQ (1998) *Environmental Quality 1996: Along the American River.* Washington, DC: USGPO.

CEQ (1999) *Environmental Quality 1997: The World Wide Web.* Washington, DC: USGPO.

Council for Scientific and Industrial Research (1997) *A Protocol for Strategic Environmental Assessment in South Africa.* Draft discussion document. Stellenbosch: CSIR.

Council for the Environment (1989) *Integrated Environmental Management in South Africa.* Pretoria: Council for the Environment.

Court, J., Wright, C. and Guthrie, A. (1996) Environmental assessment and sustainability: are we ready for the challenge? *Australian Journal of Environmental Management* 3: 42–56.

Craig & Ehrlich, Ross, H., Lane, M.B. and Northern Land Council (1996) *Indigenous Participation in Commonwealth Environmental Impact Assessment.* Canberra: Commonwealth Environment Protection Agency.

Culhane, P.J. (1987) The precision and accuracy of US environmental impact statements. *Environmental Monitoring and Assessment* 8: 218–38.

Culhane, P.J., Friesema, H.P. and Beecher, J.A. (1987) *Forecasts and Environmental Decision-making: The Content and Predictive Accuracy of Environmental Impact Statements.* Boulder, CO: Westview Press.

Curran, J.M., Wood, C.M. and Hilton, M. (1998) Environmental appraisal of UK development plans: current practice and future directions. *Environment and Planning B* 25: 411–33.

David Redmond and Associates (1999) *Compliance Monitoring Review.* Hull, Quebec: CEAA.

de Boer, J.J. (1999) Bilateral agreements for the application of the UN–ECE Convention on EIA in a Transboundary Context. *Environmental Impact Assessment Review* 19: 85–98.

de Jong, M-F. (1997) Procedural aspects of EIA: the role of the Dutch EIA Commission. In Environmental Law Network International (ed.) *International Environmental Impact Assessment.* London: Cameron May.

de Vries, Y (1996) The Netherlands experience. In de Boer, J.J. and Sadler, B. (eds) *Strategic Environmental Assessment: Environmental Assessment of Policies: Briefing Papers on Experience in Selected Countries.* MER Series 54. The Hague: Ministry of Housing, Spatial Planning and the Environment (VROM).

de Vries, Y. (1998) The Dutch approach: carrot and stick. *UVP Report* 4/98: 190–2.

de Vries, Y. and Tonk, J. (1997) Assessing draft regulations – the Dutch experience. *Environmental Assessment* 5(3): 37–8.

Department of the Environment:

DoE (1989) *Environmental Assessment: A Guide to the Procedures.* London: HMSO.

DoE (1991a) *Monitoring Environmental Assessment and Planning.* London: HMSO.

DoE (1991b) *Policy Appraisal and the Environment: A Guide for Government Departments.* London: HMSO.

DoE (1993) *The Environmental Appraisal of Development Plans: A Good Practice Guide.* London: HMSO.

DoE (1994a) *Evaluation of Environmental Information for Planning Projects: A Good Practice Guide.* London: HMSO.

DoE (1994b) *Good Practice on the Evaluation of Environmental Information for Planning Projects: Research Report.* London: HMSO.

DoE (1994c) *Planning and Pollution Control.* Planning Policy Guidance Note (PPG) 23. London: HMSO.

DoE (1995) *Preparation of Environmental Statements for Planning Projects that Require Environmental Assessment: A Good Practice Guide.* London: HMSO.

DoE (1996) *Changes in the Quality of Environmental Statements for Planning Projects.* London: HMSO.

Department of Environmental Affairs (1992) *Integrated Environmental Management Guideline Series,* vols 1–6. Pretoria: DEA.

Department of Environmental Affairs and Tourism:

DEAT (1998a) *A National Strategy for Integrated Environmental Management in South Africa.* Discussion document. Pretoria: DEAT.

DEAT (1998b) *Draft National Environmental Management Bill.* Pretoria: DEAT.

DEAT (1998c) *EIA Regulations; Implementation of Sections 21, 22 and 26 of the Environment Conservation Act.* Guideline document. Pretoria: DEAT.

DEAT (1999) *A User Guide to the National Environmental Management Act.* Pretoria: DEAT.

DEAT (2000a) *Law Reform and NEMA News* (April.) Pretoria: DEAT.

DEAT (2000b) *Strategic Environmental Assessment in South Africa.* Guideline document. Pretoria: DEAT.

Department of the Environment, Transport and the Regions:

DETR (1997a) *Consultation Paper: Determining the Need for Environmental Assessment (EC Directive 97/11/EC) on Environmental Assessment.* London: DETR.

DETR (1997b) *Consultation Paper: Implementation of EC Directive (97/11/EC) on Environmental Assessment.* London: DETR.

DETR (1997c) *Experience with the 'Policy Appraisal and the Environment' Initiative.* London: DETR.

DETR (1997d) *Mitigation Measures in Environmental Statements.* London: DETR.

DETR (1998a) *A New Deal for Trunk Roads in England: Guidance on the New Approach to Appraisal.* London: DETR.

DETR (1998b) *Policy Appraisal and the Environment: Policy Guidance.* London: DETR.

DETR (1998c) *Review of Technical Guidance on Environmental Appraisal.* London: The Stationery Office.

DETR (1998d) *Strategic Environmental Appraisal: Report of the International Seminar, Lincoln.* London: DETR.

DETR (1999a) *Development Plans.* Planning Policy Guidance (PPG) Note 12. London: The Stationery Office.

DETR (1999b) *Environmental Impact Assessment.* Circular 2/99. London: The Stationery Office.

DETR (2000a) *Environmental Impact Assessment: A Guide to Procedures.* Tonbridge: Thomas Telford.

DETR (2000b) *Strategic Environmental Assessment of Plans and Programmes: Proceedings of the Intergovernmental Policy Forum, Glasgow.* London: DETR.

Department of Transport (1993) *Environmental Assessment.* Design Manual for Roads and Bridges, vol. 11. London: HMSO.

Dickerson, W. and Montgomery, J. (1993) Substantive scientific and technical guidance for NEPA analysis: pitfalls in the real world. *The Environmental Professional* 15: 7–11.

Dipper, B., Jones, C. and Wood, C. (1998) Monitoring and post-auditing in environmental impact assessment: a review. *Journal of Environmental Planning and Management* 41: 731–47.

Dixon, J.E. (1993) The integration of EIA and planning in New Zealand: changing process and practice. *Journal of Environmental Planning and Management* 36: 239–51.

Dixon, J.E., Ericksen, N.J., Crawford, J.L. and Berke, P. (1997) Planning under a co-operative mandate: new plans for New Zealand. *Journal of Environmental Planning and Management* 40: 603–14.

Dixon, J. and Fookes, T. (1995) Environmental assessment in New Zealand: prospects and practical realities. *Australian Journal of Environmental Management* 2: 104–11.

Donnelly, A., Dalal-Clayton, B. and Hughes, R. (1998) *A Directory of Impact Assessment Guidelines*, 2nd edition. London: International Institute for Environment and Development.

Doyle, D. and Sadler, B. (1996) *Environmental Assessment in Canada: Frameworks, Procedures and Attributes of Effectiveness*. Hull, Quebec: CEAA.

Dresner, S. and Gilbert, N. (1999) Decision making processes for projects requiring EIA: case studies in six European countries. *Journal of Environmental Assessment and Policy Management* 1: 105–31.

Duthie, A.G. (2001) A review of provincial EIA administrative capacity in South Africa. *Impact Assessment and Project Appraisal* 19: 215–22.

Dutton, P. and Dutton, S. (1998) Conservation victory. *African Wildlife* 52(3): 23–7.

Eccleston, C.H. (1999) *The NEPA Planning Process: A Comprehensive Guide with Emphasis on Efficiency*. New York: Wiley.

Ecologically Sustainable Development Working Group Chairs (1992) *Intersectoral Issues Report*. Canberra: AGPS.

Elkin T.J. and Smith, P.G.R. (1988) What is a good environmental impact statement? Reviewing screening reports from Canada's national parks. *Journal of Environmental Management* 26: 71–89.

Environment Australia (1999a) *An Overview of the Environment Protection and Biodiversity Conservation Act*. Canberra: Department of the Environment and Heritage.

Environment Australia (1999b) *Environment Protection and Biodiversity Conservation Act Explanatory Memorandum*. Canberra: Department of the Environment and Heritage.

Environment Australia (2000a) *EPBC Act: Administrative Guidelines on Significance, July 2000*. Canberra: Department of the Environment and Heritage (and subsequent supplements)

Environment Australia (2000b) *EPBC Act: Frequently Asked Questions*. Canberra: Department of the Environment and Heritage.

Environment Australia (2000c) *EPBC Act: Regulations: Explanatory Statement*. Canberra: Department of the Environment and Heritage.

Environment Australia (2000d) *Environmental Assessment Processes*. Environment Protection and Biodiversity Conservation Act Fact Sheet 3. Canberra: Department of the Environment and Heritage.

Environment Australia (2001) *Environmental Impact Statement Guidelines: Proposed National Low Level Radioactive Waste Repository*. Reference 2001/151. Canberra: Department of the Environment and Heritage.

Environmental Impact Assessment Commission:

EIAC (ed.) (1996) *Environmental Impact Assessment in the Netherlands: Experiences and Views presented by and to the Commission for EIA*. Utrecht: EIAC.

EIAC (1998) *Information for Members of Working Groups of the Commission for Environmental Impact Assessment in the Netherlands*. Utrecht: EIAC.

EIAC (2000) *Annual Report (2000) in Support of Dutch Development Co-operation*. Utrecht: EIAC.

EIAC (2001a) *Annual Report for 2000*. Utrecht: EIAC. (in Dutch, English summary)

EIAC (ed.) (2001b) *Further Experiences on EIA in the Netherlands*. Utrecht: EIAC.

Environmental Law Institute (1981) *NEPA in Action: Environmental Offices in 19 Federal Agencies*. Washington, DC: Council on Environmental Quality.

Environmental Law Institute (1995a) *NEPA Deskbook*. Washington, DC: Environmental Law Institute.

Environmental Law Institute (1995b) *Rediscovering the National Environment Policy Act: Back to the Future.* Washington, DC: Environmental Law Institute.

Environmental Protection Agency:

EPA (1984) *Policy and Procedures for the Review of Federal Actions Impacting the Environment.* Washington, DC: Office of Federal Activities, EPA.

EPA (1993) *Sourcebook for the Environmental Assessment (EA) Process.* Washington, DC: EPA.

EPA (1999a) *Consideration of Cumulative Impacts in EPA Review of NEPA Documents.* Washington, DC: Office of Federal Activities, EPA.

EPA (1999b) *Considering Ecological Processes in Environmental Impact Assessments.* Washington, DC: Office of Federal Activities, EPA.

EPA (1999c) *Final Guidance for Consideration of Environmental Justice in Clean Air Act 309 Reviews.* Washington, DC: Office of Federal Activities, EPA.

Essex County Council (2000) *The Essex Guide to Environmental Impact Assessment,* revised edition. Chelmsford: Essex Planning Officers' Association, Essex County Council.

European Commission (1994) *Strategic Environmental Assessment: Existing Methodology.* Brussels: DGXI, EC.

European Commission (1995) *Strategic Environmental Assessment Legislation and Practice in the Community.* Brussels: DGXI, EC.

European Commission (1996a) *Environmental Impact Assessment: A Study on Costs and Benefits* (2 vols). Brussels: DGXI, EC.

European Commission (1996b) *Evaluation of the Performance of the EIA Process* (2 vols). Brussels: DGXI, EC.

European Commission (1997a) *A Study to Develop and Implement an Overall Strategy for EIA/SEA Research in the EU.* Brussels: DGXI, EC.

European Commission (1997b) *Case Studies on Strategic Environmental Assessment.* Brussels: DGXI, EC.

European Commission (1997c) *Concise Revision of the Report from the Commission of the Implementation of Directive 85/337/EEC on the Assessment of the Effects of Certain Public and Private Projects on the Environment.* Brussels: DGXI, EC.

European Commission (1997d) Council Directive 97/11/EC of 3 March 1997 amending Directive 85/337/EEC on the assessment of the effects of certain public and private projects on the environment. *Official Journal of the European Communities* **L73**: 5–15, 14 March 1997.

European Commission (1998) *IMPEL Report on the Interrelationship between IPPC, EIA, SEVESO Directives and EMAS Regulations.* Brussels: DGXI, EC.

European Commission (1999a) *Guidelines for the Assessment of Indirect and Cumulative Impacts as well as Impact Interactions.* Brussels: DGXI, EC.

European Commission (1999b) *Manual on Strategic Environmental Assessment of Transport Infrastructure Plans.* Brussels: DGVII, EC.

European Commission (2001a) Directive 2001/42/EC of the European Parliament and of the Council of 27 June 2001 on the assessment of the effects of certain plans and programmes on the environment. *Official Journal of the European Communities* **L197**: 30–7, 21 July 2001.

European Commission (2001b) *EIS Review,* 2nd edn. Guidance on EIA. Brussels: DG Environment, EC.

European Commission (2001c) *Screening,* 2nd edn. Guidance on EIA. Brussels: DG Environment, EC.

European Commission (2001d) *Scoping*, 2nd edn. Guidance on EIA. Brussels: DG Environment.

European Commission (2001e) Proposal for a Directive of the European Parliament and of the Council providing for public participation in respect of the drawing up of certain plans and programmes relating to the environment and amending Council Directives 85/337/EEC and 96/61/EC. COM (2000) 839 final, 18 January 2001. Brussels: DG Environment, EC.

European Commission (2001f) *SEA and Integration of the Environment into Strategic Decision-making* (3 vols). Brussels: DG Environment, EC.

Evaluation Committee on the Environmental Protection Act:

    ECW (1990) *Towards a Better Procedure to Protect the Environment: Report on the Working of the Regulation on Environmental Impact Assessment Contained in the Environmental Protection (General Provisions) Act: Summary*. Advisory Report 3. The Hague: Ministry of Housing, Spatial Planning and Environment (VROM).

    ECW (1996) *Towards a Sustainable System of Environmental Impact Assessment: Second Advisory Report on the EIA Regulations contained in the Environmental Management Act: Summary*. Advisory Report 9. The Hague: Ministry of Housing, Spatial Planning and the Environment (VROM).

Fairfax, S.K. (1978) A disaster in the environmental movement. *Science* 199: 743–8.

Fairfax, S.K. and Ingram, H.M. (1981) The United States experience. In O'Riordan, T. and Sewell, W.R.D. (eds) *Project Appraisal and Policy Review*. Chichester: Wiley.

Federal Environmental Assessment Review Office (1988) *The National Consultation Workshop on Federal Environmental Assessment Reform: Report of Proceedings*. Hull, Quebec: FEARO.

Federal Environmental Assessment Review Office (1992) *Oldman River Dam: Report of the Environmental Assessment Panel*. Report 42. Hull, Quebec: FEARO.

Federal Environmental Assessment Review Office (1993) *The Environmental Assessment Process for Policy and Program Proposals*. Hull, Quebec: FEARO.

Feldmann, L., Vanderhaegen, M. and Pirotte, C. (2001) The EU's SEA Directive: status and links to integration and sustainable development. *Environmental Impact Assessment Review* 21: 203–22.

Fenge, T. and Smith, L.G. (1986) Reforming the Federal Environmental Assessment and Review Process. *Canadian Public Policy* 12: 596–605.

Fischer, T., Wood, C.M. and Jones, C.E. (2002) Policy, plan and programme environmental assessment in England, the Netherlands and Germany: practice and prospects. *Environment and Planning B*, 29: 159–72.

Fogleman, V.M. (1990) *Guide to the National Environment Policy Act: Interpretations, Applications and Compliance*. New York: Quorum.

Fookes, T. (2000) Environmental assessment under the Resource Management Act 1991. In Memon, P.A. and Perkins, H.C. (eds) *Environmental Planning in New Zealand*, 2nd edn. Palmerston North: Dunmore Press.

Fowler, R.J. (1982*) Environmental Impact Assessment, Planning and Pollution Measures in Australia*. Canberra: Department of Home Affairs and Environment, AGPS.

Fowler, R.J. (1985) Legislative bases of environmental impact assessment. *Environmental and Planning Law Journal* 2: 200–5.

Fowler, R.J. (1996) Environmental impact assessment: what role for the Commonwealth? An overview. *Environmental and Planning Law Journal* 13: 246–59.

Frieder, J. (1997) *Approaching Sustainability: Integrated Environmental Management and New Zealand's Resource Management Act.* Wellington: New Zealand – United States Educational Foundation.

Frost, R. (1997) EIA monitoring and audit. In Weston, J (ed.) *Planning and Environmental Impact Assessment in Practice.* Harlow: Longman.

Fuggle, R.F. (1996) Integrated environmental management in South Africa: the conceptual underpinning. Keynote address, International Association for Impact Assessment Annual Conference, Durban. In Wood, C.M., Wynberg, R. and Raimondo, J. (eds) *Involving People in the Management of Change.* Fargo, ND: IAIA.

Fuggle, R.F. (1998) EIA in South Africa. Paper to International Association for Impact Assessment Conference, Christchurch. Fargo, ND: IAIA.

Fuggle, R.F. and Rabie, M.A. (1996) Postscript. In Fuggle, R.F. and Rabie, M.A. (eds) *Environmental Management in South Africa.* Cape Town: Juta.

Gasson, B. and Todeschini, F. (1997) Settlement planning and integrated environmental management compared: some lessons. Paper to International Association for Impact Assessment (South Africa Chapter) Conference, KwaMaritane. Mill Street: IAIA-SA.

Gibb Africa (1998) *Summary of Draft Environmental Impact Report for the Maputo Iron and Steel Project.* Capetown: Gibb Africa.

Gibson, R.B. (1993) Environmental assessment design: lessons from the Canadian experience. *The Environmental Professional* **15**: 12–24.

Gibson, R.B. (2000) Favouring the higher test: contribution to sustainability as the central criterion for reviews and decisions under the Canadian Environmental Assessment Act. *Journal of Environmental Law and Practice* **10**: 39–54.

Glasson, J. (1994) Life after the decision: the importance of monitoring in EIA. *Built Environment* **20**: 309–19.

Glasson, J. (1999a) Environmental impact assessment – impact on decisions. In Petts, J. (ed.) *Handbook of Environmental Impact Assessment*, vol. 1. Oxford: Blackwell.

Glasson, J. (1999b) The first 10 years of the UK EIA system: strengths, weaknesses, opportunities and threats. *Planning Practice and Research* **14**: 363–75.

Glasson, J., Therivel, R., Weston, J., Wilson, E. and Frost, R. (1997) EIA – learning from experience: changes in the quality of environmental impact statements for UK planning projects. *Journal of Environmental Planning and Management* **40**: 451–64.

Glasson, J., Therivel, R. and Chadwick, A. (1999*) Introduction to Environmental Impact Assessment*, 2nd edn. London: UCL Press.

Glazewski, J. (2000) *Environmental Law in South Africa.* Durban: Butterworths.

Goudie, S. and Kilian, D. (1996) Gender and environmental impact assessment. *Agenda* **29**: 43–54.

Government of Canada (1984) *Environmental Assessment and Review Process Guidelines Order. Canada Gazette* Part II, **118**(14): 2794–802.

Government of Canada (1992) *North America Free Trade Agreement: Canadian Environment Review.* Hull, Quebec: Environment Canada.

Government of Canada (2001) *Bill C-19: An Act to amend the Canadian Environmental Assessment Act.* Ottawa: House of Commons.

Granger, S. (1998) Application of the South African environmental regulations – a local government perspective. Paper to International Association for Impact Assessment Conference, Christchurch. Fargo, ND: IAIA.

Hagler Bailly (1999) *On-going Monitoring Program of the Canadian Environmental Assessment Act.* Hull, Quebec: Canadian Environmental Assessment Agency.

Haigh, N. (1991) *Manual of European Policy: the EC and Britain.* Harlow: Longman.

Hart, S.L. (1984) The costs of environmental review: assessment methods and trends. In Hart, S.L., Enk, G.A. and Hornick, W.F. (eds) *Improving Impact Assessment: Increasing the Relevance and Utilisation of Scientific and Technical Information.* Boulder, CO: Westview Press.

Hart, S.L. and Enk, G.A. (1980) *Green Goals and Greenbacks: State-level Environmental Programs and their Associated Costs.* Boulder, CO: Westview Press.

Hart, S.L., Enk, G.A. and Hornick, W.F. (eds) (1984) *Improving Impact Assessment: Increasing the Relevance and Utilisation of Scientific and Technical Information.* Boulder, CO: Westview Press.

Harvey, N. (1994) Timing of environmental impact assessment. *Australian Planner* **31**: 125–30.

Harvey, N. (1998) *Environmental Impact Assessment: Procedures, Practice and Prospects in Australia.* Melbourne: Oxford University Press.

Harvey, N. (2000) Strategic environmental assessment in coastal zones, especially Australia's. *Impact Assessment and Project Appraisal* **18**: 225–32.

Hazell, S. (1999) *Canada v. the Environment: Federal Environmental Assessment 1984–1998.* Toronto: Canadian Environmental Defence Fund.

Her Majesty's Government (1999) *A Better Quality of Life: A Strategy for Sustainable Development for the UK.* Cm 4345. London: The Stationery Office.

Hildebrand, S.G. and Cannon, J.B. (eds) (1993) *Environmental Analysis: The NEPA Experience.* Boca Raton, FL: Lewis.

Hill, R. (1998) *Reform of Commonwealth Environment Legislation.* Ministerial consultation paper. Canberra: Department of the Environment.

Hill, R.C. (2000) Integrated environmental management systems in the implementation of projects. *South African Journal of Science* **96**: 50–4.

Holland, K.M. (1996) The role of the courts in the making and administration of environmental policy in the United States. In Holland, K.M., Morton, F.L. and Galligan, B. (eds) *Federalism and the Environment: Environmental Policymaking in Australia, Canada, and the United States.* Westport, CN: Greenwood Press.

Holland, K.M., Martin, F.L. and Galligan, B. (eds) (1996) *Federalism and the Environment: Environmental Policy Making in Australia, Canada and the United States.* Westport, CN: Greenwood Press.

Hollick, M. (1981) Enforcement of mitigation measures resulting from environmental impact assessment. *Environmental Management* **5**: 507–13.

Hollick, M. (1986) Environmental impact assessment: an international evaluation. *Environmental Management* **10**: 157–78.

Holling, C.S. (ed.) (1978) *Adaptive Environmental Assessment and Management.* Chichester: Wiley.

Horton, S. and Memon, A. (1997) SEA: the uneven development of the environment? *Environmental Impact Assessment Review* **17**: 163–75.

House of Lords (1981) *Environmental Assessment of Projects.* Select Committee on the European Communities, 11th Report, Session 1980–81. London: HMSO.

Hughes, H.R. (1996) Simultaneous preparation and review: a new approach to environmental assessments in New Zealand. *Impact Assessment* **14**: 97–103.

Hughes, L. (1999) Environmental impact assessment in the Environment Protection and Biodiversity Act 1999 (Cth). *Environmental and Planning Law Journal* **16**: 441–67.

Hyman, E.L. and Stiftel, B. (1988) *Combining Facts and Values in Environmental Impact Assessment: Theories and Techniques.* Boulder, CO: Westview Press.

Institute of Environmental Assessment (1993) *Digest of Environmental Statements* (2 vols). London: Sweet & Maxwell.

International Union for Conservation of Nature and Natural Resources (1980) *World Conservation Strategy – Living Resource Conservation for Sustainable Development.* Switzerland: IUCN, Gland.

Jacobs, P. and Sadler, B. (eds) (1990) *Sustainable Development and Environmental Assessment: Perspectives and Planning for a Common Future.* Hull, Quebec: CEARC.

Jambrich, T., Lindsay, R. and Macartney, P. (1992) *Living with our Decisions: Commonwealth Environmental Impact Assessment Processes.* Audit Report 10, 1992–93. Canberra: Australian National Audit Office.

James, D. (1995) Environmental impact assessment: improving processes and techniques. *Australian Journal of Environmental Management* 2: 78–89.

Jeffery, M.I. (1987) Accommodating negotiation in environmental impact assessment and project approval processes. *Environmental and Planning Law Journal* 4: 244–52.

Jessee, L. (1998) The National Environmental Policy Act Net (NEPAnet) and DoE NEPA web: what they bring to environmental impact assessment. *Environmental Impact Assessment Review* 18: 73–82.

Jones, C.E. (1999) Screening, scoping and consideration of alternatives. In Petts, J. (ed.) *Handbook of Environmental Impact Assessment*, vol. 1. Oxford: Blackwell.

Jones, C.E. and Wood, C.M. (1995) The impact of environmental assessment on public inquiry decisions. *Journal of Planning and Environment Law* [1995]: 890–904.

Jones, C.E., Wood, C.M. and Dipper, B. (1998) Environmental assessment in the UK planning process. *Town Planning Review* 69: 315–39.

Jones, M.G. (1984) The evolving EIA procedure in the Netherlands. In Clark, B.D., Gilad, A., Bisset, R. and Tomlinson, P. (eds) *Perspectives on Environmental Impact Assessment.* Dordrecht: Reidel.

Joughin, J. (1997) Decision-driven environmental assessment. Paper to International Association for Impact Assessment (South Africa Chapter) Conference, KwaMaritane. Mill Street: IAIA-SA.

Kennedy, W.V. (1988) Environmental impact assessment in North America, Western Europe: what has worked where, how and why? *International Environment Reporter* 11: 257–62.

Khan, F. (1998) Public participation and environmental decision-making in South Africa – the Frankdale environmental health project. *South African Geographical Journal* 80: 73–80.

King, T.F. (1998) How the archaeologists stole culture: a gap in American environmental impact assessment practice and how to fill it. *Environmental Impact Assessment Review* 18: 117–33.

Kinhill Engineers (1994) *Public Participation in the Environmental Impact Assessment Process.* Canberra: CEPA.

Kjellerup, U. (1999) Significance determination: a rational reconstruction of decisions. *Environmental Impact Assessment Review* 19: 3–20.

Kleinschmidt, V. and Wagner, D. (eds) (1998) *Strategic Environmental Assessment in Europe.* London: Kluwer.

Kornov, L. and Thissen, W.A. (2000) Rationality in decision and policy making: implications for strategic environmental assessment. *Impact Assessment and Project Appraisal* 18: 191–200.

Kramer, L. (2000) *EC Environmental Law*, 4th edn. London: Sweet & Maxwell.

Krawetz, N.M., MacDonald, W.R. and Nichols, P. (1987) *A Framework for Effective Monitoring*. Hull, Quebec: CEARC.

Kreske, D.L. (1996) *Environmental Impact Statements: A Practical Guide for Agencies, Citizens and Consultants*. New York: Wiley.

Kreuser, P. and Hammersley, R. (1999) Assessing the assessments: British planning authorities and the review of environmental statements. *Journal of Environmental Assessment Policy and Management* 1: 369–88.

Lawrence, D.P. (1994) Designing and adapting the EIA planning process. *The Environmental Professional* 16: 2–21.

Lee, N. (1995) Environmental assessment in the European Union: a tenth anniversary. *Project Appraisal* 10: 77–90.

Lee, N. and Brown, D. (1992) Quality control in environmental assessment. *Project Appraisal* 7: 41–5.

Lee, N. and George, C (eds) (2000) *Environmental Assessment in Developing and Transitional Countries*. Chichester: Wiley.

Lee, N. and Walsh, F. (1992) Strategic environmental assessment: an overview. *Project Appraisal* 7: 126–36.

Lee, N. and Wood, C.M. (1976) *The Introduction of Environmental Impact Statements in the European Community*. ENV/197/76. Brussels: CEC.

Lee, N. and Wood, C.M. (1978a) EIA – a European perspective. *Built Environment* 4: 101–10.

Lee, N. and Wood, C.M. (1978b) Environmental impact assessment of projects in EEC countries. *Journal of Environmental Management* 6: 57–71.

Lee, N., Colley, R., Bonde, J. and Simpson, J. (1999) *Reviewing the Quality of Environmental Statements and Environmental Appraisals*. Occasional paper 55. University of Manchester: School of Planning and Landscape.

Leu, W.-S., Williams, W.P. and Bark, A.W. (1995) An evaluation of the implementation of environmental assessment by UK local authorities. *Project Appraisal* 10: 91–102.

Leu, W.-S., Williams, W.P. and Bark, A.W. (1996) Quality control mechanisms and environmental impact assessment effectiveness with special reference to the UK. *Project Appraisal* 11: 2–12.

Lundquist, L.J. (1978) The comparative study of environmental politics: from garbage to gold? *International Journal of Environmental Studies* 12: 89–97.

Lynn, S. and Wathern, P. (1991) Intervenor funding in the environmental assessment process in Canada. *Project Appraisal* 5: 169–73.

MacLaren, V.W. and Whitney, J.B. (eds) (1985) *New Directions in Environmental Impact Assessment in Canada*. Toronto: Methuen.

Mafune, I., McLean, B., Rodkin, H. and Hill, R.C. (1997) The early years of EA in South Africa: a review of case studies from 1971 to 1986. Paper to International Association for Impact Assessment (South Africa Chapter) Conference, KwaMaritane. Mill Street: IAIA-SA.

Malik, M. and Bartlett, R.V. (1993) Formal guidance for the use of science in EIA: analysis of agency procedures for implementing NEPA. *The Environmental Professional* 15: 34–45.

Mandelker, D.R. (1993) Environmental policy: the next generation. *Town Planning Review* 64: 107–17.

Mandelker, D.R. (2000) *NEPA Law and Litigation*, 2nd edn, release 8. St Paul, MN: West Publishing.

Marriott, B. (1997) *Environmental Impact Assessment: A Practical Guide.* New York: McGraw Hill.

Marsden, S. (1998a) Importance of context in measuring the effectiveness of strategic environmental assessment. *Impact Assessment and Project Appraisal* **16**: 255–66.

Marsden, S. (1998b) Why is legislative EA in Canada ineffective, and how can it be enhanced? *Environmental Impact Assessment Review* **18**: 241–65.

McCold, L.N. and Saulsbury, J.W. (1998) Defining the no-action alternative for National Environmental Policy Act analyses of continuing actions. *Environmental Impact Assessment Review* **18**: 15–37.

McNab, A. (1997) Scoping and public participation. In Weston, J. (ed.) *Planning and Environmental Impact Assessment in Practice.* Harlow: Longman.

McShane, O. (1998) *Land Use Control under the Resource Management Act.* Wellington: Ministry for the Environment.

Memon, P.A. (1993) *Keeping New Zealand Green: Recent Environmental Reforms.* Dunedin: University of Otago Press.

Minister of the Environment (2001) *Strengthening Environmental Assessment for Canadians: Report of the Minister of the Environment to the Parliament of Canada on the Review of the Canadian Environmental Assessment Act.* Hull, Quebec: Canadian Environmental Assessment Agency.

Ministry for the Environment:

MfE (1987) *Environmental Protection and Enhancement Procedures.* Wellington: MfE.

MfE (1992a) *Hinengaro Bay: A Fictional Case Study of Policy and Plan Making.* Wellington: MfE.

MfE (1992b) *Scoping of Environmental Effects: A Guide to Scoping and Public Review Methods in Environmental Assessment.* Wellington: MfE.

MfE (1993) *Section 32: A Guide to Good Practice.* Wellington: MfE.

MfE (1998) *Proposals for Amendment to the Resource Management Act.* Wellington: MfE.

MfE (1999a) *A Guide to Preparing a Basic AEE.* Wellington: MfE.

MfE (1999b) *Auditing Assessments of Environmental Effects (AEEs): A Good Practice Guide.* Wellington: MfE.

MfE (1999c) *Striking a Balance: A Practice Guide on Consultation and Communication for Project Advocates,* Wellington: MfE.

MfE (2000a) *Resource Management Act 1991: Annual Survey of Local Authorities 1998/1999.* Wellington: MfE.

MfE (2000b) *What are the Options? A Guide to Using Section 32 of the Resource Management Act.* Wellington: MfE.

MfE (2001a) *A Template for Quality Processing of Resource Consents.* Wellington: MfE.

MfE (2001b) *Effective and Enforceable Consent Conditions.* Wellington: MfE.

MfE (2001c) *The Resource Management Act and You: Getting in on the Act.* Wellington: MfE.

MfE (2001d) *Resource Management Act 1991: Annual Survey of Local Authorities 1999/2000.* Wellington: MfE.

Ministry of Housing, Spatial Planning and Environment and Ministry of Agriculture, Nature Management and Fisheries (1991) *Environmental Impact Assessment: The Netherlands: Fit for Future Life.* The Hague: VROM.

Ministry of Housing, Spatial Planning and Environment and Ministry of Agriculture, Nature Management and Fisheries (1994a) *The Quality of Environmental Impact Statements.* MER Series 47. The Hague: VROM.

Ministry of Housing, Spatial Planning and Environment and Ministry of Agriculture, Nature Management and Fisheries (1994b) *Use and Effectiveness of Environmental Impact Assessments in Decision-Making.* MER Series 49. The Hague: VROM.

Montz, B.E., and Dixon, J.E. (1993) From law to practice: EIA in New Zealand. *Environmental Impact Assessment Review* 13: 89–108.

Morel, S.A.A. (1996) Success of EIA in preventing, mitigating and compensating negative environmental impacts of projects. In *Environmental Impact Assessment in the Netherlands: Experiences and Views presented by and to the Commission for EIA.* Utrecht: Environmental Impact Assessment Commission.

Morel, S.A.A. (1998) Integrating environmental objectives in the planning of natural gas exploration drillings in sensitive areas in the Netherlands: the North Sea coastal zone and the Wadden Sea. In *New Experiences on EIA in the Netherlands: Process, Methodology, Case Studies.* Utrecht: Environmental Impact Assessment Commission.

Morgan, R.K. (1988) Reshaping environmental impact assessment in New Zealand. *Environmental Impact Assessment Review* 8: 293–306.

Morgan, R.K. (1995) Progress with implementing the environmental assessment requirements of the Resource Management Act in New Zealand. *Journal of Environmental Planning and Management* 38: 333–48.

Morgan, R.K. (1998) *Environmental Impact Assessment: A Methodological Perspective.* London: Chapman & Hall.

Morgan, R.K. (2000a) *A Structured Approach to Reviewing AEEs in New Zealand,* 2nd edn. Publication 3. Dunedin: Centre for Impact Assessment Research and Training, University of Otago.

Morgan, R.K. (2000b) The practice of environmental impact assessment in New Zealand: problems and prospects. In Memon, P.A. and Perkins, H.C. (eds) *Environmental Planning in New Zealand,* 2nd edn. Palmerston North: Dunmore Press.

Morgan, R.K. (2001) Evaluation in the AEE process: issues, processes and methods. In Lumsden, J. (ed.) *Assessment of Environmental Effects: Information, Evaluation and Outcomes.* Christchurch: Centre for Advanced Engineering, University of Canterbury.

Morgan, R.K. and Memon, P.A. (1993) *Assessing the Environmental Effects of Major Projects: A Practical Guide.* Publication 4. Dunedin: Environmental Policy and Management Research Centre, University of Otago.

Morris, P. and Therivel, R. (eds) (2001) *Methods of Environmental Impact Assessment,* 2nd edn. London: Spon.

Morrison-Saunders, A. (1996) Environmental impact assessment as a tool for ongoing environmental management. *Project Appraisal* 11: 95–104.

Morrison-Saunders, A. and Bailey, J. (2000) Transparency in environmental impact assessment decision-making: recent developments in Western Australia. *Impact Assessment and Project Appraisal* 18: 260–70.

Mostert, E. (1995) *Commissions for Environmental Impact Assessment.* Delft: Delft University Press.

Mostert, E. (1996) Subjective environmental impact assessment: causes, problems and solutions. *Impact Assessment* 14: 191–213.

Mulvihill, P.R. and Jacobs, P. (1998) Using scoping as a design process. *Environmental Impact Assessment Review* 18: 351–69.

Münchenberg, S. (1994) Judicial review and the Commonwealth Environment Protection (Impact of Proposals) Act 1974. *Environmental and Planning Law Journal* 11: 461–78.

Munro, D.A., Bryant, T.J. and Matte-Barker, A. (1986) *Learning from Experience: A State-of-the-Art Review and Evaluation of Environmental Impact Assessment Audits.* Hull, Quebec: Canadian Environmental Assessment Research Council.

National Academy of Public Administration (1998) *Managing NEPA at the Department of Energy.* Washington, DC: Office of NEPA Policy and Assistance, US Department of Energy.

Nay Htun (1988) The EIA process in Asia and the Pacific region. In Wathern, P. (ed.) *Environmental Impact Assessment: Theory and Practice.* London: Allen & Unwin.

Niekerk, F. and Arts, J. (1996) Impact assessments for infrastructure planning: towards better timing and integration. *Project Appraisal* **11**: 237–46.

Niekerk, F. and Voogd, H. (1999) Impact assessment in Dutch infrastructure planning: some Dutch dilemmas. *Environmental Impact Assessment Review* **19**: 21–36.

Nikiforuk, A. (1997) *'The Nasty Game': The Failure of Environmental Assessment in Canada.* Toronto: Walter and Duncan Gordon Foundation.

Noble, B.F. (2000a) Strategic environmental assessment: what is it? And what makes it strategic? *Journal of Environmental Assessment and Policy Management* **2**: 203–24.

Noble, B.F. (2000b) Strengthening EIA through adaptive management: a systems perspective. *Environmental Impact Assessment Review* **20**: 97–111.

Northey, R. (1994) *The 1995 Annotated Canadian Environmental Assessment Act.* Toronto: Carswell.

Offringa, P. (1997) Creating a user-friendly NEPA. In Clark, E.R. and Canter, L. (eds) *Environmental Policy and NEPA: Past, Present and Future.* Boca Raton, FL: St Lucie Press.

Ogle, L. (2000) The Environment Protection and Biodiversity Conservation Act 1999 (Cth): how workable is it? *Environment and Planning Law Journal* **17**: 468–77.

Olshansky, R.B. (1996) The California Environmental Quality Act and local planning. *Journal of the American Planning Association* **62**: 313–30.

Organisation for Economic Cooperation and Development (1992) *Good Practices for Environmental Impact Assessment of Development Projects.* Paris: Development Assistance Committee, OECD.

O'Riordan, T. (1998) Sustainability for survival in South Africa. *Global Environmental Change* **8**: 99–108.

Ortolano, L. (1993) Controls on project proponents and environmental impact assessment effectiveness. *The Environmental Professional* **15**: 352–63.

Ortolano, L. (1997) *Environmental Regulation and Impact Assessment.* New York: Wiley.

Ortolano, L., Jenkins, B. and Abracosa, R.P. (1987) Speculations on when and why EIA is effective. *Environmental Impact Assessment Review* **7**: 285–92.

Padgett, R. and Kriwoken, L.K. (2001) The Australian Environment Protection and Biodiversity Conservation Act 1999: what role for the Commonwealth in environmental impact assessment? *Australian Journal of Environmental Management* **8**: 25–36.

Parliamentary Commissioner for the Environment:

    PCE (1995) *Assessment of Environmental Effects (AEE): Administration by Three Territorial Authorities.* Wellington: Office of the PCE.

    PCE (1996a) *Administration of Compliance with Resource Consents: Report of an Investigation of Three Councils.* Wellington: Office of the PCE.

    PCE (1996b) *Public Participation under the Resource Management Act 1991: the Management of Conflict.* Wellington: Office of the PCE.

PCE (1998) *Towards Sustainable Development, the Role of the Resource Management Act 1991*. Wellington: Office of the PCE.

Partidario, M.R. (1999) Strategic environmental assessment – principles and potential. In Petts, J. (ed.) *Handbook of Environmental Impact Assessment*, vol. 1. Oxford: Blackwell.

Partidario, M.R. (2000) Elements of an SEA framework – improving the added value of SEA. *Environmental Impact Assessment Review* 20: 647–63.

Partidario, M.R. and Clark, E.R. (2000) *Perspectives on Strategic Environmental Assessment*. Boca Raton, FL: Lewis.

Peckham, B. (1997) Environmental impact assessments in South African law. *South African Journal of Environmental and Planning Law* 4: 113–33.

Pendall, R. (1998) Problems and prospects in local environmental assessment: lessons from the United States. *Journal of Environmental Planning and Management* 41: 5–23.

Peterson, E.B., Chan, Y.H., Peterson, N.M., Constable, G.S., Caton, R.B., Davis, C.S., Wallace, R.R. and Yarranton, G.A. (1987) *Cumulative Effects Assessment in Canada: an Agenda for Action and Research*. Hull, Quebec: CEARC.

Petts, J. (1999) Public participation and environmental impact assessment. In Petts, J. (ed.) *Handbook of Environmental Impact Assessment*, vol. 1. Oxford: Blackwell.

Porter, C.F. (1985) *Environmental Impact Assessment: A Practical Guide*. St Lucia: University of Queensland Press.

Preston, G.R., Robins, N. and Fuggle, R.F. (1996) Integrated environmental management. In Fuggle, R.F. and Rabie, M.A. (eds) *Environmental Management in South Africa*. Cape Town: Juta.

Read, R. (1997) Planning authority review. In Weston, J. (ed.) *Planning and Environmental Impact Assessment in Practice*. Harlow: Longman.

Rees, W.E. (1987) Introduction: a rationale for northern land-use planning. In Fenge, T. and Rees, W.E. (eds) *Hinterland or Homeland: Land-use Planning in Northern Canada*. Ottawa: Canadian Arctic Resources Committee.

Rees, W.E. (1995) Cumulative environmental assessment and global change. *Environmental Impact Assessment Review* 15: 295–309.

Regulatory Advisory Committee (2000) *Five-year Review of the Canadian Environmental Assessment Act*. Hull, Quebec: CEAA.

Renton, S. and Bailey, J. (2000) Policy development and the environment. *Impact Assessment and Project Appraisal* 18: 245–51.

Republic of South Africa (1997) *Environmental Impact Assessment Regulations (Environment Conservation Act 1989)*. Nos R 1182–1184, *Government Gazette* 387, no. 18621, 5 September 1997.

Republic of South Africa (1998) *White Paper on Environmental Management Policy for South Africa*. *Government Gazette* 395, no. 18894, 15 May 1998.

RIAS and Gartner Lee (2000) *Comparative Analysis of Impacts on Competitiveness of Environmental Assessment Requirements*. Hull, Quebec: Canadian Environmental Assessment Agency.

Richardson, B.J. and Boer, B.W. (1995) Federal public inquiries and environmental assessment. *Australian Journal of Environmental Management* 2: 90–103.

Ridl, J. (1994) 'IEM': lip service and licence? *South African Journal of Environmental and Planning Law* [1994] (1): 61–83.

Roberts, R. (1995) Public involvement: from consultation to participation. In Vanclay, F. and Bronstein, D.A. (eds) *Environmental and Social Impact Assessment*. Chichester: Wiley.

Ross, W.A. (1987) Evaluating environmental impact statements. *Journal of Environmental Management* 25: 137–47.

Ross, W.A. (1998) Cumulative effects assessment: learning from Canadian case studies. *Impact Assessment and Project Appraisal* 16: 267–76.

Ross, W.A. (2000) Reflections of an environmental assessment panel member. *Impact Assessment and Project Appraisal* 18: 91–8.

Rossouw, N., Audoin, M., Lochner, P., Heather-Clark, S. and Wiseman, K. (2000) Development of strategic environmental assessment in South Africa. *Impact Assessment and Project Appraisal* 18: 217–23.

Royal Commission on Environmental Pollution (1976) *Fifth Report: Air Pollution Control: an Integrated Approach.* Cm 6371. London: HMSO.

Royal Commission on Environmental Pollution (1988) *Twelfth Report: Best Practicable Environmental Option.* Cm 310, London: HMSO.

Russell, S. (1999) Environmental appraisal of development plans. *Town Planning Review* 70: 529–46.

Sadar, M.H. and Stolte, W.J. (1996) An overview of the Canadian experience in environmental impact assessment. *Impact Assessment* 14: 215–28.

Sadler, B. (1988) The evaluation of assessment: post-EIS research and process development. In Wathern, P. (ed.) *Environmental Impact Assessment: Theory and Practice.* London: Unwin Hyman.

Sadler, B. (1994) Mediation provisions and options in Canadian environmental assessment. *Environmental Impact Assessment Review* 13: 375–90.

Sadler, B. (1995) Canadian experience with environmental assessment: recent changes in process and practice. *Australian Journal of Environmental Management* 2: 112–29.

Sadler, B. (1996) *Environmental Assessment in a Changing World: Evaluating Practice to Improve Performance.* Final Report, International Study of the Effectiveness of Environmental Assessment. Hull, Quebec: CEAA.

Sadler, B. (1998) Ex-post evaluation of the effectiveness of environmental assessment. In Porter, A.L. and Fittipaldi, J.J. (eds) (1998) *Environmental Methods Review: Retooling Impact Assessment for the New Century.* Atlanta, GA: Army Environmental Policy Institute.

Sadler, B. and Verheem, R. (1996) *Strategic Environmental Assessment: Status, Challenges and Future Directions.* MER Series. The Hague, Ministry of Housing, Spatial Planning and the Environment (VROM).

Scanlon, J. and Dyson, M. (2001) Will practice hinder principle? Implementing the EPBC Act. *Environmental and Planning Law Journal* 18: 14–22.

Scholten, J.J. (1997) Reviewing EISs – the Netherlands experience. *Environmental Assessment* 5(1): 24–5.

Scholten, J.J. (1998) The role of EIA in land reclamation of a wetland area for urban expansion of Amsterdam. In *New Experiences on EIA in the Netherlands: Process, Methodology, Case Studies.* Utrecht: Environment Impact Assessment Commission.

Scholten, J.J. and Bonte, R.J. (1994) Independent reviewing in environmental impact assessment in the Netherlands. In *EIA Methodology in the Netherlands: Views of the Commission for EIA.* Utrecht: Environment Impact Assessment Commission.

Scholten, J.J. and van Eck, M. (1994) Reviewing environmental impact statements. In *EIA Methodology in the Netherlands: Views of the Commission for EIA.* Utrecht: Environment Impact Assessment Commission.

Scottish Executive (1999) *Environmental Impact Assessment.* Planning Advice Note 58. Edinburgh: Scottish Executive.

Sewell, W.R.D. (1981) How Canada responded: the Berger Inquiry. In O'Riordan, T. and Sewell, W.R.D. (eds) *Project Appraisal and Policy Review*. Chichester: Wiley.

Sewell, W.R.D. and Coppock, J.T. (eds) (1976) *Public Participation in Planning*. Chichester: Wiley.

Sewell, G.H. and Korrick, S. (1984) The fate of EIS projects: a retrospective study. In Hart, S.L., Enk, G.A. and Hornick, W.F. (eds) (1984) *Improving Impact Assessment: Increasing the Relevance and Utilisation of Scientific and Technical Information*. Boulder, CO: Westview Press.

Sheate, W. (1996) *Environmental Impact Assessment: Law and Policy: Making an Impact II*. London: Cameron May.

Sheate, W. (1997) The environmental impact assessment amendment Directive 97/11/EC – a small step forward? *European Environmental Law Review* 6: 235–43.

Sheate, W.R. and Macrory, R.B. (1989) Agriculture and the EC environmental assessment directive: lessons for community policy making. *Journal of Common Market Studies* 28: 68–81.

Shepherd, A. (1998) Post-project impact assessment and monitoring. In Porter, A.L. and Fittipaldi, J.J. (eds) (1998) *Environmental Methods Review: Retooling Impact Assessment for the New Century*. Atlanta, GA: Army Environmental Policy Institute.

Shillington and Burns Consultants (1999) *Public Participation in Screening and Comprehensive Studies*. Hull, Quebec: Canadian Environmental Assessment Agency.

Shopley, J.B. and Fuggle, R.F. (1984) A comprehensive review of current environmental impact assessment methods and techniques. *Journal of Environmental Management* 28: 68–81.

Sielcken, R.-J., Scholten, J.J. and van Eck, M. (1996) Review criteria employed by the Commission for EIA in the Netherlands. In *Environmental Impact Assessment in the Netherlands: Experiences and Views presented by and to the Commission for EIA*. Utrecht: Environmental Impact Assessment Commission.

Sigal, L.L. and Webb, J.W. (1989) The programmatic environmental impact statement: its purpose and use. *The Environmental Professional* 11: 14–24.

Simpson, J. (2001) Developing a review package to assess the quality of EA reports of local authority structure and local plans in the UK. *Environmental Impact Assessment Review* 21: 83–5.

Sinclair, J. and Diduck, A. (1995) Public education: an undervalued component of the environmental assessment public involvement process. *Environmental Impact Assessment Review* 15: 219–40.

Smith, G. (1996) The role of assessment of environmental effects under the Resource Management Act 1991. *Environmental and Planning Law Journal* 13: 82–102.

Smith, L.G. (1993) *Impact Assessment and Sustainable Resource Management*. Harlow: Longman.

Smythe, R.B. (1997) The historical roots of NEPA. In Clark, E.R. and Canter, L. (eds) *Environmental Policy and NEPA: Past, Present and Future*. Boca Raton, FL: St Lucie Press.

Sonntag, N.C., Everitt, R.R., Rattie, L.P., Colnett, D.L., Wolf, C.P., Truett, J.C., Dorcey, A.H.J. and Holling, C.S. (1987) *Cumulative Effects Assessment: A Context for Further Research and Development*. Hull, Quebec: Canadian Environmental Assessment Research Council.

Sowman, M., Fuggle, R. and Preston, G. (1995) A review of the evolution of environmental evaluation procedures in South Africa. *Environmental Impact Assessment Review* 15: 45–67.

Steinemann, A. (2001) Improving alternatives for environmental impact assessment. *Environmental Impact Assessment Review* **21**: 3–21.

Taylor, S. (1984) *Making Bureaucracies Think: The Environmental Impact Statement Strategy of Administrative Reform.* Stanford, CA: Stanford University Press.

ten Heuvelhof, E. and Nauta, C. (1997) The effects of environmental impact assessment in the Netherlands. *Project Appraisal* **12**: 25–30.

ten Holder, V. and Verheem, R. (1996) Strategic EIA in the Netherlands. In *Environmental Impact Assessment in the Netherlands: Experiences and Views presented by and to the Commission for EIA.* Utrecht: Environmental Impact Assessment Commission.

ten Holder, V. and Verheem, R. (1997) Strategic EIA in the Netherlands – ten years of experience *Environmental Assessment* **5**(3): 34–6.

Terence O'Rourke (1999) *Farnborough Aerodrome Environmental Statement.* Bournemouth: Terence O'Rourke plc.

Therivel, R. (1998) Strategic environmental assessment of development plans in Great Britain. *Environmental Impact Assessment Review* **18**: 39–57.

Therivel, R. and Brown, A.L. (1999) Methods of strategic environmental assessment. In Petts, J. (ed.) *Handbook of Environmental Impact Assessment,* vol. 1. Oxford: Blackwell.

Therivel, R. and Partidario, M.R. (1996) (eds) *The Practice of Strategic Environmental Assessment.* London: Earthscan.

Therivel, R., Wilson, E., Thompson, S., Heanley, D. and Pritchard, D. (1992) *Strategic Environmental Assessment.* London: Earthscan.

Thomas, I. (1998) *Environmental Impact Assessment in Australia: Theory and Practice,* 2nd edn. Sydney: Federation Press.

Thompson, M.A. (1990) Determining impact significance in EIA: a review of 24 methodologies. *Journal of Environmental Management* **30**: 235–56.

Tilleman, W.A. (1995) Public participation in the environmental impact assessment process: a comparative study of impact assessment in Canada, the United States and the European Community. *Columbia Journal of Transnational Law* **33**: 337–39.

Tomlinson, P. and Atkinson, S.F. (1987a) Environmental audits: proposed terminology. *Environmental Monitoring and Assessment* **8**: 187–98.

Tomlinson, P. and Atkinson, S.F. (1987b) Environmental audits: a literature review. *Environmental Monitoring and Assessment* **8**: 239–61.

Tzoumis, K. and Finegold, L. (2000) Looking at the quality of draft environmental impact statements over time in the United States: have ratings improved? *Environmental Impact Assessment Review* **20**: 557–78.

Ugoretz, S. (2001) Towards a new model of environmental review: preparing the US National Environmental Policy Act for new management paradigms. *Impact Assessment and Project Appraisal* **19**: 3–8.

United Nations Economic Commission for Europe:

    UNECE (1992) *Application of Environmental Impact Assessment Principles to Policies, Plans and Programmes.* Environmental series 5. Geneva: UNECE.

    UNECE (1994) *(Espoo) Convention on Environmental Impact Assessment in a Transboundary Context.* Geneva: UNECE.

    UNECE (1998) *(Aarhus) Convention on Access to Information, Public Participation in Decision-Making and Access to Justice in Environmental Matters.* Geneva: UNECE.

United Nations Environment Programme (1988) *Environmental Impact Assessment: Basic Procedures for Developing Countries.* Bangkok: UNEP, Regional Office for Asia and the Pacific.

van Breda, L.M. and Dijkema, G.P.J. (1998) EIA's contribution to environmental decision-making on large chemical plants. *Environmental Impact Assessment Review* 18: 391–410.

van Eck, M. (1994) Comparing alternatives in environmental impact assessment. In *EIA Methodology in the Netherlands: Views of the Commission for EIA*. Utrecht: Environmental Impact Assessment Commission.

van Eck, M. and Scholten, J.J. (1996) EIA – added value for decision-making. In *Environmental Impact Assessment in the Netherlands: Experiences and Views presented by and to the Commission for EIA*. Utrecht: Environmental Impact Assessment Commission.

van Eck, M., Scholten, J.J. and Morel, S.A.A. (1994) EIA methodology: scoping of alternatives. In *EIA Methodology in the Netherlands: Views of the Commission for EIA*. Utrecht: Environmental Impact Assessment Commission.

Verheem, R. (1992) Environmental assessment at the strategic level in the Netherlands. *Project Appraisal* 7: 150–6.

Verheem, R. (1996) SEA of the Dutch Ten Year Programme on Waste Management 1992–2002. In Therivel, R. and Partidario, M.R. (eds) *The Practice of Strategic Environmental Assessment*. London: Earthscan.

Verheem, R. (1999) Dutch SEA practice and the proposed EU SEA Directive. *Evaluation Environnementale des Plans et Programme. Amenagement et Nature* 134: 184–95.

Verheem, R. and Tonk, J., (2000) Strategic environmental assessment: one concept, multiple forms. *Impact Assessment and Project Appraisal* 18: 177–82.

Verheem, R., Odijk, M. and Scholten, J. (1998) The environment: from add-on to guiding principle. In *New Experiences on EIA in the Netherlands: Process, Methodology, Case Studies*. Utrecht: Environmental Impact Assessment Commission.

Verocai Moreira, I. (1988) EIA in Latin America. In Wathern, P. (ed.) *Environmental Impact Assessment: Theory and Practice*. London: Unwin Hyman.

Vig, N. and Kraft, M. (eds) (1984) *Environmental Policy in the 1980s: Reagan's New Agenda*. Washington, DC: CQ Press.

von Moltke, K. (1984) Impact assessment in the United States and Europe. In Clark, B.D., Gilad, A., Bisset, R. and Tomlinson, P. (eds) *Perspectives on Environmental Impact Assessment*. Dordrecht: Reidel.

Wandesforde-Smith, G. (1979) Environmental impact assessment in the European Community. *Zeitschrift für Umwelt Politik* 1: 35–76.

Wandesforde-Smith, G. (1989) Environmental impact assessment, entrepreneurship, and policy change. In Bartlett, R.V. (ed.) *Policy through Impact Assessment*. New York: Greenwood Press.

Wandesforde-Smith, G. and Kerbavaz, J. (1988) The co-evaluation of politics and policy: elections, entrepreneurship and EIA in the United States. In Wathern, P. (ed.) *Environmental Impact Assessment: Theory and Practice*. London: Unwin Hyman.

Wathern, P. (1988) The EIA directive of the European Community. In Wathern, P. (ed.) *Environmental Impact Assessment: Theory and Practice*. London: Unwin Hyman.

Wathern, P. (1989) Implementing supranational policy: environmental impact assessment in the United Kingdom. In Bartlett, R.V. (ed.) *Policy through Impact Assessment*. New York: Greenwood Press.

Weaver, A. (1996) Life after EIA: the IAIAsa response. Paper to International Association for Impact Assessment (South Africa Chapter) Conference, Cape Town. Mill Street: IAIA-SA.

Weaver, A., Rossouw, N. and Grobler, D. (1999) Scoping and 'issues focussed' environmental impact assessment in South Africa. *African Journal of Environmental Assessment and Management* 1: 1–11.

Weaver, A., Morant, P., Ashton, P. and Kruger, F. (2000) Environmental impact assessment. In Sampson, I.R. (ed.) *The Guide to Environmental Auditing in South Africa.* Durban: Butterworths.

Weiner, K.S. (1997) Basic purposes and policies of the NEPA regulations. In Clark, E.R. and Canter, L. (eds) *Environmental Policy and NEPA: Past, Present and Future.* Boca Raton, FL: St Lucie Press.

Welles, H. (1997) The CEQ NEPA Effectiveness Study: learning from our past and shaping our future. In Clark, E.R. and Canter, L. (eds) *Environmental Policy and NEPA: Past, Present and Future.* Boca Raton, FL: St Lucie Press.

Wells, C. and Fookes, T. (1988) *Impact Assessment in Resource Management.* Resource Management Law Reform working paper 20. Wellington: Ministry for the Environment.

Weston, J. (1997) EIA and public inquiries. In Weston, J. (ed.) *Planning and Environmental Impact Assessment in Practice.* Harlow: Longman.

Weston, J. (2000) EIA, decision-making theory and screening and scoping in UK practice. *Journal of Environmental Planning and Management* 43: 185–203.

Weston, J. and Prenton-Jones, P. (1997) Improving EIA for intensive livestock projects. *Journal of Environmental Planning and Management* 40: 527–33.

Weston, S.M.C. (1991) *The Canadian Federal Environmental Assessment and Review Process: An Analysis of the Initial Assessment Phase.* Hull, Quebec: Canadian Environmental Assessment Research Council.

Whyte, A.V. (ed.) (1995) *Environment, Reconstruction, and Development.* Building a new South Africa, vol. 4. Ottawa: International Development Research Centre.

Williams, D.A.R. (1997) *Environmental and Resource Management Law in New Zealand.* Wellington: Butterworths.

Wilson, L. (1998) A practical method for environmental impact assessment audits. *Environmental Impact Assessment Review* 18: 59–71.

Wiseman, K. (2000) Environmental assessment and planning in South Africa: the SEA connection. In Partidario, M.R. and Clark, E.R. (eds) *Perspectives on Strategic Environmental Assessment.* Boca Raton, FL: Lewis.

Wood, C.M. (1988a) EIA in plan-making. In Wathern, P. (ed.) *Environmental Impact Assessment: Theory and Practice.* London: Unwin Hyman.

Wood, C.M. (1988b) The genesis and implementation of environmental impact assessment in Europe. In Clark, M. and Herington, J. (eds) *The Role of Environmental Assessment in the Planning Process.* London: Mansell.

Wood, C.M. (1989) *Planning Pollution Prevention.* Oxford: Heinemann Newnes.

Wood, C.M. (1993) Environmental impact assessment in Victoria: Australian discretion rules EA! *Journal of Environmental Management* 39: 281–95.

Wood, C.M. (1999a) Environmental planning. In Cullingworth, J.B. (ed.) *British Planning: Fifty Years of Urban and Regional Policy.* London: Athlone Press.

Wood, C.M. (1999b) Pastiche or postiche? Environmental impact assessment in South Africa. *South African Geographical Journal* 81: 52–9.

Wood, C.M. and Bailey, J. (1994) Predominance and independence in environmental impact assessment: the Western Australia model. *Environmental Impact Assessment Review* 14: 37–59.

Wood, C.M. and Coppell, L. (1999) An evaluation of the Hong Kong environmental assessment system. *Impact Assessment and Project Appraisal* 17: 21–31.

Wood, C.M. and Djeddour, M. (1992) Strategic environmental assessment: EA of policies, plans and programmes. *Impact Assessment Bulletin* **10**: 3–22.

Wood, C.M. and Jones, C.E. (1997) The effect of environmental assessment on UK local planning authority decisions. *Urban Studies* **34**: 1237–57.

Wood, C.M. and McDonic, G. (1989) Environmental assessment: challenge and opportunity. *The Planner* **75** (11): 12–18.

Wood, C., Dipper, B. and Jones, C. (2000) Auditing the assessment of the environmental impacts of planning projects. *Journal of Environmental Planning and Management* **43**: 23–47.

Wood, G. (1999) Assessing techniques of assessment: post-development auditing of noise predictive schemas in environmental impact assessment. *Impact Assessment and Project Appraisal* **17**: 217–26.

Wood, G. (2000) Is what you see what you get? Post-development auditing of methods used for predicting the zone of visual influence in EIA. *Environmental Impact Assessment Review* **20**: 537–56.

Wood, G. and Bellanger, C. (1998) *Directory of Environmental Impact Statements.* July 1988–April 1998, working paper 179. Oxford: Oxford Brookes University.

World Bank (1987) *Environment, Growth and Development.* Development Committee paper 14. Washington, DC: World Bank.

World Bank (1996) *Analysis of Alternatives in Environmental Assessment. Environmental Assessment Sourcebook* Update no. 17. Washington, DC: World Bank.

World Bank (1999) *Operational Policy, Bank Procedure and Good Practice* 4.01: *Environmental Assessment.* Washington, DC: World Bank.

World Commission on Environment and Development (1987) *Our Common Future.* Oxford: Oxford University Press.

Yost, N.C. (1990) NEPA's promise partially fulfilled. *Environmental Law* **20**: 533–49.

Index